T0192613

A COURSE IN MATHEMATICAL ANALYSIS
Volume II: Metric and Topological Spaces, Functions of a Vector Variable

The three volumes of *A Course in Mathematical Analysis* provide a full and detailed account of all those elements of real and complex analysis that an undergraduate mathematics student can expect to encounter in the first two or three years of study. Containing hundreds of exercises, examples and applications, these books will become an invaluable resource for both students and instructors.

Volume I focuses on the analysis of real-valued functions of a real variable. This second volume goes on to consider metric and topological spaces. Topics such as completeness, compactness and connectedness are developed, with emphasis on their applications to analysis. This leads to the theory of functions of several variables: differentiation is developed in a coordinate free way, while integration (the Riemann integral) is established for functions defined on subsets of Euclidean space. Differential manifolds in Euclidean space are introduced in a final chapter, which includes an account of Lagrange multipliers and a detailed proof of the divergence theorem. Volume III covers complex analysis and the theory of measure and integration.

D. J. H. GARLING is Emeritus Reader in Mathematical Analysis at the University of Cambridge and Fellow of St. John's College, Cambridge. He has fifty years' experience of teaching undergraduate students in most areas of pure mathematics, but particularly in analysis.

A COURSE IN
MATHEMATICAL ANALYSIS

Volume II
Metric and Topological Spaces,
Functions of a Vector Variable

D. J. H. GARLING

Emeritus Reader in Mathematical Analysis,
University of Cambridge, and
Fellow of St John's College, Cambridge

CAMBRIDGE
UNIVERSITY PRESS

CAMBRIDGE
UNIVERSITY PRESS

University Printing House, Cambridge CB2 8BS, United Kingdom

One Liberty Plaza, 20th Floor, New York, NY 10006, USA

477 Williamstown Road, Port Melbourne, VIC 3207, Australia

314-321, 3rd Floor, Plot 3, Splendor Forum, Jasola District Centre, New Delhi - 110025, India

79 Anson Road, #06-04/06, Singapore 079906

Cambridge University Press is part of the University of Cambridge.

It furthers the University's mission by disseminating knowledge in the pursuit of education, learning and research at the highest international levels of excellence.

www.cambridge.org
Information on this title: www.cambridge.org/9781107675322

© D. J. H. Garling 2013

This publication is in copyright. Subject to statutory exception and to the provisions of relevant collective licensing agreements, no reproduction of any part may take place without the written permission of Cambridge University Press.

First published 2013

A catalogue record for this publication is available from the British Library

Library of Congress Cataloging in Publication data
Garling, D. J. H.
Metric and topological spaces, functions of a vector variable / D. J. H. Garling.
pages   cm. – (A course in mathematical analysis; volume 2)
Includes bibliographical references and index.
ISBN 978-1-107-03203-3 (hardback) – ISBN 978-1-107-67532-2 (paperback)
1. Metric spaces   2. Topological spaces.   3. Vector valued functions.   I. Title.
QA611.28.G37   2013
514′.325–dc23   2012044992

ISBN   978-1-107-03203-3   Hardback
ISBN   978-1-107-67532-2   Paperback

Cambridge University Press has no responsibility for the persistence or accuracy of URLs for external or third-party internet websites referred to in this publication, and does not guarantee that any content on such websites is, or will remain, accurate or appropriate.

Contents

Introduction

This book is the second volume of a full and detailed course in the elements of real and complex analysis that mathematical undergraduates may expect to meet. Indeed, it was initially based on those parts of analysis that undergraduates at Cambridge University meet, or used to meet, in their first two years. There is however always a temptation to go a bit further, and this is a temptation that I have not resisted. Thus I have included accounts of Baire's category theorem, and the Arzelà–Ascoli theorem, which are taught in the third year, and the Mazur–Ulam theorem, which, as far as I know, has never been taught. As a consequence, there are certain sections that can be omitted on a first reading. These are indicated by asterisks.

Volume I was concerned with analysis on the real line. In Part Three, the analysis is extended to a more general setting. We introduce and consider metric and topological spaces, and normed spaces. In fact, metric and metrizable spaces are sufficient for all subsequent needs, but many of the properties that we investigate are topological properties, and it is well worth understanding what this means. The study of topological spaces can degenerate into the construction of pathological examples; once again, temptation is not resisted, and Section 11.6 contains a collection of these. This section can be omitted at a first reading (and indeed at any subsequent reading). Baire's category theorem is proved in Section 12.6; it is remarkable that a theorem with a rather easy proof can lead to so many strong conclusions, but this is another section that can be omitted at a first reading. The notion of compactness, which is a fundamental topological idea, is studied in some detail. Tychonoff's theorem on the compactness of the product of compact spaces, which involves the axiom of choice, is too hard to include here: a proof is given in Appendix D.

In Part Four, we come back down to earth. The principal concern is the differentiation and integration of functions of several variables. Differentiation is interesting and reasonably straightforward, and we consider functions defined on a normed space; this shows that the results do not depend on any particular choice of coordinate system. Integration is another matter. To begin with it seems that the ideas of Riemann integration developed in Part Two carry over easily to higher dimensions, but serious problems arise as soon as a non-linear change of variables is considered. It is however possible to establish results that suffice in a great number of contexts. For example, the change of variables results are used in Volume III, where we introduce the Lebesgue measure, and the corresponding theory of integration. These results on differentiation and integration are applied in Chapter 19, where we consider subspaces of a Euclidean space which are differential manifolds – subspaces which locally look like Euclidean space.

This volume requires the knowledge of some elementary results in linear algebra; these are described and established in Appendix B.

The text includes plenty of exercises. Some are straightforward, some are searching, and some contain results needed later. All help to develop an understanding of the theory: do them!

I am extremely grateful to Zhuo Min 'Harold' Lim who read the proofs, and found embarrassingly many errors. Any remaining errors are mine alone. Corrections and further comments can be found on a web page on my personal home page at www.dpmms.cam.ac.uk.

Part Three

Metric and topological spaces

11

Metric spaces and normed spaces

11.1 Metric spaces: examples

In Volume I, we established properties of real analysis, starting from the properties of the ordered field $\mathbf{R}$ of real numbers. Although the fundamental properties of $\mathbf{R}$ depend upon the order structure of $\mathbf{R}$, most of the ideas and results of the real analysis that we considered (such as the limit of a sequence, or the continuity of a function) can be expressed in terms of the distance $d(x, y) = |x - y|$ defined in Section 3.1. The concept of distance occurs in many other areas of analysis, and this is what we now investigate.

A *metric space* is a pair (X, d), where X is a set and d is a function from the product $X \times X$ to the set $\mathbf{R}^+$ of non-negative real numbers, which satisfies

1. $d(x, y) = d(y, x)$ for all $x, y \in X$ *(symmetry)*;
2. $d(x, z) \leq d(x, y) + d(y, z)$ for all $x, y, z \in X$ *(the triangle inequality)*;
3. $d(x, y) = 0$ if and only if $x = y$.

d is called a *metric*, and $d(x, y)$ is the *distance* from x to y. The conditions are very natural: the distance from x to y is the same as the distance from y to x; the distance from x to y *via* z is at least as far as any more direct route, and any two distinct points of X are a positive distance apart.

Let us give a few examples, to get us started.

Example 11.1.1 $\mathbf{R}$, with the metric $d(x, y) = |x - y|$, is a metric space, as is $\mathbf{C}$, with the metric $d(z, w) = |z - w|$.

This was established in Volume I, in Propositions 3.1.2 and 3.7.2. These metrics are called the *usual metrics*. If we consider $\mathbf{R}$ or $\mathbf{C}$ as a metric space, without specifying the metric, we assume that we are considering the usual metric.

Example 11.1.2   The Euclidean metric.

We can extend the ideas of the previous example to higher dimensions. We need an inequality.

Proposition 11.1.3 (Cauchy's inequality)   *If $x, y \in \mathbf{R}^d$ then*

$$\sum_{j=1}^{d} x_j y_j \leq \left(\sum_{j=1}^{d} x_j^2 \right)^{\frac{1}{2}} \left(\sum_{j=1}^{d} y_j^2 \right)^{\frac{1}{2}}.$$

Equality holds if and only if $x_i y_j = x_j y_i$ for $1 \leq i, j \leq d$.

Proof   We give the proof given by Cauchy in 1821, using Lagrange's identity:

$$\left(\sum_{j=1}^{d} x_j y_j \right)^2 + \sum_{\{(i,j):i<j\}} (x_i y_j - x_j y_i)^2 = \left(\sum_{j=1}^{d} x_j^2 \right) \left(\sum_{j=1}^{d} y_j^2 \right),$$

which follows by expanding the terms. This clearly establishes the inequality, and also shows that equality holds if and only if $x_i y_j = x_j y_i$ for $1 \leq i, j \leq d$. $\square$

Corollary 11.1.4   *If $x, y \in \mathbf{R}^d$, let $d(x, y) = (\sum_{j=1}^{d} (x_j - y_j)^2)^{1/2}$. Then d is a metric on $\mathbf{R}^d$.*

Proof   Conditions (i) and (iii) are clearly satisfied. We must establish the triangle inequality. First we use Cauchy's inequality to show that if $a, b \in \mathbf{R}^2$, then $d(a + b, 0) \leq d(a, 0) + d(b, 0)$:

$$d(a + b, 0)^2 = \sum_{j=1}^{d} (a_j + b_j)^2$$

$$= \sum_{j=1}^{d} a_j^2 + 2 \sum_{j=1}^{d} a_j b_j + \sum_{j=1}^{d} b_j^2$$

$$\leq d(a, 0)^2 + 2 d(a, 0).d(b, 0) + d(b, 0)^2 = (d(a, 0) + d(b, 0))^2.$$

Note that it follows from the definitions that d is translation invariant: $d(a, b) = d(a + c, b + c)$. In particular, $d(a, b) = d(a - b, 0)$. If $x, y, z \in \mathbf{R}^d$, set $a = x - y$ and $b = y - z$, so that $a + b = x - z$. Then

$$d(x, z) = d(a + b, 0) \leq d(a, 0) + d(b, 0) = d(x, y) + d(y, z).$$

$\square$

The metric d is called the *Euclidean metric*, or *standard metric*, on $\mathbf{R}^d$. When $d = 2$ or 3 it is the usual measure of distance.

We can also consider complex sequences.

Corollary 11.1.5 *If $z, w \in \mathbf{C}^d$, let $d(z, w) = (\sum_{j=1}^d |z_j - w_j|^2)^{1/2}$. Then d is a metric on $\mathbf{C}^d$.*

Proof Again, conditions (i) and (iii) are clearly satisfied, and we must establish the triangle inequality. First we show that if $z, w \in \mathbf{C}^d$ then $d(z + w, 0) \le d(z, 0) + d(w, 0)$: using the inequality of the previous corollary,

$$
d(z + w, 0) = \left(\sum_{j=1}^d |z_j + w_j|^2 \right)^{1/2} \le \left(\sum_{j=1}^d (|z_j| + |w_j|)^2 \right)^{1/2}
$$

$$
\le \left(\sum_{j=1}^d |z_j|^2 \right)^{1/2} + \left(\sum_{j=1}^d |w_j|^2 \right)^{1/2} = d(z, 0) + d(w, 0).
$$

Again d is translation invariant, so that $d(r, s) = d(r - s, 0)$. If $r, s, t \in \mathbf{C}^d$ let $z = r - s$ and $w = s - t$, so that $z + w = r - t$ and

$$
d(r, t) = d(z + w, 0) \le d(z, 0) + d(w, 0) = d(r, s) + d(s, t).
$$

$\square$

The metric d is called the *standard metric* on $\mathbf{C}^d$.

We shall study these metrics in more detail, later.

Example 11.1.6 The discrete metric.

Let X be any set. We define $d(x, y) = 1$ if $x \ne y$ and $d(x, y) = 0$ if $x = y$. Then d is a metric on X, the *discrete* metric. If $x \in X$, there are no other points of X close to x; this means, as we shall see, that analysis on X is rather trivial.

Example 11.1.7 The subspace metric.

If (X, d) is a metric space, and Y is a subset of X, then the restriction of d to $Y \times Y$ is a metric on Y. This metric is the *subspace metric* on Y, and Y, with the subspace metric, is called a *metric subspace* of (X, d).

The subspace metric is a special case of the following. Suppose that (X, d) is a metric space and that f is an injective mapping of a set Y into X. If we set $\rho(y, y') = d(f(y), f(y'))$ then it is immediately obvious that ρ is a metric on Y. For example, we can give $(-\pi/2, \pi/2)$ the usual metric, as a subset of $\mathbf{R}$. The mapping $j = \tan^{-1}$ is a bijection of $\mathbf{R}$ onto $(-\pi/2, \pi/2)$. Thus if we set

$$\rho(y, y') = |j(y) - j(y')| = \left|\tan^{-1}(y) - \tan^{-1}(y')\right|,$$

then ρ is a metric on $\mathbf{R}$.

Example 11.1.8   A metric on the extended real line $\overline{\mathbf{R}}$.

We can extend the mapping j of the previous example to $\overline{\mathbf{R}}$ by setting $j(-\infty) = -\pi/2$ and $j(+\infty) = -\pi/2$. Then j is a bijection of $\overline{\mathbf{R}}$ onto $[-\pi/2, \pi/2]$, and we can again define a metric on $\overline{\mathbf{R}}$ by setting $\rho(y, y') = |j(y) - j(y')|$. Thus

$$\rho(y, \infty) = \pi/2 - \tan^{-1}(y),$$
$$\rho(-\infty, y) = \tan^{-1}(y) + \pi/2$$
$$\text{and } \rho(-\infty, \infty) = \pi.$$

Example 11.1.9   A metric on $\overline{\mathbf{N}} = \mathbf{N} \cup \{\infty\}$.

Here is a similar construction. If $n \in \mathbf{N}$, let $f(n) = 1/n$, and let $f(+\infty) = 0$. f is an injective map of $\overline{\mathbf{N}}$ onto a closed and bounded subset of $\mathbf{R}$. Define $\rho(x, x') = |f(x) - f(x')|$. This defines a metric on $\overline{\mathbf{N}}$:

$$\rho(m, n) = |1/m - 1/n| \text{ and } \rho(m, \infty) = 1/m.$$

Example 11.1.10   The uniform metric.

There are many cases where we define a metric on a space of functions. Here is the first and most important example. First we need some definitions. Suppose that B is a non-empty subset of a metric space (X, d). The *diameter* diam (B) of B is defined to be $\sup\{d(b, b') : b, b' \in B\}$. The set B is *bounded* if diam $(B) < \infty$. If $f : S \to (X, d)$ is a mapping and A is a non-empty subset of S, we define the *oscillation* $\Omega(f, A)$ of f on A to be the diameter of $f(A)$: $\Omega(f, A) = \sup\{d(f(a), f(a')) : a, a' \in A\}$. The function f is *bounded* if $\Omega(f, S) = \text{diam}(f(S)) < \infty$.

Proposition 11.1.11   *Let* $B(S, X) = B_X(S)$ *denote the set of all bounded mappings* f *from a non-empty set* S *to a metric space* (X, d). *If* $f, g \in B_X(S)$, *let* $d_\infty(f, g) = \sup\{d(f(s), g(s)) : s \in S\}$. *Then* d_∞ *is a metric on* $B_X(S)$.

Proof   First we show that $d_\infty(f, g)$ is finite. Let $s_0 \in S$. If $s \in S$ then, by the triangle inequality,

$$d(f(s), g(s)) \le d(f(s), f(s_0)) + d(f(s_0), g(s_0)) + d(g(s_0), g(s))$$
$$\le \Omega(f, S) + d(f(s_0), g(s_0)) + \Omega(g, S).$$

Taking the supremum,

$$d_\infty(f, g) \le \Omega(f, S) + d(f(s_0), g(s_0)) + \Omega(g, S) < \infty.$$

Conditions (i) and (iii) are clearly satisfied, and it remains to establish the triangle inequality. Suppose that $f, g, h \in B_X(S)$ and that $s \in S$. Then

$$d(f(s), h(s)) \le d(f(s), g(s)) + d(g(s), h(s)) \le d_\infty(f, g) + d_\infty(g, h).$$

Taking the supremum, $d_\infty(f, h) \le d_\infty(f, g) + d_\infty(g, h)$.   □

This metric is called the *uniform metric*.

Example 11.1.12   Pseudometrics.

We shall occasionally need to consider functions p for which the third condition in the definition of a metric is replaced by the weaker condition
(3') if $x = y$ then $p(x, y) = 0$.
In other words, we allow distinct points to be zero p-distance apart. Such a function is called a *pseudometric*. It is easy to relate a pseudometric to a metric on a quotient space.

Proposition 11.1.13   *Suppose that p is a pseudometric on a set X. The relation on X defined by setting $x \sim y$ if $d(x, y) = 0$ is an equivalence relation on X. Let q be the quotient mapping from X onto the quotient space $X/\sim$. Then there exists a metric d on $X/\sim$ such that $d(q(x), q(y)) = p(x, y)$ for $x, y \in X$.*

Proof   The fact that $\sim$ is an equivalence relation on X is an immediate consequence of the symmetry property and the triangle inequality. We need a lemma, which will be useful elsewhere.

Lemma 11.1.14   *Suppose that p is a pseudometric, or a metric, on a set X, and that $a, a', b, b' \in X$. Then*

$$|p(a, b) - p(a', b')| \le p(a, a') + p(b, b').$$

Proof   Using the triangle inequality twice,

$$p(a, b) \le p(a, a') + p(a', b) \le p(a, a') + p(a', b') + p(b', b),$$

so that

$$p(a, b) - p(a', b') \leq p(a, a') + p(b, b').$$

Similarly

$$p(a', b') - p(a, b) \leq p(a, a') + p(b, b'),$$

which gives the result.   □

We now return to the proof of the proposition. If $a \sim a'$ and $b \sim b'$ then it follows from the lemma that $p(a, b) = p(a', b')$. Thus if we define $d(q(a), q(b)) = p(a, b)$, this is well-defined. Symmetry and the triangle inequality for d now follow immediately from the corresponding properties of p. Finally if $d(q(a), q(b)) = 0$ then $p(a, b) = 0$, so that $a \sim b$ and $q(a) = q(b)$.   □

We shall meet more examples of metric spaces later.

Exercises

11.1.1 If $x, y \in [0, 2\pi)$, let $d(x, y) = \min(|x - y|, 2\pi - |x - y|)$. Show that d is a metric on $[0, 2\pi)$. Define $f : [0, 2\pi) \to \mathbf{R}^2$ by setting $f(x) = (\cos x, \sin x)$. Let $\rho(f(x), f(y)) = d(x, y)$. Show that ρ is a metric on $f([0, 2\pi))$, the *arc-length metric*.

11.1.2 Suppose that p is a prime number. If r is a non-zero rational number, it can be written uniquely as $r = p^{v(r)} s/t$, where $v(r) \in \mathbf{Z}$ and s/t is a fraction in lowest terms, with neither s nor t having p as a divisor. Thus if $p = 3$ then $v(6/7) = 1$ and $v(5/18) = -2$. Let $n(r) = p^{-v(r)}$. If $r, r' \in \mathbf{Q}$, set $d_p(r, r') = n(r - r')$ for $r \neq r'$ and $d_p(r, r') = 0$ if $r = r'$. Show that d is a metric on $\mathbf{Q}$. This metric, the *p-adic metric*, is useful in number theory, but we shall not consider it further.

11.1.3 As far as I know, the next example is just a curiosity. Consider $\mathbf{R}^d$ with its usual metric d. If $x = \alpha y$ for some $\alpha \in \mathbf{R}$ (that is, x and y lie on a real straight line through the origin) set $\rho(x, y) = d(x, y)$; otherwise, set $\rho(x, y) = d(x, 0) + d(0, y)$. Show that ρ is a metric on $\mathbf{R}^d$.

11.1.4 Let P_n be the power set of $\{1, \ldots, n\}$; the set of subsets of $\{1, \ldots, n\}$. Let $h(A, B) = |A \Delta B|$, where $A \Delta B$ is the symmetric difference of A and B. Show that h is a metric on P_n (the *Hamming metric*).

11.1.5 Let $P(\mathbf{N})$ be the set of subsets of $\mathbf{N}$. If A and B are distinct subsets of $\mathbf{N}$, let $d(A, B) = 2^{-j}$, where $j = \inf(A \Delta B)$, and let $d(A, A) = 0$. Show that d is a metric on $P(\mathbf{N})$ and that

$$d(A, C) \leq \max(d(A, B), d(B, C)) \text{ for } A, B, C \in P(\mathbf{N}).$$

11.1.6 Let f be the real-valued function on the extended real line $\overline{\mathbf{R}}$ defined by $f(x) = x/\sqrt{1+x^2}$ if $x \in \mathbf{R}$, $f(+\infty) = 1$ and $f(-\infty) = -1$. If $x, y \in \overline{\mathbf{R}}$, let $d(x, y) = |f(x) - f(y)|$. Show that d is a metric on $\overline{\mathbf{R}}$. Show that a sequence $(x_n)_{n=1}^{\infty}$ of real numbers converges to $+\infty$ as $n \to \infty$ if and only if $d(x_n, +\infty) \to 0$ as $n \to \infty$.

11.2 Normed spaces

Many of the metric spaces that we shall consider are real or complex vector spaces, and it is natural to consider metrics which relate to the algebraic structure. We shall assume knowledge of the basic algebraic properties of vector spaces and linear mappings; these are described in Appendix B. Suppose that E is a real or complex vector space. It is then natural to consider those metrics d which are *translation-invariant*: that is, $d(x + a, y + a) = d(x, y)$ for all $x, y, a \in E$. Note that this implies that

$$d(x, y) = d(x - y, 0) = d(0, x - y) = d(-x, -y).$$

It is also natural to require that if we multiply by a scalar then the distance is scaled in an appropriate way: $d(\lambda x, \lambda y) = |\lambda| d(x, y)$ for all $x, y \in E$ and all scalars λ. It is easy to characterize such metrics.

A real-valued function $x \to \|x\|$ on a real or complex vector space E is a *norm* if

1. $\|x + y\| \le \|x\| + \|y\|$ (p is *subadditive*),
2. $\|\alpha x\| = |\alpha| \, \|x\|$ for every scalar α and
3. $\|x\| = 0$ if and only if $x = 0$,

for α a scalar and x, y vectors in E. $(E, \|.\|)$ is then called a *normed space*. Note that $\|x\| = \|-x\|$ and that $\|0\| = \|0.0\| = 0 \, \|0\| = 0$. A norm is necessarily non-negative, since $0 = \|0\| \le \|x\| + \|-x\| = 2 \, \|x\|$.

A subset C of a real or complex vector space is *convex* if whenever $x, y \in C$ and $0 \le t \le 1$ then $(1 - t)x + ty \in C$. A real-valued function on a convex subset C of a real or complex vector space E is *convex* if whenever $x, y \in C$ and $0 \le t \le 1$ then

$$f((1 - t)x + ty) \le (1 - t)f(x) + tf(y).$$

Proposition 11.2.1 *If $\|.\|$ is a norm on a real or complex vector space E, then $\|.\|$ is a convex function on E.*

Proof Suppose that $x, y \in E$ and $0 \le t \le 1$. Then

$$\|(1 - t)x + ty\| \le \|(1 - t)x\| + \|ty\| = (1 - t) \, \|x\| + t \, \|y\|. \qquad \square$$

Corollary 11.2.2   *The sets* $U = \{x : \|x\| < 1\}$ *and* $B = \{x : \|x\| \leq 1\}$ *are convex subsets of* E.

Theorem 11.2.3   *Suppose that* d *is a metric on a real or complex vector space* E. *Then the following are equivalent:*

(i) d is translation-invariant and satisfies $d(\lambda x, \lambda y) = |\lambda| d(x, y)$ for all $x, y \in E$ and all scalars λ;

(ii) there exists a norm $\|.\|$ on E such that $d(x, y) = \|x - y\|$.

Proof   If d is a translation-invariant metric with the desired scaling properties, and we set $\|x\| = d(x, 0)$, then $\|x\| = 0$ if and only if $x = 0$,

$$\|x + y\| = d(x + y, 0) = d(x, -y) \leq d(x, 0) + d(0, -y)$$
$$= d(x, 0) + d(y, 0) = \|x\| + \|y\|,$$

and

$$\|\lambda x\| = d(\lambda x, 0) = d(\lambda x, \lambda 0) = |\lambda| d(x, 0) = |\lambda| \|x\|.$$

Thus (i) implies (ii).

Conversely, suppose that $\|.\|$ is a norm on E, and that we set $d(x, y) = \|x - y\|$. First we show that d is a metric on E:

$$d(x, y) = \|x - y\| = \|y - x\| = d(y, x),$$

$d(x, y) = 0$ if and only if $\|x - y\| = 0$, if and only if $x - y = 0$, if and only if $x = y$, and

$$d(x, z) = \|x - z\| = \|(x - y) + (y - z)\|$$
$$\leq \|x - y\| + \|y - z\| = d(x, y) + d(y, z),$$

so that the triangle inequality holds. Further,

$$d(x + a, y + a) = \|(x + a) - (y + a)\| = \|x - y\| = d(x, y)$$

so that d is translation invariant, and

$$d(\lambda x, \lambda y) = \|\lambda x - \lambda y\| = \|\lambda(x - y)\| = |\lambda| \|x - y\| = |\lambda| d(x, y).$$

Thus (ii) implies (i). $\square$

A vector x in a normed space $(E, \|.\|)$ with $\|x\| = 1$ is called a *unit vector*. If y is a non-zero vector in E, then $y = \lambda y_1$, where $\lambda = \|y\|$ and y_1 is the unit vector $y/\|y\|$.

Let us give some examples of norms.

Theorem 11.2.4 *The function* $\|x\|_2 = (\sum_{j=1}^d x_j^2)^{1/2}$ *is a norm on* $\mathbf{R}^d$ *(the* Euclidean *norm).*

Proof   Clearly $\|x\|_2 = 0$ if and only if $x = 0$ and $\|\lambda x\|_2 = |\lambda|\,\|x\|_2$. Let d be the Euclidean metric on $\mathbf{R}^d$. Then, as in Corollary 11.1.4,

$$\|x + y\|_2 = d(x + y, 0) \le d(x, 0) + d(y, 0) = \|x\|_2 + \|y\|_2,$$

so that $\|.\|_2$ is subadditive. □

In the same way, we have the following.

Theorem 11.2.5 *The function* $\|z\|_2 = (\sum_{j=1}^d |z_j|^2)^{1/2}$ *is a norm on* $\mathbf{C}^d$.

The norm $\|.\|_2$ on $\mathbf{C}^d$ is again called the *Euclidean* norm.

We shall generalize these two examples in the next section.

Suppose that $(E, \|.\|)$ is a normed space. Then, since

$$\|x - y\| \le \|x\| + \|y\| \quad \text{and} \quad \|y\| \le \|y - x\| + \|x\|,$$

a subset B of E is bounded if and only if $\sup\{\|b\| : b \in B\} < \infty$; we say that B is *norm bounded*, or *bounded in norm*.

Thus if S is a set and $(E, \|.\|)$ is a normed space, then the set $B_E(S) = \{f : S \to E : f(S) \text{ is bounded}\}$ is equal to the set $\{f : S \to E : \sup\{\|f(s)\| : s \in S\} < \infty\}$. Further, $B_E(S)$ is a vector space, when addition and scalar multiplication are defined pointwise:

$$(f + g)(s) = f(s) + g(s) \quad \text{and} \quad (\lambda f)(s) = \lambda(f(s)).$$

Arguing as in Proposition 11.1.11,

$$\|f\|_\infty = d_\infty(f, 0) = \sup\{\|f(s)\| : s \in S\},$$

is a norm on $B_E(S)$, and $d_\infty(f, g) = \|f - g\|_\infty$, for $f, g \in B_E(S)$. The norm $\|.\|_\infty$ is called the *uniform norm*. We denote the normed space $(B_E(S), \|.\|_\infty)$ by $l_\infty(S, E)$. When $E = \mathbf{R}$ or $\mathbf{C}$, and the context is clear, we write $l_\infty(S)$ for $l_\infty(S, \mathbf{R})$ or $l_\infty(S, \mathbf{C})$. We denote $l_\infty(\mathbf{N})$ (or $l_\infty(\mathbf{Z}^+)$) by l_∞, and $l_\infty(\{1, \ldots, d\})$ by l_∞^d.

As with pseudometrics, we occasionally need to consider functions on a vector space which do not satisfy the third condition for a norm. A *seminorm* p on a vector space E is a real-valued function which satisfied the first two conditions for a norm, but not necessarily the third. Note that, as for a norm, $p(x) = p(-x)$, $p(0) = 0$ and $p(x) \ge 0$ for all $x \in E$.

Proposition 11.2.6   *If p is a seminorm on a vector space E, then the set $\{x \in E : p(x) = 0\}$ is a linear subspace N of E.*

Proof   If λ is a scalar and $x, y \in N$ then $p(\lambda x) = |\lambda| p(x) = 0$ and $0 \leq p(x + y) \leq p(x) + p(y) = 0$. $\square$

If p is a seminorm on a vector space E, then the function $\pi(x, y) = p(x - y)$ is a pseudometric on E. If $\sim$ is the equivalence relation which this defines, then $x \sim y$ if and only if $x - y \in N$. Thus the quotient space $E/\sim$ is the quotient vector space E/N, and q is a linear mapping of E onto E/N. If we set $\|q(x)\| = p(x)$, then $\|.\|$ is a norm on E/N which defines the metric of Proposition 11.1.13.

Exercises

11.2.1 Suppose that f is a real-valued function on a convex subset C of a real or complex vector space E. Show that the following are equivalent.
   (a) f is a convex function on C.
   (b) The set $\{(c, t) \in C \times \mathbf{R} : f(c) < t\}$ is a convex subset of $E \times \mathbf{R}$.
   (c) The set $\{(c, t) \in C \times \mathbf{R} : f(c) \leq t\}$ is a convex subset of $E \times \mathbf{R}$.

11.2.2 If $x = (x_1, \ldots, x_d) \in \mathbf{R}^d$, let $\|x\|_1 = \sum_{j=1}^{d} |x_j|$. Show that $\|.\|_1$ is a norm on $\mathbf{R}^d$. The normed space $(\mathbf{R}^d, \|.\|_1)$ is denoted by $l_1^d(\mathbf{R})$. Prove a similar result in the complex case.

11.2.3 Let $l_1(\mathbf{R})$ denote the set of real sequences $(a_n)_{n=1}^{\infty}$ for which $\sum_{n=1}^{\infty} |a_n|$ is finite. Show that $l_1(\mathbf{R})$ is a real vector space (with the algebraic operations defined pointwise) and that the function $\|a\|_1 = \sum_{n=1}^{\infty} |a_n|$ is a norm on $l_1(\mathbf{R})$.

11.3 Inner-product spaces

In 1885, Hermann Schwarz gave another proof of Cauchy's inequality, this time for two-dimensional integrals. Schwarz's proof is quite different from Cauchy's, and extends to a more general and more abstract setting. This provides some important examples of normed spaces.

Suppose that V is a real vector space. An *inner product* on V is a real-valued function $(x, y) \to \langle x, y \rangle$ on $V \times V$ which satisfies the following:
   (i) (bilinearity)

$$\langle \alpha_1 x_1 + \alpha_2 x_2, y \rangle = \alpha_1 \langle x_1, y \rangle + \alpha_2 \langle x_2, y \rangle,$$

$$\langle x, \beta_1 y_1 + \beta_2 y_2 \rangle = \beta_1 \langle x, y_1 \rangle + \beta_2 \langle x, y_2 \rangle,$$

for all x, x_1, x_2, y, y_1, y_2 in V and all real $\alpha_1, \alpha_2, \beta_1, \beta_2$;

(ii) (symmetry)

$$\langle y, x \rangle = \langle x, y \rangle \ \text{ for all } x, y \text{ in } V;$$

(iii) (positive definiteness)

$$\langle x, x \rangle > 0 \text{ for all non-zero } x \text{ in } V.$$

For example, if $V = \mathbf{R}^d$, we define the *usual* inner product, by setting $\langle z, w \rangle = \sum_{i=1}^d z_i w_i$ for $z = (z_i), w = (w_i)$.

A function which satisfies (i) and (ii) is called a *symmetric bilinear form*.

Similarly, an *inner product* on a complex vector space V is a function $(x, y) \to \langle x, y \rangle$ from $V \times V$ to the complex numbers $\mathbf{C}$ which satisfies the following:

(i$'$) (sesquilinearity)

$$\langle \alpha_1 x_1 + \alpha_2 x_2, y \rangle = \alpha_1 \langle x_1, y \rangle + \alpha_2 \langle x_2, y \rangle,$$
$$\langle x, \beta_1 y_1 + \beta_2 y_2 \rangle = \overline{\beta_1} \langle x, y_1 \rangle + \overline{\beta_2} \langle x, y_2 \rangle,$$

for all x, x_1, x_2, y, y_1, y_2 in V and all complex $\alpha_1, \alpha_2, \beta_1, \beta_2$ (note that complex conjugation is applied to the second term; theoretical physicists do it the other way round);

(ii$'$) (the Hermitian condition)

$$\langle y, x \rangle = \overline{\langle x, y \rangle} \ \text{ for all } x, y \text{ in } V;$$

(iii$'$) (positive definiteness)

$$\langle x, x \rangle > 0 \text{ for all non-zero } x \text{ in } V.$$

A function which satisfies (i$'$) and (ii$'$) is called a *Hermitian bilinear form*. Note that it follows from (ii$'$) that

$$\langle x, y \rangle + \langle y, x \rangle = 2\Re \langle x, y \rangle \ \text{ and } \ \langle x, y \rangle - \langle y, x \rangle = 2i\Im \langle x, y \rangle.$$

For example, if $V = \mathbf{C}^d$, we define the *usual* inner product, by setting $\langle z, w \rangle = \sum_{i=1}^d z_i \overline{w_i}$ for $z = (z_i), w = (w_i)$. As another example, the space $C[a, b]$ of continuous (real or) complex functions on the closed interval $[a, b]$ is an inner-product space when the inner product is defined by

$$\langle f, g \rangle = \int_a^b f(x)\overline{g(x)} \, dx.$$

A (real or) complex vector space V equipped with an inner product is called an *inner-product space*. If x is a vector in V, we set $\|x\| = \langle x, x \rangle^{\frac{1}{2}}$. We shall show that $\|.\|$ is a norm on V. Certainly $\|x\| = 0$ if and only if $x = 0$, and $\|\lambda x\| = |\lambda| \, \|x\|$.

In what follows, we shall consider the complex case: the real case is easier, since we do not need to consider complex conjugation.

Proposition 11.3.1 (The Cauchy–Schwarz inequality) *If x and y are vectors in an inner-product space V then*

$$| \langle x, y \rangle | \leq \|x\| \cdot \|y\|,$$

with equality if and only if x and y are linearly dependent.

Proof This depends upon the quadratic nature of the inner product. The inequality is trivially true if $\langle x, y \rangle = 0$. If $\|x\| = 0$, then $x = 0$ and $\langle x, y \rangle = 0$, so that the inequality is true, and the same holds if $\|y\| = 0$.

Otherwise, if $\lambda \in \mathbf{C}$ then

$$0 \leq \|x + \lambda y\|^2 = \langle x + \lambda y, x + \lambda y \rangle$$
$$= \langle x, x \rangle + \overline{\lambda} \langle x, y \rangle + \lambda \langle y, x \rangle + |\lambda|^2 \langle y, y \rangle.$$

Put

$$\lambda = -\frac{\langle x, y \rangle}{| \langle x, y \rangle |} \cdot \frac{\|x\|}{\|y\|}.$$

It follows that

$$0 \leq \|x\|^2 - 2\frac{| \langle x, y \rangle |^2}{| \langle x, y \rangle |} \cdot \frac{\|x\|}{\|y\|} + \frac{\|x\|^2}{\|y\|^2} \|y\|^2 = 2\left(\|x\|^2 - | \langle x, y \rangle | \cdot \frac{\|x\|}{\|y\|} \right),$$

so that $| \langle x, y \rangle | \leq \|x\| \cdot \|y\|$.

If $x = 0$ or $y = 0$, then equality holds, and x and y are linearly dependent.

Otherwise, if equality holds, then $\|x + \lambda y\| = 0$, so that $x + \lambda y = 0$, and x and y are linearly dependent. Conversely, if x and y are linearly dependent, then $x = \alpha y$ for some scalar α, and so

$$| \langle x, y \rangle | = |\alpha| \, \|y\|^2 = \|x\| \cdot \|y\|.$$

$\square$

Note that we obtain Cauchy's inequality by applying this result to $\mathbf{R}^d$ or $\mathbf{C}^d$, with their usual inner products.

Corollary 11.3.2 $\|x + y\| \leq \|x\| + \|y\|$, *with equality if and only if either $y = 0$ or $x = \alpha y$, with $\alpha \geq 0$.*

Proof   We have

$$\|x + y\|^2 = \|x\|^2 + \langle x, y \rangle + \langle y, x \rangle + \|y\|^2$$
$$\leq \|x\|^2 + 2\|x\| \cdot \|y\| + \|y\|^2 = (\|x\| + \|y\|)^2.$$

Equality holds if and only if $\Re \langle x, y \rangle = \|x\| \cdot \|y\|$, which is equivalent to the condition stated.   □

Thus $\|.\|$ is a norm on V.

Note also that the inner product is determined by the norm: in the real case, we have the *polarization formulae*

$$\langle x, y \rangle = \tfrac{1}{2}(\|x + y\|^2 - \|x\|^2 - \|y\|^2) = \tfrac{1}{4}(\|x + y\|^2 - \|x - y\|^2),$$

and in the complex case we have the *polarization formula*

$$\langle x, y \rangle = \tfrac{1}{4}\left(\sum_{j=0}^{3} i^j \|x + i^j y\|^2 \right).$$

We also have the following.

Proposition 11.3.3 (The parallelogram law)   *If x and y are vectors in an inner-product space V, then*

$$\|x + y\|^2 + \|x - y\|^2 = 2\|x\|^2 + 2\|y\|^2.$$

Proof   For

$$\|x + y\|^2 + \|x - y\|^2 = (\langle x, x \rangle + \langle x, y \rangle + \langle y, x \rangle + \langle y, y \rangle) +$$
$$+ (\langle x, x \rangle - \langle x, y \rangle - \langle y, x \rangle + \langle y, y \rangle)$$
$$= 2\|x\|^2 + 2\|y\|^2.$$

□

Many of the geometric and metric properties of inner-product spaces can be expressed in terms of orthogonality. Vectors x and y in an inner-product space V are said to be *orthogonal* if $\langle x, y \rangle = 0$; if so, we write $x \perp y$. For real spaces, this property can be expressed metrically, in terms of the norm.

Proposition 11.3.4   *If x and y are vectors in a real inner-product space V then $x \perp y$ if and only if $\|x + y\|^2 = \|x\|^2 + \|y\|^2$.*

Proof   For $\|x + y\|^2 = \|x\|^2 + \|y\|^2 + 2\langle x, y \rangle$.   □

On the other hand, if x is a vector in a complex inner-product space, $\|x + ix\|^2 = 2\|x\|^2 = \|x\|^2 + \|ix\|^2$, while $\langle x, ix \rangle = -i\|x\|^2$.

If A is a subset of an inner-product space V, we set

$$A^\perp = \{x \in V : \langle a, x \rangle = 0 \text{ for all } a \in A\}.$$

We write $x^\perp$ for $\{x\}^\perp$. $A^\perp$ is the *annihilator* of A: it has the following properties.

Proposition 11.3.5 *Suppose that A and B are subsets of an inner-product space V.*

1. $A^\perp = \{x \in V : \langle x, a \rangle = 0 \text{ for all } a \in A\}$.
2. $A^\perp$ *is a linear subspace of V.*
3. *If $A \subseteq B$ then $B^\perp \subseteq A^\perp$.*
4. $A \subseteq A^{\perp\perp}$.
5. $A^\perp = A^{\perp\perp\perp}$.
6. $A \cap A^\perp = \{0\}$.

Proof   These all follow easily from the definitions. For example, to prove 5, $A^\perp \subseteq (A^\perp)^{\perp\perp}$, by (iv), while $(A^{\perp\perp})^\perp \subseteq A^\perp$, by (iii), since $A \subseteq A^{\perp\perp}$.   □

Suppose that x is a unit vector in V, and that $z \in V$. Let $\lambda = \langle z, x \rangle$ and let $y = z - \lambda x$. Then $\langle y, x \rangle = \langle z, x \rangle - \langle z, x \rangle \langle x, x \rangle = 0$. Thus $z = \lambda x + y$, where $\lambda x \in \mathrm{span}\,(x)$ and $y \in x^\perp$. If $z = \mu x + w$, with $w \in x^\perp$, then $\langle z, x \rangle = \mu$, so that $\mu = \lambda$ and $w = y$; the decomposition is unique.

Exercises

11.3.1 Let l_2 denote the set of real sequences $(a_n)_{n=1}^\infty$ for which $\sum_{n=1}^\infty |a_n|^2$ is finite. Show that l_2 is a vector space (with the algebraic operations defined pointwise), that if $a, b \in l_2$ then $\sum_{n=1}^\infty a_n b_n$ converges absolutely, and that the function $(a, b) \to \langle a, b \rangle = \sum_{n=1}^\infty a_n b_n$ is an inner product on l_2.

11.3.2 Establish corresponding results for complex sequences.

11.3.3 Let x, y and z be elements of a real inner-product space, such that $\|x - z\| = \|x - y\| + \|y - z\|$. Show that there exists $0 \leq \lambda \leq 1$ such that $y = (1 - \lambda)x + \lambda z$.

11.3.4 A *pre-inner-product space* is a vector space E with a symmetric bilinear (Hermitian sesquilinear) form $\langle ., . \rangle$ which is *positive semi-definite*: $\langle x, x \rangle \geq 0$ for all $x \in E$. Show that $N = \{x \in E : \langle x, x \rangle = 0\}$ is a linear subspace of E, and that if $q : E \to E/N$ is the quotient mapping then there exists an inner product $\langle .,. \rangle_N$ on E/N such that $\langle q(x), q(y) \rangle_N = \langle x, y \rangle$, for $x, y \in E$.

11.4 Euclidean and unitary spaces

We now restrict attention to finite-dimensional spaces; a finite-dimensional real inner-product space is called a *Euclidean space* and a finite-dimensional complex inner-product space is called a *unitary space*. Throughout this section, V will denote a Euclidean or unitary space of dimension n. The key idea is that of *Gram–Schmidt orthonormalization*.

Theorem 11.4.1 *Suppose that* $(x_1, \ldots, x_d)$ *is a basis for a Euclidean or unitary space* V. *Then there exists a basis* $(e_1, \ldots, e_d)$ *for* V *with the following properties.*

(i) $\|e_j\| = 1$ *for* $1 \le j \le d$ *and* $\langle e_j, e_i \rangle = 0$ *for* $1 \le i < j \le d$.
(ii) *If* $W_j = \mathrm{span}\,(x_1, \ldots, x_j)$, *then* $W_j = \mathrm{span}\,(e_1, \ldots, e_j)$, *for* $1 \le j \le d$.

Proof   The proof is by an iterative construction. We set $e_1 = x_1 / \|x_1\|$. Suppose that we have constructed $e_1, \ldots, e_{j-1}$, satisfying the conclusions of the theorem. Let $f_j = x_j - \sum_{i=1}^{j-1} \langle x_j, e_i \rangle e_i$. Since $x_j \notin W_{j-1}$, $f_j \ne 0$. Let $e_j = f_j / \|f_j\|$. Then $\|e_j\| = 1$, and

$$\mathrm{span}\,(e_1, \ldots e_j) = \mathrm{span}\,(W_{j-1}, e_j) = \mathrm{span}\,(W_{j-1}, x_j) = W_j.$$

Thus $(e_1, \ldots, e_n)$ is a basis for W_j. If $1 \le k < j$ then

$$\langle f_j, e_k \rangle = \langle x_j, e_k \rangle - \sum_{i=1}^{j-1} \langle x_j, e_i \rangle \langle e_i, e_k \rangle = \langle x_j, e_k \rangle - \langle x_j, e_k \rangle = 0,$$

so that $\langle e_j, e_k \rangle = 0$. Thus $(e_1, \ldots, e_j)$ is a basis for W_j. In particular, $(e_1, \ldots, e_d)$ is a basis for V with the required properties.   □

The construction made in the proof is known as Gram–Schmidt orthonormalization. A basis $(e_1, \ldots, e_d)$ which satisfies condition (i) of the theorem is called an *orthonormal basis*. More generally, if $(e_1, \ldots, e_k)$ is a sequence in an inner-product space V which satisfies condition (i) (with k replacing d), then $(e_1, \ldots, e_k)$ is called an *orthonormal sequence*. Note that if $(e_1, \ldots, e_k)$ is an orthonormal sequence and $\sum_{j=1}^{k} x_j e_j = 0$ then $x_i = \left\langle \sum_{j=1}^{k} x_j e_j, e_i \right\rangle = 0$ for $1 \le i \le k$; an orthonormal sequence of vectors is linearly independent.

If $(e_1, \ldots, e_d)$ is an orthonormal basis for V, and $x = \sum_{j=1}^d x_j e_j \in V$ then $\langle x, e_i \rangle = x_i$ for $1 \le i \le d$, so that

$$x = \sum_{j=1}^d \langle x, e_j \rangle \, e_j.$$

Thus if $x, y \in V$ then

$$\langle x, y \rangle = \sum_{j=1}^d \langle x, e_j \rangle \, \langle e_j, y \rangle = \sum_{j=1}^d \langle x, e_j \rangle \, \overline{\langle y, e_j \rangle}.$$

In particular,

$$\|x\|^2 = \sum_{j=1}^d | \langle x, e_j \rangle |^2.$$

Corollary 11.4.2 *If W is a k-dimensional linear subspace of a Euclidean or unitary space V then $V = W \oplus W^\perp$, and there exists an orthonormal basis $(e_1, \ldots, e_d)$ of V such that $(e_1, \ldots, e_k)$ is a basis for W and $(e_{k+1}, \ldots, e_d)$ is a basis for $W^\perp$.*

Proof   Let $(x_1, \ldots, x_k)$ be a basis for W. Extend it to a basis $(x_1, \ldots, x_d)$ for V, and apply Gram–Schmidt orthonormalization to obtain an orthonormal basis $(e_1, \ldots, e_d)$ for V. Then $(e_1, \ldots, e_k)$ is an orthonormal basis for W, and $\mathrm{span}\,(e_{k+1}, \ldots e_d) \subseteq W^\perp$. On the other hand, if $x = \sum_{j=1}^d \langle x, e_j \rangle e_j \in W^\perp$ then $\langle x, e_j \rangle = 0$ for $1 \le j \le k$, so that $x = \sum_{j=k+1}^d \langle x, e_j \rangle e_j \in \mathrm{span}\,(e_{k+1}, \ldots e_d)$. Thus $(e_{k+1}, \ldots, e_d)$ is an orthonormal basis for $W^\perp$. Since $W \cap W^\perp = \{0\}$, it follows that $V = W \oplus W^\perp$. $\square$

If $x \in V$ we can write x uniquely as $y + z$, with $y \in W$ and $z \in W^\perp$. Let us set $P_W(x) = y$. P_W is a linear mapping of V onto W, and $P_W^2 = P_W$. P_W is called the *orthogonal projection* of V onto W. Note that $P_{W^\perp} = I - P_W$.

Although it is easy, the next result is important. It shows that an orthogonal projection is a 'nearest point' mapping; since it is linear, it relates the linear structure to metric properties.

Proposition 11.4.3 *If W is a linear subspace of a Euclidean or unitary space V and $x \in V$ then $P_W(x)$ is the nearest point in W to x, and is the unique point in W with this property: $\|x - P_W(x)\| \le \|x - w\|$ for $w \in W$, and if $\|x - P_W(x)\| = \|x - w\|$ then $w = P_W(x)$.*

Proof   Let $(e_1, \ldots, e_d)$ be an orthonormal basis for V which satisfies the conditions of Corollary 11.4.2. If $x = \sum_{j=1}^d x_j e_j \in V$ and $w = \sum_{j=1}^k w_j e_j \in W$

then

$$\|x - w\|^2 = \sum_{j=1}^{k} |x_j - w_j|^2 + \sum_{j=k+1}^{d} |x_j|^2,$$

and this is minimized if and only if $w_j = x_j$ for $1 \le j \le k$, in which case $w = P_W(x)$. $\qquad\square$

We shall extend Corollary 11.4.2 and Proposition 11.4.3 to certain inner product spaces in Section 14.3.

Exercise

11.4.1 Suppose that $(x_1, \ldots, x_n)$ is a basis for a Euclidean or unitary space V, and that $(e_1, \ldots, e_n)$ and $(f_1, \ldots, f_n)$ satisfy the conclusions of Theorem 11.4.1. Show that there are scalars $\lambda_1, \ldots, \lambda_n$ of unit modulus such that $f_j = \lambda_j e_j$ for $1 \le j \le n$.

11.5 Isometries

A mapping f from a metric space (X, d) to a metric space (Y, ρ) is an *isometry* if it preserves distances: that is, if $\rho(f(x), f(x')) = d(x, x')$ for all $x, x' \in X$.

If T is a linear mapping from a normed space $(E, \|.\|_E)$ into a normed space $(F, \|.\|_F)$ which is an isometry, then T is called a *linear isometry*. The mapping T is an isometry if and only if $\|T(x)\|_F = \|x\|_E$ for all $x \in E$. The condition is necessary, since

$$\|T(x)\|_F = \|T(x) - T(0)\|_F = \|x - 0\|_E = \|x\|_E.$$

It is sufficient, since

$$\|T(x) - T(y)\|_F = \|T(x - y)\|_F = \|x - y\|_E.$$

An isometry preserves the metric geometry of (X, d). Let us give some examples.

Example 11.5.1 An isometry of $\bar{N}$ into $\mathbf{R}$.

Let $\bar{N} = N \cup \{\infty\}$ be given the metric ρ defined in Example 11.1.9 in Section 11.1. Let $f(n) = 1/n$ and let $f(\infty) = 0$. Then f is an isometry of $\bar{N}$ into $\mathbf{R}$, with its usual metric.

Example 11.5.2 The mapping $(x, y) \to x + iy$ is a linear isometry of $\mathbf{R}^2$ onto $\mathbf{C}$, when $\mathbf{C}$ is considered as a real vector space.

Example 11.5.3   The conjugation mapping.

The *conjugation mapping* $z \to \bar{z}$ is an isometry of $\mathbf{C}$ onto itself. It is a linear mapping when $\mathbf{C}$ is considered as a real vector space, but is not linear when $\mathbf{C}$ is considered as a complex vector space

Example 11.5.4   Rotations of $\mathbf{R}^2$.

Since

$$(x \cos t - y \sin t)^2 + (x \sin t + y \cos t)^2 = x^2 + y^2,$$

the linear mapping r_t from $\mathbf{R}^2 \to \mathbf{R}^2$ defined by

$$r_t(x, y) = (x \cos t - y \sin t, x \sin t + y \cos t)$$

is a linear isometry of $\mathbf{R}^2$ onto $\mathbf{R}^2$. It is a *rotation* of $\mathbf{R}^2$. It is a bijection, with inverse r_{-t}.

Example 11.5.5   Translations of a normed space.

Suppose that $(E, \|.\|)$ is a normed space. If $a \in E$, let $T_a(x) = x + a$; T_a is a *translation*. It is an isometry of $(E, \|.\|)$ onto itself, since

$$\|T_a(x) - T_a(y)\| = \|(x + a) - (y + a)\| = \|x - y\|.$$

Example 11.5.6   If $(E, \|.\|)$ is a normed space, and λ is a scalar with $|\lambda| = 1$ then the mapping $x \to \lambda x$ is a linear isometry of E onto itself.

Example 11.5.7   A linear isometry of $l_1^2(\mathbf{R})$ onto $l_\infty^2(\mathbf{R})$.

If $x, y \in \mathbf{R}$ then $\max(|x + y|, |x - y|) = |x| + |y|$. Thus the linear mapping $T : l_1^2(\mathbf{R}) \to l_\infty^2(\mathbf{R})$ defined by $T((x, y)) = (x + y, x - y)$ is an isometry of $l_1^2(\mathbf{R})$ onto $l_\infty^2(\mathbf{R})$.

Example 11.5.8   Reflections of a real inner-product space.

Suppose that x is a non-zero vector in a real inner-product space V. If $z \in V$, we can write z uniquely as $z = \lambda x + y$, where $\lambda \in \mathbf{R}$ and $y \in x^\perp$. Let $\rho_x(z) = -\lambda x + y$, so that $\rho_x(x) = -x$ and $\rho_x(z) = z$ if and only if $z \in x^\perp$. Then ρ_x is a linear mapping of V onto V, and is an *involution*: ρ_x^2 is the identity mapping. It is an isometry, since

$$\|\rho_x(z)\|^2 = \lambda^2 \|x\|^2 + \|y\|^2 = \|z\|^2.$$

It is called the *simple reflection in the direction x, with mirror $x^\perp$*.

Suppose that x and y are distinct vectors in a real inner-product space V, with $\|x\| = \|y\|$. Then

$$\langle x + y, x - y \rangle = \langle x, x \rangle - \langle x, y \rangle + \langle y, x \rangle - \langle y, y \rangle = \|x\|^2 - \|y\|^2 = 0,$$

so that $(x + y) \perp (x - y)$. Thus

$$\rho_{x-y}(x) = \rho_{x-y}(\tfrac{1}{2}(x - y)) + \rho_{x-y}(\tfrac{1}{2}(x + y)) = \tfrac{1}{2}(y - x) + \tfrac{1}{2}(x + y) = y,$$

and $\rho_{x-y}(y) = x$.

Example 11.5.9 Linear isometries between inner-product spaces.

Proposition 11.5.10 *Suppose that $S : V \to W$ is a linear mapping from a real inner-product space V to a real inner-product space W. Then S is an isometry if and only if $\langle S(x), S(y) \rangle = \langle x, y \rangle$ for $x, y \in V$.*

Proof   If S is an isometry, then

$$\langle S(x), S(y) \rangle = \tfrac{1}{2}(\|S(x)\|^2 + \|S(y)\|^2 - \|S(x) - S(y)\|^2)$$
$$= \tfrac{1}{2}(\|x\|^2 + \|y\|^2 - \|x - y\|^2) = \langle x, y \rangle.$$

The condition is sufficient, since $\|S(x)\| = \sqrt{\langle x, x \rangle}$,   □

Thus if $(e_1, \dots, e_k)$ is an orthonormal sequence in V and S is an isometry then $(S(e_1), \dots, S(e_k))$ is an orthonormal sequence in W.

Corollary 11.5.11 *If $(e_1, \dots, e_d)$ is an orthonormal basis for a Euclidean space V and T is a linear mapping from V into an inner-product space W then T is an isometry if and only if $(T(e_1), \dots, T(e_d))$ is an orthonormal sequence in W.*

Proof   As we have just observed, the condition is necessary. If it is satisfied and $x = \sum_{j=1}^{d} x_j e_j \in V$ then

$$\|T(x)\|^2 = \left\| \sum_{j=1}^{d} x_j T(e_j) \right\|^2 = \sum_{j=1}^{d} x_j^2 = \|x\|^2 .$$

□

If $(e_1, \dots, e_k)$ is an orthonormal sequence in a real (or complex) inner-product space V, then the mapping T from $\mathbf{R}^k$ (or $\mathbf{C}^k$) into V defined by

$T(x) = T((x_1, \ldots, x_k)) = \sum_{j=1}^k x_j e_j$ is an isometry, since

$$\|T(x)\|^2 = \sum_{j=1}^k |x_j|^2 = \|x\|.$$

Any two Euclidean spaces of the same dimension are linearly isometric: if V and W are Euclidean spaces of dimension k, then there exists a linear mapping of V onto W which is an isometry. Let $(e_1, \ldots, e_k)$ be an orthonormal basis for V, and let $(f_1, \ldots, f_k)$ be an orthonormal basis for W. Let

$$J(x_1 e_1 + \cdots + x_k e_k) = x_1 f_1 + \cdots + x_k f_k.$$

Then J is a linear isometry of V onto W.

Example 11.5.12 An isometry of a metric space (X, d) into $l_\infty(X)$.

This example will be useful to us later. Let (X, d) be a metric space, with X non-empty, and let $l_\infty(X) = l_\infty(X, \mathbf{R})$ be the normed space of bounded real-valued functions on X introduced in Section 11.2. Let x_0 be an element of X. If $x \in X$, let

$$f_x(y) = d(x, y) - d(x_0, y) \text{ for } y \in X.$$

Since $d(x_0, y) \leq d(x_0, x) + d(x, y)$ and $d(x, y) \leq d(x_0, x) + d(x_0, y)$, by the triangle inequality, it follows that $|f_x(y)| \leq d(x_0, x)$, so that $f_x \in l_\infty(X)$, and $\|f_x\|_\infty \leq d(x_0, x)$. We claim that the mapping $x \to f_x : X \to l_\infty(X)$ is an isometry. Since

$$f_x(y) - f_{x'}(y) = d(x, y) - d(x', y) \leq d(x, x'),$$
$$\text{and } f_{x'}(y) - f_x(y) = d(x', y) - d(x, y) \leq d(x, x')$$

it follows that $|f_x(y) - f_{x'}(y)| \leq d(x, x')$ for all $y \in X$. Hence $\|f_x - f_{x'}\|_\infty \leq d(x, x')$. On the other hand,

$$f_x(x') - f_{x'}(x') = d(x, x') - d(x', x') = d(x, x'),$$

and so $\|f_x - f_{x'}\|_\infty = d(x, x')$.

Two metric spaces (X, d) and (Y, ρ) are said to be *congruent* if there is an isometry of X onto Y. (They are said to be *similar* if there exists $\alpha > 0$ and a mapping f of X onto Y such that $\rho(f(x), f(x')) = \alpha d(x, x')$.)

The composition of two isometries is an isometry, and the inverse of a bijective isometry is an isometry. Thus the set of bijective isometries of a metric space (X, d) onto itself forms a group under composition, the group of *metric symmetries* of (X, d). This group gives valuable information about the metric space.

Exercises

Let l_1^d denote $\mathbf{R}^d$ with norm $\|x\|_1 = \sum_{j=1}^d |x_j|$.

11.5.1 If $A \in P_d$, let I_A be its indicator function: $I_A(j) = 1$ if $j \in A$ and $I_A(j) = 0$ otherwise. Show that the mapping $A \to I_A$ is an isometry of P_d, with its Hamming metric, into l_1^d.

11.5.2 Let $e_j = (0, \ldots 0, 1, 0, \ldots 0)$, with 1 in the jth place.

    (a) Show that if x is a unit vector in l_1^d and if $\max(\|x + y\|_1, \|x - y\|_1) > 1$ for all $y \neq 0$ then $x = \pm e_j$ for some $1 \leq j \leq d$.

    (b) Let f be an isometry of l_1^d with $f(0) = 0$. Show that there exists a permutation σ of $\{1, \ldots, d\}$ and a choice of signs $(\epsilon_1, \ldots, \epsilon_d)$ (that is, $\epsilon_j = \pm 1$ for $1 \leq j \leq d$) such that $f(e_j) = \epsilon_j e_{\sigma(j)}$, $f(-e_j) = -\epsilon_j e_{\sigma(j)}$ for $1 \leq j \leq d$.

    (c) Show that f is linear, so that $f(x) = \sum_{j=1}^d \epsilon_j x_j e_{\sigma(j)}$ for $x \in l_1^d$.

11.5.3 By considering vectors of the form $(\epsilon_1, \ldots, \epsilon_d)$, where $\epsilon_j = \pm 1$ for $1 \leq j \leq d$, show that if a mapping f of l_∞^d into itself is an isometry and if $f(0) = 0$ then there exists a permutation σ of $\{1, \ldots, d\}$ and a choice of signs $(\epsilon_1, \ldots, \epsilon_d)$ (that is, $\epsilon_j = \pm 1$ for $1 \leq j \leq d$) such that $f(e_j) = \epsilon_j e_{\sigma(j)}$, $f(-e_j) = -\epsilon_j e_{\sigma(j)}$ for $1 \leq j \leq d$. Show that f is linear.

11.6 *The Mazur–Ulam theorem*

(This section can be omitted on a first reading.)

Suppose that $(E, \|.\|_E)$ and $(F, \|.\|_F)$ are real normed spaces and that $J : E \to F$ is an isometry. Let $L = T_{-J(0)} \circ J$, where $T_{-J(0)}$ is the translation of F mapping $J(0)$ to 0. Thus $L(x) = J(x) - J(0)$, so that L is an isometry of E into F, with $L(0) = 0$. Our principal aim is to show that if L is surjective, then it must be linear. This extends the results of Exercises 11.5.2 and 11.5.3 of the previous section.

Theorem 11.6.1 (The Mazur–Ulam theorem)   *If $L : E \to F$ is an isometry of a real normed space $(E, \|.\|_E)$ onto a real normed space $(F, \|.\|_F)$ with $L(0) = 0$, then L is a linear mapping.*

In order to prove this, we introduce some ideas concerning the geometry of metric spaces, of interest in their own right. First, suppose that x, y, z are elements of a metric space. We say that y is *between* x and z if $d(x, y) + d(y, z) = d(x, z)$, and we say that y is *halfway between* x and z if $d(x, y) =$

$d(y, z) = \frac{1}{2}d(x, z)$. We denote the set of points halfway between x and z by $H(x, z)$. The set $H(x, z)$ may be

- empty, for example if (X, d) has the discrete metric,
- a singleton set, for example if (X, d) is the real line $\mathbf{R}$, with the usual metric, when $H(x, z) = \{\frac{1}{2}(x + z)\}$,
- or may contain more than one point: let $(X, d) = l^2_\infty(\mathbf{R})$, $x = (-1, 0)$, and $z = (1, 0)$; then $H(x, z) = \{(0, y) : -1 \le y \le 1\}$.

If $(E, \|.\|_E)$ is a normed space, then $\frac{1}{2}(x + z) \in H(x, z)$, but $H(x, z)$ may contain other points, as the last example shows. The set $H(x, z)$ is always bounded, since if $y, y' \in H(x, z)$ then $d(y, y') \le d(y, x) + d(x, y') = d(x, z)$.

Suppose that A is a bounded subset of a metric space (X, d). Can we find a special point in A which is the centre of A, in some metric sense? In general, the answer must be 'no', since, for example, in a metric space with the discrete metric, there is no obvious special point. In certain cases, however, the answer is 'yes'. First, let

$$\kappa(A) = \{x \in A : d(x, y) \le \tfrac{1}{2}\mathrm{diam}\,(A) \text{ for all } y \in A\};$$

$\kappa(A)$ is the *central core* of A. Again, $\kappa(A)$ may be empty, may consist of one point (which would then be the centre of A) or may consist of more than one point; for example, if

$$A = \{(x, y) \in l^2_\infty(\mathbf{R}) : -1 \le x \le 1, -\tfrac{1}{2} \le y \le \tfrac{1}{2}\},$$

then $\kappa(A) = \{(0, y) : -\frac{1}{2} \le y \le \frac{1}{2}\}$. Note though that $\mathrm{diam}\,(\kappa(A)) \le \frac{1}{2}\mathrm{diam}\,A$. This suggests that we iterate the procedure: we set $\kappa_1(A) = \kappa(A)$, and if $\kappa_n(A) \ne \emptyset$ we set $\kappa_{n+1}(A) = \kappa(\kappa_n(A))$. There are then three possible outcomes:

- $\kappa_n(A) = \emptyset$ for some $n \in \mathbf{N}$;
- $\kappa_n(A) \ne \emptyset$ for all $n \in \mathbf{N}$, but $\cap_{n=1}^\infty (\kappa_n(A)) = \emptyset$;
- $\kappa_n(A) \ne \emptyset$ for all $n \in \mathbf{N}$, and $\cap_{n=1}^\infty (\kappa_n(A)) \ne \emptyset$.

If either of the first two cases occurs, then A does not have a centre. In the third case, $\mathrm{diam}\,\cap_{n=1}^\infty (\kappa_n(A)) \le \mathrm{diam}\,\kappa_n(A) \le \mathrm{diam}\,(A)/2^n$, for all $n \in \mathbf{N}$, so that $\mathrm{diam}\,\cap_{n=1}^\infty (\kappa_n(A)) = 0$, and $\cap_{n=1}^\infty \kappa_n(A) = \{c(A)\}$, a singleton set. Then we call $c(A)$ the *centre* of A.

Let us give an example. A subset A of a real vector space E is *symmetric* if $A = -A$: that is, if $x \in A$ then $-x \in A$.

Proposition 11.6.2 *If A is a bounded symmetric subset of a normed space $(E, \|.\|)$ and if $0 \in A$ then 0 is the centre of A.*

Proof   Let us consider $\kappa(A)$. We show that $\kappa(A)$ is symmetric and that $0 \in \kappa(A)$. First, if $y \in \kappa(A)$ and $x \in A$ then

$$d(-y, x) = \|-y - x\| = \|y + x\| = \|y - (-x)\| = d(y, -x) \le \tfrac{1}{2}\mathrm{diam}\,(A),$$

so that $-y \in \kappa(A)$ and $\kappa(A)$ is symmetric. Secondly, if $x \in A$ then

$$d(x, 0) = \|x\| = \tfrac{1}{2}\|2x\| = \tfrac{1}{2}\|x - (-x)\| = \tfrac{1}{2}d(x, -x) \le \tfrac{1}{2}\mathrm{diam}\,(A),$$

so that $0 \in \kappa(A)$. We can therefore iterate the procedure: $\kappa_n(A)$ is symmetric, and $0 \in \kappa_n(A)$, and so it follows that 0 is the centre of A. $\square$

Corollary 11.6.3   *If $x, z \in E$ then $\tfrac{1}{2}(x + z)$ is the centre of $H(x, z)$.*

Proof   First consider the case where $z = -x$. Then $0 \in H(x, -x)$, and if $y \in H(x, -x)$ then

$$\|(-y) - x\| = \|y - (-x)\| = \tfrac{1}{2}\|x - (-x)\| = \|x\|.$$

Similarly, $\|(-y) - (-x)\| = \tfrac{1}{2}\|x - (-x)\|$. Thus $H(x, -x)$ is symmetric, and 0 is its centre. In the general case, let $y = \tfrac{1}{2}(x + z)$. Since translation is an isometry,

$$H(x, z) = T_y(H(x - y, z - y)) = T_y(H(x - y, -(x - y))),$$

so that $y = T_y(0)$ is the centre of $H(x, z)$. $\square$

The importance of this is that the centre is defined purely in terms of the metric, and not in terms of the vector space structure of E.

Proof   of Theorem 11.6.1. Note that if A is a bounded subset of E, with centre $c(A)$, then, since L is a surjective isometry, $L(A)$ is a bounded subset of F, with centre $c(L(A)) = L(c(A))$. If $x, z \in E$, then $L(\tfrac{1}{2}(x + z)) = \tfrac{1}{2}(L(x) + L(z))$, since $\tfrac{1}{2}(x + z)$ is the centre of $H(x, z)$ and $\tfrac{1}{2}(L(x) + L(z))$ is the centre of $H(L(x), L(z))$. In particular, considering $2x$ and 0, $L(x) = \tfrac{1}{2}L(2x)$, and considering $2x$ and $2z$, $L(x + z) = \tfrac{1}{2}(L(2x) + L(2z)) = L(x) + L(z)$. Thus L is additive.

From this, we deduce the fact that $L(\lambda x) = \lambda L(x)$, for $\lambda \in \mathbf{R}$ and $x \in E$, in a few easy stages. First, an easy induction argument shows that $L((n+1)x) = L(nx) + L(x) = nL(x) + L(x) = (n+1)L(x)$, for $n \in \mathbf{N}$; thus the result holds for $\lambda \in \mathbf{Z}^+$. Secondly, $L(nx) + L(-nx) = L(0) = 0$, so that the result holds for $\lambda \in \mathbf{Z}$. Thirdly, if $m \in \mathbf{Z}$ and $n \in \mathbf{N}$ then $L((m/2^{n-1})x) = 2L((m/2^n)x)$, so that $L((m/2^n)x) = \tfrac{1}{2}L((m/2^{n-1})x)$, and another induction argument shows that $L((m/2^n)x) = (m/2^n)L(x)$. Thus the result holds for all dyadic rationals (numbers of the form $m/2^n$, with $m \in \mathbf{Z}$ and $n \in \mathbf{N}$).

Finally, suppose that $\lambda \in \mathbf{R}$ and that $\epsilon > 0$. Then there exists a dyadic rational $r = m/2^n$ such that $|\lambda - r| < \epsilon/2(\|x\|_E + 1)$. Thus $\|\lambda x - rx\|_E = |\lambda - r|\,\|x\|_E < \epsilon/2$. Since L is an isometry, $\|L(\lambda x) - L(rx)\|_F < \epsilon/2$, and also

$$\|rL(x) - \lambda L(x)\|_F = |\lambda - r|\,\|L(x)\|_F = |\lambda - r|\,\|x\|_E < \epsilon/2.$$

Since $L(rx) = rL(x)$,

$$\|L(\lambda x) - \lambda L(x)\|_F$$
$$\leq \|L(\lambda x) - L(rx)\|_F + \|L(rx) - rL(x)\|_F + \|rL(x) - \lambda L(x)\|_F < \epsilon.$$

Since this holds for all $\epsilon > 0$, $\|L(\lambda x) - \lambda L(x)\|_F = 0$, and so $L(\lambda x) = \lambda L(x)$. $\qquad\qquad\square$

The condition that L is surjective cannot be dropped. It follows from the mean-value theorem that if $x < y$ then there exists $x < z < y$ such that $\sin x - \sin y = (x - y)\cos z$, so that $|\sin x - \sin y| \leq y - x$. Thus the mapping $L : \mathbf{R} \to l_\infty^2(\mathbf{R})$ defined by $L(t) = (t, \sin t)$ is an isometry of $\mathbf{R}$ into $l_\infty^2(\mathbf{R})$ with $L(0) = (0, 0)$ which is clearly not linear.

There is however one important circumstance in which the surjective condition can be dropped. A normed space $(E, \|.\|_E)$ is *strictly convex* if whenever $x, y \in E$, $\|x\|_E = \|z\|_E = 1$ and $x \neq y$ then $\|\frac{1}{2}(x + y)\| < 1$.

Proposition 11.6.4   *If $(E, \|.\|_E)$ is strictly convex and $x, z \in E$ then the set $H(x, z)$ of points halfway between x and z is the singleton set $\{\frac{1}{2}(x + z)\}$.*

Proof   First consider the case where $z = -x$, and $\|x\| = 1$. If $u \in H(x, -x)$ then $\|x + u\| = \|x - u\| = \|x\| = 1$. Since

$$x = \tfrac{1}{2}((x + u) + (x - u)),$$

it follows from strict convexity that $u = 0$, so that $H(x, -x) = \{0\}$. Then, by scaling, $H(x, -x) = \{0\}$, for all $x \in E$. Finally, $H(x, z) = \{\frac{1}{2}(x + z)\}$, by translation. $\qquad\qquad\square$

Corollary 11.6.5   *If $L : E \to F$ is an isometry of a real normed space $(E, \|.\|_E)$ into a strictly convex real normed space $(F, \|.\|_F)$ with $L(0) = 0$, then L is a linear mapping, and $(E, \|.\|_E)$ is also strictly convex.*

Proof   If $x, z \in E$, then $\frac{1}{2}(x + z)$ is the centre of $H(x, z)$, and so $\frac{1}{2}(L(x) + L(z))$ must be the centre of $H(L(x), L(z))$ in $L(E)$. But the only possible centre of $H(L(x), L(z))$ in $L(E)$ is $\frac{1}{2}(L(x) + L(z))$. Thus $L(\frac{1}{2}(x + z)) = \frac{1}{2}(L(x) + L(z))$ and $\frac{1}{2}(L(x) + L(z)) \in L(E)$. An argument exactly like the one given in Theorem 11.6.1 then shows that $L(x + z) = L(x) + L(z)$, and that $L(\lambda x) = \lambda L(x)$, for all $x \in E$ and $\lambda \in \mathbf{R}$.

Thus L is an isometric linear mapping of E onto a linear subspace of F. Since a linear subspace of a strictly convex normed space is strictly convex, it follows that $(E, \|.\|_E)$ is strictly convex. $\qquad\square$

Many important normed spaces are strictly convex, and indeed have stronger metric convexity properties.

Proposition 11.6.6   *An inner-product space E is strictly convex.*

Proof   This follows easily from the parallelogram law. If $\|x\| = \|y\| = 1$ and $x \neq y$ then it follows from the parallelogram law that

$$ 4 = 2(\|x\|^2 + \|y\|^2) = \|x + y\|^2 + \|x - y\|^2, $$

so that $\left\|\tfrac{1}{2}(x + y)\right\|^2 = 1 - \tfrac{1}{4}\|x - y\|^2 < 1$, and $\left\|\tfrac{1}{2}(x + y)\right\| < 1$. $\qquad\square$

Corollary 11.6.7   *If $L : E \to F$ is an isometry of a real normed space $(E, \|.\|_E)$ into a real inner-product space $(F, \|.\|_F)$ with $L(0) = 0$, then L is a linear mapping, and $(E, \|.\|_E)$ is also an inner-product space.*

Exercise

11.6.1   Let $J : \mathbf{R} \to l_1^2(\mathbf{R})$ be defined as $J(t) = \tfrac{1}{2}(t - 1/(t^2 + 1), t + 1/(t^2 + 1))$. Show that J is an isometry. Why does this not contradict the Mazur–Ulam theorem?

11.7 The orthogonal group O_d

We now consider the group O_d of linear isometries of $\mathbf{R}^d$, with its Euclidean metric; this is the *orthogonal group*, and its elements are called *orthogonal mappings*. As an example, a simple reflection ρ_x in the direction x is an orthogonal mapping. It follows from the polarization formula that a linear mapping T is orthogonal if and only if $\langle T(x), T(y) \rangle = \langle x, y \rangle$, for all $x, y \in \mathbf{R}^d$. Let $(e_1, \ldots, e_d)$ be the standard basis of $\mathbf{R}^d$. It then follows that a linear mapping S from $\mathbf{R}^d$ to itself is orthogonal if and only if $(S(e_1), \ldots, S(e_d))$ is also an orthogonal basis for $\mathbf{R}^d$.

This can be expressed in terms of the matrix representing S. If S is represented by the matrix (s_{ij}) in the usual way (so that $S(e_j) = \sum_{i=1}^d s_{ij} e_i$ for $1 \leq j \leq d$), then S is orthogonal if and only if

$$ \sum_{i=1}^d s_{ij}^2 = 1 \text{ for } 1 \leq j \leq d, \text{ and } \sum_{i=1}^d s_{ij} s_{ik} = 0 \text{ for } 1 \leq j < k \leq d. $$

Such a matrix is called an *orthogonal matrix*.

Theorem 11.7.1 *Suppose that $T \in O_d$. Then T can be written as the product of at most d simple reflections.*

Proof Let $S = T - I$. We prove the result by induction on the rank $r(S)$ of S. We show that if $r(S) = r$ then T is the product of at most r simple reflections. If $r(S) = 0$ then $T = I$, which is the product of no simple reflections. Suppose that the result holds if $r(S) \leq r$, where $r < d$. Suppose that $T \in O_d$ and that $r(S) = r + 1$. Let N be the null-space of S: $N = \{x \in \mathbf{R}^d : T(x) = x\}$. By the rank-nullity formula, $\dim(N) = d - r - 1$. Let x be a unit vector in $N^\perp$, so that $S(x) \neq 0$. We consider the simple reflection $\rho_{S(x)}$.

If $y \in N$ then

$$\langle y, S(x) \rangle = \langle y, T(x) \rangle - \langle y, x \rangle = \langle T(y), T(x) \rangle - \langle y, x \rangle = 0,$$

so that $\rho_{S(x)}(y) = y$. Also

$$\langle S(x), T(x) + x \rangle = \langle T(x) - x, T(x) + x \rangle$$
$$= \langle T(x), T(x) \rangle - \langle x, T(x) \rangle + \langle T(x), x \rangle - \langle x, x \rangle = 0,$$

so that $T(x) + x \in (S(x))^\perp$. Hence $\rho_{S(x)}(T(x) + x) = T(x) + x$. But $\rho_{S(x)}(T(x) - x) = -T(x) + x$, and so $\rho_{S(x)}(T(x)) = x$. Let $U = \rho_{S(x)} \circ T$. Then $U \in O_d$, $U(x) = x$ and $U(y) = y$ for $y \in N$. Let $M = \mathrm{span}\,(N, x)$, so that $\dim(M) = d - r$. Then $(U - I)(z) = 0$ for $z \in M$, and so $r(U - I) \leq r$. By the inductive hypothesis, U is the product of at most r simple reflections. Since $T = \rho_{S(x)} \circ U$, T is the product of at most $r + 1$ simple reflections. □

Here are some more easy examples. Suppose that σ is a permutation of $\{1, \ldots, n\}$. If $x \in \mathbf{R}^d$ let $T_\sigma(x) = (x_{\sigma(1)}, \ldots, x_{\sigma(d)})$. Then $T_\sigma \in O_d$; it is a permutation operator. Note that $T_\sigma^{-1} = T_{\sigma^{-1}}$. Suppose that $0 \leq t \leq 2\pi$. If $x \in \mathbf{R}^d$, let

$$R_t(x) = (x_1 \cos t - x_2 \sin t, x_1 \sin t + x_2 \cos t, x_3, \ldots, x_d).$$

Then $R_t \in O_d$; it is an *elementary rotation*. Note that $R_t^{-1} = R_{2\pi - t}$.

Theorem 11.7.2 *For $d \geq 2$, let G_d be the subgroup of O_d generated by the permutation operators and the elementary rotations. Then $G_d = O_d$.*

Proof We leave this as an exercise for the reader. □

Exercises

11.7.1 Let $\tau_{i,j} \in \Sigma_n$ be the permutation of $\{1, \ldots, n\}$ which transposes i and j: $\tau_{i,j}(i) = j$, $\tau_{i,j}(j) = i$ and $\tau_{i,j}(k) = k$ otherwise. Show that Σ_n is generated by the transpositions $\{\tau_{1,j} : 2 \le j \le n\}$.

11.7.2 Suppose that $T \in O_d$, and that $r(T - I) = r$. Show that T cannot be written as the product of fewer than r simple reflections.

11.7.3 Interpret the equation $\langle T(x) + x, T(x) - x \rangle = 0$ that occurs in Theorem 11.7.1 geometrically.

11.7.4 Prove Theorem 11.7.2.

12

Convergence, continuity and topology

12.1 Convergence of sequences in a metric space

We now turn to analysis on metric spaces. The definitions and results that we shall consider are straightforward generalizations of the corresponding definitions and results for the real line. The same is true of the proofs; in most cases, they will be completely straightforward modifications of proofs of results in Volume I. We shall however present the material in a slightly different order.

Suppose that (X, d) is a metric space, that $(a_n)_{n=1}^{\infty}$ is a sequence of elements of X, and that $l \in X$. We say that a_n *converges to* l, or *tends to* l, as n *tends to infinity*, and write $a_n \to l$ as $n \to \infty$, if whenever $\epsilon > 0$ there exists n_0 (which usually depends on ϵ) such that $d(a_n, l) < \epsilon$ for $n \geq n_0$. In other words, the real-valued sequence $(d(a_n, l))_{n=1}^{\infty}$ tends to 0 as $n \to \infty$.

Suppose that $x \in X$ and $\epsilon > 0$. The *open ϵ-neighbourhood* $N_\epsilon(x)$ is defined to be the set of all elements of X distant less than ϵ from x:

$$N_\epsilon(x) = \{y \in X : d(y, x) < \epsilon\}.$$

We can express convergence in terms of open ϵ-neighbourhoods: $a_n \to l$ as $n \to \infty$ if and only if for each $\epsilon > 0$ there exists n_0 such that $a_n \in N_\epsilon(l)$ for $n \geq n_0$.

We have the following consequence of Lemma 11.1.14.

Proposition 12.1.1 *If $a_n \to l$ and $b_n \to m$ as $n \to \infty$ then*

$$d(a_n, b_n) \to d(l, m) \ \text{as } n \to \infty.$$

Proof   For $|d(a_n, b_n) - d(l, m)| \leq d(a_n, l) + d(b_n, m)$, and

$$d(a_n, l) + d(b_n, m) \to 0 \text{ as } n \to \infty.$$

$\square$

When they exist, limits are unique.

Corollary 12.1.2   *If $a_n \to l$ as $n \to \infty$ and $a_n \to m$ as $n \to \infty$, then $l = m$.*

Proof   Put $b_n = a_n$. Then $d(a_n, b_n) = 0$, so that $d(l, m) = 0$ and $l = m$.

$\square$

A subsequence of a convergent sequence converges to the same limit.

Proposition 12.1.3   *If $a_n \to l$ as $n \to \infty$ and if $(a_{n_k})_{k=0}^{\infty}$ is a subsequence, then $a_{n_k} \to l$ as $k \to \infty$.*

Proof   Given $\epsilon > 0$ there exists N such that $d(a_n, l) < \epsilon$ for $n \geq N$, and there exists k_0 such that $n_k > N$ for $k \geq k_0$. Thus if $k \geq k_0$ then $d(a_{n_k}, l) < \epsilon$.

$\square$

Let us give two examples. First, suppose that

$$(x^{(n)})_{n=0}^{\infty} = ((x_1^{(n)}, \dots, x_d^{(n)}))_{n=0}^{\infty}$$

is a sequence in $\mathbf{R}^d$, and that $x = (x_1, \dots, x_d) \in \mathbf{R}^d$. We consider the Euclidean norm $\|.\|_2$ and Euclidean metric on $\mathbf{R}^d$. If $1 \leq j \leq d$ then $|x_j^{(n)} - x_j| \leq \|x^{(n)} - x\|_2$, so that if $x^{(n)} \to x$ as $n \to \infty$ then $x_j^{(n)} \to x_j$ as $n \to \infty$. Conversely, suppose that $x_j^{(n)} \to x_j$ as $n \to \infty$ for $1 \leq j \leq d$. Given $\epsilon > 0$ and $1 \leq j \leq d$ there exists $n_j \in \mathbf{N}$ such that $|x_j^{(n)} - x_j| < \epsilon/\sqrt{d}$ for $n \geq n_j$. Let $N = \max\{n_j : 1 \leq j \leq d\}$. If $n \geq N$ then

$$\left\| x^{(n)} - x \right\|_2^2 = \sum_{j=1}^{d} |x_j^{(n)} - x_j|^2 \leq \epsilon^2,$$

so that $x^{(n)} \to x$ as $n \to \infty$. Thus a sequence in $\mathbf{R}^d$ converges in the Euclidean metric if and only if each sequence of coordinates converges: convergence in the Euclidean metric is the same as *coordinate-wise convergence*.

The second example is extremely important. Suppose that S is a set, that (X, d) is a metric space, that $(f_n)_{n=0}^{\infty}$ is a sequence in the space $B_X(S)$

of bounded functions on S taking values in X, and that $f \in B_X(S)$. We consider the uniform metric d_∞ on $B_X(S)$. If $s \in S$, then

$$d(f_n(s), f(s)) \leq \sup\{d(f_n(t), f(t)) : t \in S\} = d_\infty(f_n, f),$$

so that if $f_n \to f$ in the uniform metric as $n \to \infty$ then $f_n(x) \to f(x)$ as $n \to \infty$: $f_n \to f$ pointwise. But the convergence is stronger than that: given $\epsilon > 0$ there exists n_0 such that

$$d_\infty(f_n, f) = \sup\{|f_n(s) - f(s)| : s \in S\} < \epsilon \text{ for } n \geq n_0.$$

Thus there exists an n_0 *independent of* s such that $|f_n(s) - f(s)| < \epsilon$ for all $n \geq n_0$ and all $s \in S$. We say that $f_n \to f$ *uniformly* on X as $n \to \infty$. The distinction between uniform convergence and pointwise convergence is most important. It is reassuring that the uniform convergence of bounded functions can be characterized in terms of a metric (and in terms of a norm, when X is a normed space).

It is however useful to have a slightly more general definition. Suppose that $(f_n)_{n=1}^\infty$ is a sequence of functions on S, taking values in (X, d), and that f is a function on S with values in X. Then we say that f_n converges *uniformly* on S to f as $n \to \infty$ if $\sup_{t \in S} d(f_n(t), f(t)) \to 0$ as $n \to \infty$; in other words, we do not restrict attention to functions bounded on S.

Let us give some easy but important results about uniform convergence. We shall generalize them later.

Theorem 12.1.4 *If $(f_n)_{n=1}^\infty$ is a sequence of continuous real-valued functions on a subset A of $\mathbf{R}$, and if f_n converges uniformly on A to f as $n \to \infty$, then f is continuous.*

Proof Suppose that $t_0 \in A$ and that $\epsilon > 0$. There exists $N \in \mathbf{N}$ such that $|f_n(t) - f(t)| < \epsilon/3$ for all $t \in A$ and $n \geq N$. Since f_N is continuous at t_0, there exists $\delta > 0$ such that if $t \in A$ and $|t - t_0| < \delta$ then $|f_N(t) - f_N(t_0)| < \epsilon/3$. If $t \in A$ and $|t - t_0| < \delta$ then

$$|f(t) - f(t_0)| \leq |f(t) - f_N(t)| + |f_N(t) - f_N(t_0)| + |f_N(t_0) - f(t_0)| < \epsilon,$$

so that f is continuous at t_0. $\square$

Theorem 12.1.5 *Suppose that $(f_n)_{n=1}^\infty$ is a sequence of Riemann integrable functions on a bounded interval $[a, b]$ and that $f_n \to f$ uniformly. Then f is Riemann integrable, and $\int_a^b f_n(x)\,dx \to \int_a^b f(x)\,dx$ as $n \to \infty$.*

Proof We use the following criterion, established in Corollary 8.3.5 of Volume I.

Lemma 12.1.6 *A bounded function f on an interval $[a, b]$ is Riemann integrable if and only if given $\epsilon > 0$ there exists a dissection $D = \{a = x_0 < \cdots < x_k = b\}$ of $[a, b]$ and a partition $G \cup B$ of $\{1, \ldots, k\}$ such that*

$$\Omega(f, I_j) \leq \epsilon \text{ for } j \in G \quad \text{and} \quad \sum_{j \in B} l(I_j) < \epsilon,$$

where $I_1, \ldots I_k$ are the intervals of the dissection, and

$$\Omega(f, I_j) = \sup_{x, y \in I_j} |f(x) - f(y)|$$

is the oscillation of f on I_j.

First we show that f is bounded. There exists $N \in \mathbf{N}$ such that $d_\infty(f, f_N) < 1$. If $t \in [a, b]$ then

$$|f(t)| \leq |f(t) - f_N(t)| + |f_N(t)| \leq 1 + \sup_{s \in [a,b]} |f_N(s)|,$$

so that f is bounded.

Next we show that f is Riemann integrable. Suppose that $\epsilon > 0$. There exists $M \in \mathbf{N}$ such that $d_\infty(f, f_n) < \epsilon/3$ for $n \geq M$. By the lemma, there exists a dissection $D = \{a = x_0 < \cdots < x_k = b\}$ of $[a, b]$ and a partition $G \cup B$ of $\{1, \ldots, k\}$ such that

$$\Omega(f_M, I_j) \leq \epsilon/3 \text{ for } j \in G \quad \text{and} \quad \sum_{j \in B} l(I_j) < \epsilon,$$

where $I_1, \ldots I_k$ are the intervals of the dissection. If $j \in G$ and $s, t \in I_j$ then

$$|f(s) - f(t)| \leq |f(s) - f_M(s)| + |f_M(s) - f_M(t)| + |f_M(t) - f(t)| < \epsilon,$$

so that $\Omega(f, I_j) \leq \epsilon$. Thus f satisfies the conditions of the lemma and so it is Riemann integrable. If $n \geq M$ then

$$\left| \int_a^b f(t)\, dt - \int_a^b f_n(t)\, dt \right| \leq \int_a^b |f(t) - f_n(t)|\, dt \leq \epsilon(b - a),$$

so that $\int_a^b f_n(t)\, dt \to \int_a^b f(t)\, dt$ as $n \to \infty$. □

Corollary 12.1.7 *Suppose that $(f_n)_{n=1}^\infty$ is a sequence of continuously differentiable real-valued functions on an open interval (a, b) of $\mathbf{R}$, and that the sequence $(f_n')_{n=1}^\infty$ of derivatives converges uniformly on (a, b) to g as $n \to \infty$. Suppose also that there exists $c \in (a, b)$ such that $f_n(c) \to l$ as $n \to \infty$. Then* *there exists a continuously differentiable function f on (a, b) such that f_n converges uniformly on (a, b) to f as $n \to \infty$, and such that $f' = g$.*

Proof If $t \in (a, b)$, let $f(t) = l + \int_c^t g(s)\, ds$. Since g is continuous, by Theorem 12.1.4, we can apply the fundamental theorem of calculus: f is differentiable, with derivative g. Suppose that $\epsilon > 0$. There exists $N \in \mathbf{N}$ such that

$$|f_n(c) - f(c)| < \frac{\epsilon}{2} \text{ and } |f_n'(s) - g(s)| < \frac{\epsilon}{2(b-a)}$$

for $n \geq N$ and $s \in (a, b)$. If $t \in (a, b)$ then

$$f(t) - f_n(t) = (f(c) - f_n(c)) + \int_c^t (f_n'(s) - g(s))\, ds,$$

so that if $n \geq N$ then

$$|f(t) - f_n(t)| \leq |f(c) - f_n(c)| + \int_c^t |f_n'(s) - g(s)|\, ds$$

$$< \frac{\epsilon}{2} + \frac{|t - c|\epsilon}{2(b-a)} < \epsilon,$$

so that f_n converges uniformly on (a, b) to f as $n \to \infty$.   □

These proofs are rather easy. If we drop the continuity conditions, the proofs are considerably harder.

Theorem 12.1.8 *Suppose that $(f_n)_{n=1}^{\infty}$ is a sequence of differentiable real-valued functions on a bounded open interval (a, b). Suppose that*

(a) there exists $c \in (a, b)$ such that $f_n(c)$ converges, to l say, as $n \to \infty$, and

(b) the sequence $(f_n')_{n=1}^{\infty}$ of derivatives converges uniformly on (a, b) to a function g.

Then there exists a continuous function f on (a, b) such that

(i) $f_n \to f$ uniformly on (a, b), and

(ii) f is differentiable on (a, b), and $f'(x) = g(x)$ for all $x \in (a, b)$.

Proof Suppose that $\epsilon > 0$. There exists $N \in \mathbf{N}$ such that

$$|f_m'(x) - f_n'(x)| < \frac{\epsilon}{4 + 2(b-a)} \text{ for } x \in (a, b) \text{ and } m, n \geq N,$$

and $|f_m(c) - f_n(c)| < \epsilon/2$ for $m, n \geq N$. By the mean-value theorem,

$$|(f_m(x) - f_n(x)) - (f_m(c) - f_n(c))| \leq |x - c| \sup_{y \in [c, x]} |f_m'(y) - f_n'(y)| < \frac{\epsilon}{2}$$

for $m, n \geq N$, and so

$$|f_m(x) - f_n(x)| < \epsilon \text{ for } x \in (a, b) \text{ and } m, n \geq N.$$

Thus, for each $x \in (a, b)$, $(f_m(x))_{m=1}^{\infty}$ is a Cauchy sequence, convergent to $f(x)$, say, and

$$|f(x) - f_n(x)| \leq \epsilon \text{ for } x \in (a, b) \text{ and } n \geq N :$$

$f_n \to f$ uniformly on (a, b). Thus f is a continuous function on $[a, b]$, by Theorem 12.1.4.

Suppose that $x \in (a, b)$, that $h \neq 0$ and that $x + h \in (a, b)$. Let

$$d_n(h) = f_n(x + h) - f_n(x) \text{ for } n \in \mathbf{N},$$
$$d(h) = f(x + h) - f(x);$$

then $d_n(h) \to d(h)$ as $n \to \infty$. If $m \geq N$ then, by the mean-value theorem,

$$|d_m(h) - d_N(h)| \leq |h| \sup_{y \in [x, x+h]} |f_m'(y) - f_N'(y)| < \epsilon|h|/4,$$

so that

$$|(d_m(h) - hf_m'(x)) - (d_N(h) - hf_N'(x))|$$
$$\leq |d_m(h) - d_N(h)| + |h(f_m'(x) - f_N'(x))|$$
$$\leq \epsilon|h|/4 + \epsilon|h|/4 = \epsilon|h|/2.$$

Letting $m \to \infty$,

$$|(d(h) - hg(x)) - (d_N(h) - hf_N'(x))| \leq \epsilon|h|/2.$$

But there exists $\delta > 0$ such that $(x - \delta, x + \delta) \subset (a, b)$ and such that if $|h| < \delta$ then

$$|d_N(h) - hf_N'(x)| = |f_N(x + h) - f_N(x) - hf_N'(x)| < \epsilon|h|/2,$$

and so

$$|d(h) - hg(x)| = |f(x + h) - f(x) - hg(x)| < \epsilon|h| \text{ for } |h| < \delta.$$

Thus f is differentiable at x, with derivative $g(x)$. □

In the case where d is a metric on a vector space E (and in particular, when d is given by a norm), we can also consider the convergence of series. Suppose that $(a_n)_{n=0}^{\infty}$ is a sequence in E. Let $s_n = \sum_{j=0}^{n} a_j$, for $n \in \mathbf{N}$, and suppose that $s \in E$. Then the sum $\sum_{n=0}^{\infty} a_n$ *converges* to s if $s_n \to s$ as $n \to \infty$.

Exercises

12.1.1 Suppose that a set S is given the discrete metric d. Show that a sequence $(x_n)_{n=1}^{\infty}$ converges to a point of S if and only if it is eventually constant; there exists $N \in \mathbf{N}$ such that $x_n = x_N$ for all $n \geq N$.

12.1.2 Suppose that $(x_n)_{n=1}^{\infty}$ is a sequence in a metric space which has the property that if $(y_k)_{k=1}^{\infty} = (x_{n_k})_{k=1}^{\infty}$ is a subsequence of $(x_n)_{n=1}^{\infty}$ then there is a subsequence $(z_j)_{j=1}^{\infty} = (y_{k_j})_{j=1}^{\infty}$ of $(y_k)_{k=1}^{\infty}$ which converges to x_1. Show that $x_n \to x_1$ as $n \to \infty$.

12.1.3 Suppose that $(x_n)_{n=1}^{\infty}$ and $(y_n)_{n=1}^{\infty}$ are sequences in a normed space and that α and β are scalars. Show that if $\sum_{n=1}^{\infty} x_n$ converges to s and $\sum_{n=1}^{\infty} y_n$ converges to t then $\sum_{n=1}^{\infty} (\alpha x_n + \beta y_n)$ converges to $\alpha s + \beta t$.

12.1.4 Suppose that $\sum_{n=1}^{\infty} x_n$ is a convergent series in a normed space. Show that $\|x_n\| \to 0$ as $n \to \infty$.

12.1.5 Give an example of a sequence $(f_n)_{n=1}^{\infty}$ of continuous real-valued functions on $[0,1]$ which converges pointwise to a continuous function f on $[0,1]$, but which does not converge uniformly to f.

12.1.6 Give an example of a sequence $(f_n)_{n=1}^{\infty}$ of continuous real-valued functions on $[0,1]$ which decreases pointwise to a bounded function f on $[0,1]$, but which does not converge uniformly to f.

12.1.7 Give an example of a sequence $(f_n)_{n=1}^{\infty}$ of bounded continuous real-valued functions on $[0,\infty)$ which decreases pointwise to a bounded continuous function f on $[0,\infty)$, but which does not converge uniformly to f.

12.1.8 Let h be the *hat function*:

$$h(t) = \begin{cases} 2t & \text{if } 0 \leq t \leq \frac{1}{2}, \\ 2 - 2t & \text{if } \frac{1}{2} \leq t \leq 1, \\ 0 & \text{otherwise.} \end{cases}$$

Let

$$h_n(t) = \sum_{k=1}^{n} \frac{1}{k} \left(\sum_{j=1}^{2^{k-1}} h\left(2^n \left(t - \frac{2j-1}{2^k} \right) \right) \right).$$

Sketch h_1, h_2 and h_3. Show that h_n converges pointwise to 0, but that there exists no proper interval $[c,d]$ in $[0,1]$ on which it converges uniformly to 0.

12.1.9 Formulate versions of Theorems 12.1.8 and 12.1.5 for infinite sums of functions.

12.1.10 Show that Theorem 12.1.5 does not hold for improper integrals over $[0, \infty)$.

12.2 Convergence and continuity of mappings

Suppose that A is a subset of a metric space (X, d). An element b of X is called a *limit point* or *accumulation point* of A if whenever $\epsilon > 0$ there exists $a \in A$ (which may depend upon ϵ) with $0 < d(a, b) < \epsilon$. Thus b is a limit point of A if there are points of A, different from b, which are arbitrarily close to b. An element a of A is an *isolated point* of A if it is not a limit point of A; that is, there exists $\epsilon > 0$ such that $N_\epsilon(a) \cap A = \{a\}$.

If $a \in X$ and $\epsilon > 0$ then the *punctured ϵ-neighbourhood* $N_\epsilon^*(a)$ of a is defined as

$$N_\epsilon^*(a) = \{x \in X : 0 < d(x, a) < \epsilon\} = N_\epsilon(a) \setminus \{a\}.$$

Thus b is a limit point of A if and only if $N_\epsilon^*(b) \cap A \neq \emptyset$, for each $\epsilon > 0$.

Suppose that (X, d) and (Y, ρ) are metric spaces and that f is a mapping from a subset A of X into Y. Suppose that b is a limit point of A (which may or may not be an element of A) and that $l \in Y$. We say that $f(x)$ *converges to a limit l*, or *tends to l*, as x tends to b if whenever $\epsilon > 0$ there exists $\delta > 0$ (which usually depends on ϵ) such that if $x \in A$ and $0 < d(x, b) < \delta$, then $\rho(f(x), l) < \epsilon$. That is to say, as x gets close to b, $f(x)$ gets close to l. Note that in the case where $b \in A$, we do not consider the value of $f(b)$, but only the values of f at points nearby. We say that *l is the limit of f as x tends to b*, write '$f(x) \to l$ as $x \to b$' and write $l = \lim_{x \to b} f(x)$.

We can express the convergence of f in terms of punctured ϵ-neigh-bourhoods; $f(x) \to l$ as $x \to b$ if and only if for each $\epsilon > 0$ there exists $\delta > 0$ such that if $x \in A \cap N_\delta^*(b)$ then $f(x) \in N_\epsilon(l)$ – that is, $f(N_\delta^*(b) \cap A) \subseteq N_\epsilon(l)$.

Proposition 12.2.1 *Suppose that f is a mapping from a subset A of a metric space (X, d) into a metric space (Y, ρ), and that b is a limit point of A.*

(i) If $f(x) \to l$ as $x \to b$ and $f(x) \to m$ as $x \to b$, then $l = m$.

(ii) $f(x) \to l$ as $x \to b$ if and only if whenever $(a_n)_{n=0}^\infty$ is a sequence in $A \setminus \{b\}$ which tends to b as $n \to \infty$ then $f(a_n) \to l$ as $n \to \infty$.

Proof (i) Suppose that $\epsilon > 0$. There exists $\delta > 0$ such that if $x \in N_\delta^*(b) \cap A$ then $\rho(f(x), l) < \epsilon$ and $\rho(f(x), m) < \epsilon$. Since $N_\delta^*(b) \cap A$ is not empty, there exists $x_0 \in N_\delta^*(b) \cap A$. Then, using the triangle inequality,

$$\rho(l, m) \leq \rho(l, f(x_0)) + \rho(f(x_0), m) < \epsilon + \epsilon = 2\epsilon.$$

Since this holds for all $\epsilon > 0$, $\rho(l, m) = 0$ and $l = m$.

(ii) Suppose that $f(x) \to l$ as $x \to b$ and that $(a_n)_{n=0}^{\infty}$ is a sequence in $A \setminus \{b\}$ which tends to b as $n \to \infty$. Given $\epsilon > 0$, there exists $\delta > 0$ such that if $x \in N_{\delta}^{*}(b) \cap A$ then $\rho(f(x), l) < \epsilon$. There then exists n_0 such that $d(a_n, b) < \delta$ for $n \geq n_0$. Then $\rho(f(a_n), l) < \epsilon$ for $n \geq n_0$, so that $f(a_n) \to l$ as $n \to \infty$.

Suppose that $f(x)$ does not converge to l as $x \to b$. Then there exists $\epsilon > 0$ for which we can find no suitable $\delta > 0$. Thus for each $n \in \mathbf{N}$ there exists $x_n \in N_{1/n}^{*}(b) \cap A$ with $\rho(f(x_n), l) \geq \epsilon$. Then $x_n \to b$ as $n \to \infty$ and $f(x_n)$ does not converge to l as $n \to \infty$. $\qquad\square$

Suppose now that f is a mapping from a metric space (X, d) into a metric space (Y, ρ), and that $a \in X$. We say that f is *continuous* at a if whenever $\epsilon > 0$ there exists $\delta > 0$ (which usually depends on ϵ) such that if $d(x, a) < \delta$ then $\rho(f(x), f(a)) < \epsilon$. That is to say, as x gets close to a, $f(x)$ gets close to $f(a)$. If f is not continuous at a, we say that f has a *discontinuity* at a.

We can express the continuity of f in terms of ϵ-neighbourhoods; f is continuous at a if and only if for each $\epsilon > 0$ there exists $\delta > 0$ such that $f(N_{\delta}(x)) \subseteq N_{\epsilon}(f(a))$.

Compare this definition with the definition of convergence. First, we only consider functions defined on X. This is not a real restriction; suppose that f is a mapping from a subset A of a metric space (X, d) into a metric space (Y, ρ), and that $a \in A$. We say that f is continuous on A at a if $f : A \to Y$ is continuous at a when A is given the subspace metric. Secondly, a need not be a limit point of X. If it is a limit point, then f is continuous at a if and only if $f(x) \to f(a)$ as $x \to a$. If a is not a limit point, then there exists $\delta > 0$ such that $N_{\delta}(a) = \{a\}$, so that if $x \in N_{\delta}(a)$ then $f(x) = f(a)$, and f is continuous at a.

We have the following immediate consequence of Proposition 12.2.1.

Proposition 12.2.2 *Suppose that f is a mapping from a metric space (X, d) into a metric space (Y, ρ), and that $a \in X$. Then f is continuous at a if and only if whenever $(a_n)_{n=0}^{\infty}$ is a sequence in X which tends to a as $n \to \infty$ then $f(a_n) \to f(a)$ as $n \to \infty$.*

Suppose that f is a real- or complex-valued function on a metric space (X, d). Unless it is explicitly stated otherwise, when we consider convergence and continuity properties of f, we give $\mathbf{R}$ or $\mathbf{C}$ its usual metric.

Theorem 12.2.3 *Suppose that f and g are functions on a metric space (X, d) taking values in a normed space $(E, \|.\|_E)$, that λ is a scalar-valued function on (X, d) and that $a \in X$.*

(i) If f is continuous at a then there exists $\delta > 0$ such that f is bounded on $N_\delta(a)$.

(ii) If $f(a) = 0$, f is continuous at a, and $\lambda(x)$ is bounded on $N_\delta(a)$ for some $\delta > 0$, then λf is continuous at a.

(iii) If f and g are continuous at a then $f + g$ is continuous at a.

(iv) If f and λ are continuous at a then λf is continuous at a.

(v) If $\lambda(x) \neq 0$ for $x \in X$, and if λ is continuous at a, then $1/\lambda$ is continuous at a.

Proof   These results correspond closely to results for functions of a real variable (Volume I, Theorem 6.3.1). We prove (i), (iv) and (v), and leave the others as exercises for the reader.

(i) There exists $\delta > 0$ such that $\|f(x) - f(a)\|_E \leq 1$ for $x \in N_\delta(a)$. Then

$$\|f(x)\|_E \leq \|f(x) - f(a)\|_E + \|f(a)\|_E \leq 1 + \|f(a)\|_E, \text{ for } x \in N_\delta(a).$$

(iv) Suppose that $\epsilon > 0$. Let $M = \max(\|f(a)\|_E, |\lambda(a)|)$, and let $\eta = \min(\epsilon/(2M + 1), 1)$. There exists $\delta > 0$ such that if $x \in N_\delta(a)$ then $\|(f(x) - f(a)\|_E \leq \eta$ and $|\lambda(x) - \lambda(a)| < \eta$. If $x \in N_\delta(a)$, then $\|f(x)\|_E \leq \|f(x) - f(a)\|_E + \|f(a)\|_E \leq \eta + M$, so that

$$\begin{aligned}
\|\lambda(x)f(x) - \lambda(a)f(a)\|_E &= \|(\lambda(x) - \lambda(a))f(x) + \lambda(a)(f(x) - f(a))\|_E \\
&\leq |\lambda(x) - \lambda(a)| . \|f(x)\|_E + |\lambda(a)| . \|f(x) - f(a)\|_E \\
&\leq \eta(\eta + M) + M\eta \leq \epsilon.
\end{aligned}$$

(v) Suppose that $\epsilon > 0$. Let $\eta = |\lambda(a)|^2 \epsilon/2$. There exists $\delta > 0$ such that $|\lambda(x) - \lambda(a)| < \max(|\lambda(a)|/2, \eta)$ for $x \in N_\delta(a)$. If $x \in N_\delta(a)$, then $|\lambda(x)| \geq |\lambda(a)|/2$, and so

$$\left| \frac{1}{\lambda(x)} - \frac{1}{\lambda(a)} \right| = \left| \frac{\lambda(a) - \lambda(x)}{\lambda(x)\lambda(a)} \right| \leq \frac{2\eta}{|\lambda(a)|^2} = \epsilon.$$

$\square$

Proposition 12.2.4   *(The sandwich principle) Suppose that f, g and h are real-valued functions on a metric space (X, d), and that there exists $\eta > 0$ such that $f(x) \leq g(x) \leq h(x)$ for all $x \in N_\eta(a)$, and that $f(a) = g(a) = h(a)$. If f and h are continuous at a, then so is g.*

Proof   This follows easily from the fact that

$$|g(x) - g(a)| \leq \max(|f(x) - f(a)|, |h(x) - h(a)|).$$

$\square$

Continuity behaves well under composition.

Theorem 12.2.5   *Suppose that f is a mapping from a metric space (X, d) into a metric space (Y, ρ) and that g is a mapping from Y into a metric space (Z, σ). If f is continuous at $a \in A$ and g is continuous at $f(a)$, then $g \circ f$ is continuous at a.*

Proof   Suppose that $\epsilon > 0$. Then there exists $\eta > 0$ such that

$$g(N_\eta(f(a))) \subseteq N_\epsilon(g(f(a))).$$

Similarly there exists $\delta > 0$ such that $f(N_\delta(a)) \subseteq N_\eta(f(a))$. Then $g(f(N_\delta(a))) \subseteq g(N_\eta(f(a))) \subseteq N_\epsilon(g(f(a)))$. $\square$

This proof is almost trivial: the result has great theoretical importance and practical usefulness.

Continuity is a local phenomenon. Nevertheless, there are many important cases where f is continuous at every point of X. In this case we say that f is *continuous on X*, or, more simply, that f is continuous. Let us give some easy examples.

1. An isometry from a metric space (X, d) into a metric space (Y, ρ) is continuous on X: given $\epsilon > 0$, take $\delta = \epsilon$.
2. In particular, if A is a subset of a metric space (X, d) and A is given the subspace metric, then the inclusion mapping $i : A \to X$ is continuous. If $f : (X, d) \to (Y, \rho)$ is continuous, then so is the restriction $f \circ i : A \to Y$ of f to A.
3. More generally, if f is a mapping from a metric space (X, d) to a metric space (Y, ρ) and if $x \in X$ then f is a *Lipschitz mapping, with constant K*, at x if $\rho(f(x), f(x')) \leq K d(x, x')$ for all $x' \in X$. If there exists $K > 0$ such that $\rho(f(x), f(x')) \leq K d(x, x')$, for all $x, x' \in X$, then f is a *Lipschitz mapping on X, with constant K*. A Lipschitz mapping at x is continuous at x (given $\epsilon > 0$, take $\delta = \epsilon/K$).
4. If f is a constant mapping from a metric space (X, d) into a metric space (Y, ρ) – that is, $f(x) = f(y)$ for any $x, y \in X$ – then f is continuous: given $\epsilon > 0$, any $\delta > 0$ will do.
5. If d is the discrete metric on a set X then every point of X is isolated, and any mapping $f : (X, d) \to (Y, \rho)$ is continuous.

6. Suppose that A is a non-empty subset of a metric space (X, d). If $x \in X$, let $d(x, A) = \inf\{d(x, a) : a \in A\}$. The mapping $x \to d(x, A)$ is a mapping from X to $\mathbf{R}$. We show that it is a Lipschitz mapping with constant 1, and is therefore continuous. Suppose that $x, y \in X$. If $\epsilon > 0$, there exists $a \in A$ with $d(x, a) < d(x, A) + \epsilon$. Then

$$d(y, A) \leq d(y, a) \leq d(y, x) + d(x, a) < d(y, x) + d(x, A) + \epsilon.$$

Since ϵ is arbitrary, $d(y, A) - d(x, A) \leq d(y, x)$. In the same way, $d(x, A) - d(y, A) \leq d(x, y) = d(y, x)$, and so $|d(x, A) - d(y, A)| \leq d(x, y)$.

7. In particular, if $(E, \|.\|)$ is a normed space, then the mapping $x \to \|x\|$ is a Lipschitz mapping from E to $\mathbf{R}$ with constant 1. For $\|x\| = d(x, \{0\})$.

8. The function tan is a continuous bijection from $(-\pi/2, \pi/2)$ onto $\mathbf{R}$, when both are given the usual metric, and the inverse mapping $\tan^{-1}$ is also continuous. A bijective continuous mapping f from a metric space (X, d) onto a metric space (Y, ρ) whose inverse is also continuous is called a *homeomorphism*.

9. We can give $\mathbf{R}^d$ the Euclidean metric d_2. We can also consider $\mathbf{R}^d$ as $B(\{1, 2, \ldots, d\})$, and give it the uniform metric d_∞. Then

$$d_\infty(x, y) = \max_{1 \leq j \leq d} |x_j - y_j| \leq \left(\sum_{j=1}^{d} |x_j - y_j|^2 \right)^{1/2}$$

$$= d_2(x, y) \leq d^{1/2} d_\infty(x, y),$$

so that the identity mapping $i : (\mathbf{R}^d, d_2) \to (\mathbf{R}^d, d_\infty)$ is a homeomorphism. If ρ_1 and ρ_2 are two metrics on a set X for which the identity mapping $i : (X, \rho_1) \to (X, \rho_2)$ is a homeomorphism, then the metrics are said to be *equivalent*. If, as in the present case, i and i^{-1} are Lipschitz mappings, then the metrics are said to be *Lipschitz equivalent*.

10. Let $C[0, 1]$ be the vector space of (real or) complex continuous functions on $[0, 1]$. We can give $C[0, 1]$ the uniform metric: $d_\infty(f, g) = \sup\{|f(x) - g(x)| : x \in [0, 1]\}$. We can also give it the metric defined by the inner product: $d_2(f, g) = (\int_0^1 |f(x) - g(x)|^2, dx)^{1/2}$. Since $d_2(f, g) \leq d_\infty(f, g)$, the identity mapping $i : (C[0, 1], d_\infty) \to (C[0, 1], d_2)$ is a Lipschitz mapping, with constant 1. On the other hand, if we set $f_n(x) = x^n$, then $\|f_n\|_2 = d_2(f_n, 0) = 1/(2n + 1)^{1/2}$, so that $d_2(f_n, 0) \to 0$ as $n \to \infty$, while $\|f_n\|_\infty = d_\infty(f_n, 0) = 1$, so that f_n does not converge to 0 in the uniform metric as $n \to \infty$. Thus the inverse mapping $i^{-1} : (C[0, 1], d_2) \to (C[0, 1], d_\infty)$ is not continuous.

We have the following generalization of Theorem 12.1.4.

Theorem 12.2.6   *If $(f_n)_{n=1}^{\infty}$ is a sequence of continuous functions from a metric space (X, d) into a metric space (Y, ρ) and if f_n converges uniformly on X to f as $n \to \infty$, then f is continuous.*

Proof   Verify that a proof is given by making obvious notational changes to the proof of Theorem 12.1.4.   □

Exercises

12.2.1 Show that any two metrics on a finite set are Lipschitz equivalent.

12.2.2 Let $(\overline{\mathbf{N}}, \rho)$ be the metric space defined in Example 11.1.9. Suppose that $(x_n)_{n=1}^{\infty}$ is a sequence in a metric space (X, d), and that $x \in X$. Set $f(n) = x_n$, $f(+\infty) = x$. Show that $x_n \to x$ as $n \to \infty$ if and only if $f : (\overline{\mathbf{N}}, \rho) \to (X, d)$ is continuous.

12.2.3 Suppose that (X, d), (Y, ρ) and (Z, σ) are metric spaces, that f is a continuous surjective mapping of (X, d) onto (Y, ρ) and that $g : (Y, \rho) \to (Z, \sigma)$ is continuous. Show that if $g \circ f$ is a homeomorphism of (X, d) onto (Z, σ) then f is a homeomorphism of (X, d) onto (Y, ρ) and g is a homeomorphism of (Y, ρ) onto (Z, σ).

12.2.4 Show that the punctured unit sphere $\{x \in \mathbf{R}^d : \|x\| = 1\} \setminus \{(1, 0, \dots, 0)\}$ of $\mathbf{R}^d$, with its usual metric, is homeomorphic to $\mathbf{R}^{d-1}$.

12.2.5 Give an example of three metric subspaces A, B and C of $\mathbf{R}$ such that $A \subset B \subset C$, A and C are homeomorphic, and B and C are not homeomorphic.

12.3 The topology of a metric space

This section contains many definitions: we start with a few.

Suppose that A is a subset of a metric space (X, d). Recall that a point $b \in X$ is a *limit point*, or *accumulation point*, of A if and only if $N_\epsilon^*(b) \cap A \neq \emptyset$, for each $\epsilon > 0$. We now make another definition, similar enough to be confusing. An element b of X is called a *closure point* of A if $N_\epsilon(b) \cap A \neq \emptyset$, for each $\epsilon > 0$. That is to say, whenever $\epsilon > 0$ there exists $a \in A$ (which may depend upon ϵ) with $d(b, a) < \epsilon$. Thus b is a closure point of A if there are points of A arbitrarily close to b. If $b \in A$, then b is a closure point of A, since $d(b, b) = 0 < \epsilon$ for all $\epsilon > 0$.

The set of limit points of A is called the *derived set* of A, and is denoted by A', and the set of closure points of A is called the *closure* of A, and is denoted by $\overline{A}$. A set A is *perfect* if $A = A'$ and is *closed* if $A = \overline{A}$.

An element a of A is an *isolated point* of A if there exists $\epsilon > 0$ such that $N_\epsilon(a) \cap A = \{a\}$.

Proposition 12.3.1   *Suppose that A is a subset of a metric space (X, d). Let $i(A)$ be the set of isolated points of A. Then A' and $i(A)$ are disjoint, and $\overline{A} = A' \cup i(A)$.*

Proof   This follows immediately from the definitions.   □

The set A is a subset of $\overline{A}$, since each point of A is a closure point of A, but any isolated point of A is not in A'. A set is perfect if and only if it is closed, and has no isolated points.

We can characterize limit points and closure points of A in terms of convergent sequences.

Proposition 12.3.2   *Suppose that A is a subset of a metric space (X, d) and that $b \in X$.*

(i) b is a limit point of A if and only if there exists a sequence $(a_j)_{j=1}^{\infty}$ in $A \setminus \{b\}$ such that $a_j \to b$ as $j \to \infty$.

(ii) b is a closure point of A if and only if there exists a sequence $(a_j)_{j=1}^{\infty}$ in A such that $a_j \to b$ as $j \to \infty$.

Proof   (i) Suppose that there exists a sequence $(a_j)_{j=0}^{\infty}$ in $A \setminus \{b\}$ such that $a_j \to b$ as $j \to \infty$. Suppose that $\epsilon > 0$. There exists j_0 such that $d(b, a_j) < \epsilon$ for $j \geq j_0$. Then $a_{j_0} \in N_\epsilon^*(b)$. Conversely, if b is a limit point of A then for each $j \in \mathbf{N}$ there exists $a_j \in A \setminus \{b\}$ with $0 < d(b, a_j) < 1/j$. Then $a_j \to b$ as $j \to \infty$.

(ii) The proof is exactly similar.   □

Proposition 12.3.2 (ii) says that A is closed if and only if A is *closed under taking limits*.

The closure of a bounded set is bounded.

Proposition 12.3.3   *If A is a non-empty bounded subset of a metric space (X, d), then $\operatorname{diam} \overline{A} = \operatorname{diam} A$.*

Proof   Certainly $\operatorname{diam} \overline{A} \geq \operatorname{diam} A$. Suppose that $\epsilon > 0$. If $x, y \in \overline{A}$ there exist $a, b \in A$ with $d(x, a) < \epsilon/2$ and $d(y, b) < \epsilon/2$. Then, by the triangle inequality,

$$d(x, y) \leq d(x, a) + d(a, b) + d(b, y) \leq d(a, b) + \epsilon \leq \operatorname{diam} A + \epsilon,$$

so that $\operatorname{diam} \overline{A} \leq \operatorname{diam} A + \epsilon$. Since ϵ is arbitrary, the result follows.   □

A subset A of a metric space (X, d) is *dense in X* if $\overline{A} = X$. For example, the rationals are dense in $\mathbf{R}$.

Proposition 12.3.4 *Suppose that A and B are subsets of X.*

(i) If $A \subseteq B$ then $\overline{A} \subseteq \overline{B}$.

(ii) $\overline{A}$ is closed.

(iii) $\overline{A}$ is the smallest closed set containing A: if C is closed and $A \subseteq C$ then $\overline{A} \subseteq C$.

Proof (i) follows trivially from the definition of closure.

(ii) Suppose that b is a closure point of $\overline{A}$ and suppose that $\epsilon > 0$. Then there exists $c \in \overline{A}$ such that $d(b,c) < \epsilon/2$, and there exists $a \in A$ with $d(c,a) < \epsilon/2$. Thus $d(b,a) < \epsilon$, by the triangle inequality, and so $b \in \overline{A}$.

(iii) By (i), $\overline{A} \subseteq \overline{C} = C$. $\qquad\qquad\qquad\qquad\qquad\qquad\qquad\square$

Suppose that Y is a metric subspace of a metric space (X,d). How are the closed subsets of Y related to the closed subsets of X?

Theorem 12.3.5 *Suppose that Y is a metric subspace of a metric space (X,d) and that $A \subseteq Y$. Let $\overline{A}^Y$ denote the closure of A in Y, and $\overline{A}^X$ the closure in X.*

(i) $\overline{A}^Y = \overline{A}^X \cap Y$.

(ii) A is closed in Y if and only if there exists a closed set B in X such that $A = B \cap Y$.

Proof (i) Certainly $\overline{A}^Y \subseteq \overline{A}^X$, so that $\overline{A}^Y \subseteq \overline{A}^X \cap Y$. On the other hand, if $y \in \overline{A}^X \cap Y$ there exists a sequence $(a_n)_{n=1}^\infty$ in A such that $a_n \to y$ as $n \to \infty$. Thus $y \in \overline{A}^Y$.

(ii) If A is closed in Y, then $A = \overline{A}^Y = \overline{A}^X \cap Y$, so that we can take $B = \overline{A}^X$. Conversely if B is closed in X and $A = B \cap Y$, then $\overline{A}^X \subseteq B$, so that $\overline{A}^Y = \overline{A}^X \cap Y \subseteq B \cap Y = A$, and A is closed in Y. $\qquad\square$

Here are some fundamental properties of the collection of closed subsets of a metric space (X,d).

Proposition 12.3.6 *(i) The empty set $\emptyset$ and X are closed.*

(ii) If $\mathcal{A}$ is a set of closed subsets of X then $\cap_{A \in \mathcal{A}} A$ is closed.

(iii) If $\{A_1, \ldots, A_n\}$ is a finite set of closed subsets of X then $\cup_{j=1}^n A_j$ is closed.

Proof (i) The empty set is closed, since there is nothing to go wrong, and X is trivially closed.

(ii) Suppose that b is a closure point of $\cap_{A \in \mathcal{A}} A$, and that $A \in \mathcal{A}$. If $\epsilon > 0$ then there exists $a \in \cap_{A \in \mathcal{A}} A$ with $d(b,a) < \epsilon$. But then $a \in A$. Since this holds for all $\epsilon > 0$, $b \in \overline{A} = A$. Since this holds for all $A \in \mathcal{A}$, $b \in \cap_{A \in \mathcal{A}} A$, and so $\cap_{A \in \mathcal{A}} A$ is closed.

(iii) Suppose that $b \notin \cup_{j=1}^{n} A_j$. Then for each j, $b \notin A_j = \overline{A}_j$, and so there exists $\epsilon_j > 0$ such that if $d(b,c) < \epsilon_j$ then $c \notin A_j$. Let $\epsilon = \min\{\epsilon_j : 1 \le j \le n\}$. Then $\epsilon > 0$ and if $d(b,c) < \epsilon$ then $c \notin \cup_{j=1}^{n} A_j$. Thus b is not a closure point of $\cup_{j=1}^{n} A_j$, and so $\cup_{j=1}^{n} A_j$ is closed. $\qquad\square$

Here is an important example.

Theorem 12.3.7 *Suppose that (X,d) and (Y,ρ) are metric spaces. Let $C_b(Y,X)$ denote the set of all bounded continuous mappings of Y into X. Then $C_b(Y,X)$ is a closed subset of the space $B_X(Y)$ of all bounded mappings of Y into X, when $B_X(Y)$ is given the uniform metric d_∞.*

Proof If f is in the closure of $C_b(Y,X)$, then, by Proposition 12.3.2 there exists a sequence $(f_n)_{n=1}^{\infty}$ in $C_b(Y,X)$ which converges uniformly to f. It then follows from Theorem 12.2.6 that f is continuous. $\qquad\square$

We now introduce some more definitions. Suppose that A is a subset of a metric space (X,d).

- An element a of A is an *interior point* of A if there exists $\epsilon > 0$ such that $N_\epsilon(a) \subseteq A$. In other words, all the points sufficiently close to a are in A; we can move a little way from a without leaving A.
- The *interior* A° of A is the set of interior points of A.
- A subset U of X is *open* if $U = U^\circ$. In other words U is open if and only if whenever $u \in U$ there exists $\epsilon > 0$ such that if $d(u,v) < \epsilon$ then $v \in U$. The collection of open subsets of (X,d) is called the *topology* of (X,d).

Proposition 12.3.8 *If (X,d) is a metric space, if $x \in X$ and if $\epsilon > 0$ then the open ϵ-neighbourhood $N_\epsilon(a)$ is open.*

Proof Suppose that $y \in N_\epsilon(x)$, so that $d(y,x) < \epsilon$. Let $\delta = \epsilon - d(y,x)$. If $z \in N_\delta(y)$ then, by the triangle inequality,

$$d(z,x) \le d(z,y) + d(y,x) < \delta + d(y,x) = \epsilon,$$

so that $N_\delta(y) \subseteq N_\epsilon(x)$. $\qquad\square$

If $(E, \|.\|)$ is a normed space then the ϵ-neighbourhood

$$N_\epsilon(0) = \{x \in E : \|x\| < \epsilon\}$$

is called the *open ϵ-ball*; in particular, $N_1(0) = \{x \in E : \|x\| < 1\}$ is called the *open unit ball*. The scaling property of the norm implies that $N_\epsilon(0) = \epsilon N_1(0)$. Although a normed space has plenty of closed linear subspaces, it has only one open linear subspace.

Proposition 12.3.9　*If F is an open linear subspace of a normed space $(E, \|.\|)$ then $F = E$.*

Proof　Since F is a linear subspace, $0 \in F$. Since F is open, there exists $\epsilon > 0$ such that $N_\epsilon(0) \subseteq F$. Suppose that $x \in E \setminus \{0\}$; let $y = \epsilon x / (\|x\| + 1)$. Then $0 < \|y\| < \epsilon$, so that $y \in N_\epsilon(0)$, and so $y \in F$. Since F is a linear subspace of E, $x = (\|x\| + 1)y/\epsilon \in F$. This is true for all $x \in E$, so that $F = E$.　□

'Interior' and 'closure', 'open' and 'closed', are closely related, as the next proposition shows.

Proposition 12.3.10　*Suppose that A and B are subsets of a metric space (X, d), and that $C(A) = X \setminus A$ is the complement of A in X.*
　(i) If $A \subseteq B$ then $A^\circ \subseteq B^\circ$.
　(ii) $C(A^\circ) = \overline{C(A)}$.
　(iii) A is open if and only if $C(A)$ is closed.
　(iv) A° is open.
　(v) A° is the largest open set contained in A: if U is open and $U \subseteq A$ then $U \subseteq A^\circ$.

Proof　(i) follows directly from the definition.
　(ii) If $b \notin A^\circ$ then $N_\epsilon(b) \cap C(A) \neq \emptyset$ for all $\epsilon > 0$, and so $b \in \overline{C(A)}$. Conversely, if $b \in \overline{C(A)}$ then $N_\epsilon(b) \cap C(A) \neq \emptyset$ for all $\epsilon > 0$, and so $b \notin A^\circ$.
　(iii) If A is open then $C(A) = C(A^\circ) = \overline{C(A)}$, by (ii), and so $C(A)$ is closed. If $C(A)$ is closed then $C(A^\circ) = \overline{C(A)} = C(A)$, so that $A^\circ = A$.
　(iv) $C(A^\circ) = \overline{C(A)}$ is closed, so that A° is open, by (iii).
　(v) By (i), $U = U^\circ \subseteq A^\circ$.　□

Corollary 12.3.11　*Suppose that Y is a metric subspace of a metric space (X, d) and that $A \subseteq Y$. Then A is open in Y if and only if there exists an open set B in X such that $A = B \cap Y$.*

Proof　Take complements.　□

Corollary 12.3.12　*(i) The empty set $\emptyset$ and X are open.*
　(ii) If $\mathcal{A}$ is a set of open subsets of X then $\cup_{A \in \mathcal{A}} A$ is open.
　(iii) If $\{A_1, \ldots, A_n\}$ is a finite set of open subsets of X then $\cap_{j=1}^n A_j$ is open.

Proof　Take complements.　□

Two final definitions: if A is a subset of a metric space (X, d) then the *frontier* or *boundary* ∂A of A is the set $\overline{A} \setminus A^\circ$. Since $\partial A = \overline{A} \cap \overline{C(A)}$, ∂A is closed. $x \in \partial A$ if and only if every open ϵ-neighbourhood of x contains an element of A and an element of $C(A)$.

A metric space is *separable* if it has a countable dense subset. Thus $\mathbf{R}$, with its usual metric, is a separable metric space.

There are interesting metric spaces which are not separable:

Proposition 12.3.13 *If (X, d) is a metric space with at least two points and if S is an infinite set, then the space $B_X(S)$ of bounded mappings from $S \to X$, with the uniform metric, is not separable.*

Proof We use the fact that $P(S)$ is uncountable; this was proved in Volume I, Corollary 2.3.10. Suppose that x_0 and x_1 are distinct points of X, and let $d = d(x_0, x_1)$. For each subset A of X, define the mapping $f_A : S \to X$ by setting

$$f_A(s) = x_1 \text{ if } s \in A \text{ and } f_A(s) = x_0 \text{ if } x \notin A.$$

Then f_A is bounded. If A and B are distinct subsets of S, then there exists $s \in S$ such that s is in exactly one of A and B, and so $d_\infty(f_A, f_B) = d$. Thus $N_{d/2}(f_A) \cap N_{d/2}(f_B) = \emptyset$. Suppose that G is a dense subset of $B_X(S)$. Let $H = \{g \in G : g \in N_{d/2}(f_A) \text{ for some } A \in P(X)\}$. If $g \in H$, then there exists a unique $A \in P(S)$ for which $g \in N_{d/2}(f_A)$: let this be $c(g)$. Then c is a mapping of H into $P(S)$. It is surjective, since if $A \in P(S)$ there exists $g \in G$ with $d_\infty(g, f_A) < d/2$, by the density of G, so that $c(g) = A$. Since $P(S)$ is uncountable, so is H, and since $H \subseteq G$, G is uncountable. Thus $B_X(S)$ is not separable. $\qquad \square$

Exercises

12.3.1 Show that a finite subset of a metric space (X, d) is closed.

12.3.2 Suppose that $(a_j)_{j=1}^\infty$ is a sequence in a metric space (X, d) which converges to a. Show that the set $S = \{a_j : j \in \mathbf{N}\} \cup \{a\}$ is closed.

12.3.3 Give an example of an open ϵ-neighbourhood $N_\epsilon(x)$ in a metric space (X, d) whose closure is not equal to $M_\epsilon(x) = \{y \in X : d(y, x) \leq \epsilon\}$.

12.3.4 Let $(E, \|.\|)$ be a normed space. Let $U = \{x \in E : \|x\| < 1\}$. Show that $\overline{U} = \{x \in E : \|x\| \leq 1\}$.

12.3.5 Let $D = \{(i, j) : 1 \leq i \leq j \leq d\}$. A *real quadratic form* on $\mathbf{R}^d$ is a function of the form $q_a(x) = \sum_{(i,j) \in D} a_{ij} x_i x_j$, where $a \in \mathbf{R}^D$. It is *positive definite* if $q_a(x) > 0$ for all $x \neq 0$. Show that

$$\{a \in \mathbf{R}^D : q_a \text{ is positive definite}\}$$

is open in $\mathbf{R}^D$.

12.3.6 Give an example of two subsets A and B of $\mathbf{R}$, for which there exist continuous bijections $f : A \to B$ and $g : B \to A$, but which are not homeomorphic.

12.3.7 Show that the interior of the boundary of a subset of a metric space is empty.

12.3.8 Suppose that (X, d) is a metric space and that $a \in X$. Show that the following are equivalent

   (i) a is an isolated point of (X, d).

   (ii) $\{a\}$ is open.

   (iii) Any real-valued function on X is continuous at a.

   (iv) If $x_n \to a$ as $n \to \infty$ then there exists $N \in \mathbf{N}$ such that $x_n = a$ for $n \geq N$.

12.3.9 Suppose that A is a subset of a normed space $(E, \|.\|)$, that $x \in E$ and that λ is a scalar. Show that $\overline{x + A} = x + \overline{A}$, that $\overline{\lambda A} = \lambda \overline{A}$ and that $(x + A)^{\circ} = x + A^{\circ}$. Under what circumstances is $(\lambda A)^{\circ} = \lambda A^{\circ}$? Show that $\overline{A}$ and A° are convex if A is convex, and that $\overline{-A} = -\overline{A}$ and $(-A)^{\circ} = -A^{\circ}$.

12.3.10 A collection $\mathcal{B}$ of open subsets of a metric space (X, d) is a *base* or *basis for the topology* if every open subset of X is a union of sets in $\mathcal{B}$. Show that the collection of open intervals of $\mathbf{R}$ with rational endpoints is a basis for the usual topology of $\mathbf{R}$.

12.3.11 Suppose that (X, d) is a perfect metric space, and that S is a dense subset of X. Show that if F is a finite subset of S then $S \backslash F$ is dense in X. Show (using the axiom of dependent choice) that there exists an infinite subset J of S such that $S \setminus J$ is dense in X.

12.3.12 Show that a separable metric space has a countable basis for the topology.

12.3.13 Show that a metric space with a countable basis for the topology is separable.

12.3.14 Show that a metric subspace of a separable metric space is separable.

12.3.15 Show that a set S, with the discrete metric, has a countable basis for the topology if and only if it is countable.

12.3.16 Use the three preceding exercises to give another proof that if X is infinite then $B(X)$, with the uniform metric, is not separable.

12.3.17 Let $c_0 = \{x \in l_\infty : x_n \to 0 \text{ as } n \to \infty\}$. Show that c_0 is a separable closed linear subspace of l_∞.

12.3.18 Suppose that $\mathcal{U}$ is a set of open subsets of a separable metric space, any two of which are disjoint. Show that $\mathcal{U}$ is countable.

12.3.19 Suppose that f is a real-valued function on a metric space (X, d). f has a *strict local maximum* at x if there exists $\epsilon > 0$ such that if

$0 < d(x, y) < \epsilon$ then $f(y) < f(x)$. Show that if (X, d) is separable then the set of strict local maxima is countable. (Consider a countable basis $\mathcal{B}$ for the topology, and consider the sets of $\mathcal{B}$ on which f is bounded above, and attains its supremum at a unique point).

12.3.20 Suppose that f is a mapping from a metric space (X, d) into a metric space (Y, ρ). f has a *removable discontinuity* at a if a is a limit point of X, if $l = \lim_{x \to a} f(x)$ exists, and $l \neq f(a)$. Show that if (X, d) is separable then f has only countably many removable discontinuities.

12.3.21 A metric d, which like the p-adic metric, satisfies

$$d(x, z) \le \max(d(x, y), d(y, z)) \text{ for } x, y, z \in X,$$

is called an *ultrametric*. Verify that the p-adic metric on $\mathbf{Q}$ is an ultrametric.

12.3.22 Show that if one considers the three distances between three points of an ultrametric space then either they are all equal or two are equal, and greater than the third.

12.3.23 Show that an open ϵ-neighbourhood in an ultrametric space is closed.

12.3.24 Suppose that $N_\epsilon(x)$ is an open ϵ-neighbourhood in an ultrametric space and that $y \in N_\epsilon(x)$. Show that $N_\epsilon(x) = N_\epsilon(y)$.

12.4 Topological properties of metric spaces

Recall that the *topology* of a metric space is the collection of open subsets. Many, but by no means all, of the properties of a metric space (X, d) and of mappings from (X, d) into a metric space (Y, ρ), can be defined in terms of the topologies of (X, d) and (Y, ρ). These are called *topological* properties. Thus

- convergent sequences;
- closure point, closure, closed set, dense set, separability;
- interior, frontier or boundary;
- limit point, isolated point, derived set, perfect set

are all topological notions. On the other hand, the notion of an ϵ-neighbourhood is not a topological one. But if we define a *neighbourhood* of a point x of a metric space (X, d) to be a set which contains $N_\epsilon(x)$ for some $\epsilon > 0$, then the notion of neighbourhood is a topological one, since N is a neighbourhood of x if and only if x is in the interior of N. A collection $\mathcal{N}$ of subsets of X is called a *base of neighbourhoods* of x if each $N \in \mathcal{N}$ is a neighbourhoood of x, and if each neighbourhood of x contains an element of $\mathcal{N}$. Thus the ϵ-neighbourhoods of x form a base of neighbourhoods of x, and the set $\{N_{1/n}(x) : n \in \mathbf{N}\}$ is a countable base of neighbourhoods of x. Let $M_\epsilon(x) = \{y \in X : d(y, x) \le \epsilon\}$. Then $M_\epsilon(x)$ is a neighbourhood of x, since $N_\epsilon(x) \subseteq M_\epsilon(x)$. The set $M_\epsilon(x)$ is closed, and is called the *closed ϵ-neighbourhood* of x. If $0 < \eta < \epsilon$ then $M_\eta(x) \subseteq N_\epsilon(x)$, so that the set $\{M_\epsilon(x) : \epsilon > 0\}$ of closed neighbourhoods is also a base of neighbourhoods of x.

Continuity is also a topological property. Let us make this explicit.

Theorem 12.4.1 *Suppose that f is a mapping from a metric space (X, d) into a metric space (Y, ρ) and that $a \in X$.*

(a) f is continuous at a if and only if whenever N is a neighbourhood of $f(a)$ in Y then $f^{-1}(N)$ is a neighbourhood of a.

(b) The following are equivalent.

(i) f is continuous on X.

(ii) If U is an open subset of Y then $f^{-1}(U)$ is open in X.

(iii) If F is a closed subset of Y then $f^{-1}(F)$ is closed in X.

(iv) If A is a subset of X then $f(\overline{A}) \subseteq \overline{f(A)}$.

Proof (a) Suppose that f is continuous at a and that N is a neighbourhood of $f(a)$. Then there exists $\epsilon > 0$ such that $N_\epsilon(f(a)) \subseteq N$. Since f is continuous at a there exists $\delta > 0$ such that if $d(x, a) < \delta$ then $\rho(f(x), f(a)) < \epsilon$. This says that $N_\delta(a) \subseteq f^{-1}(N_\epsilon(f(a)))$, so that $N_\delta(a) \subseteq f^{-1}(N)$, and $f^{-1}(N)$ is a neighbourhood of a.

Conversely, suppose the condition is satisfied. If $\epsilon > 0$ then $N_\epsilon(f(a))$ is a neighbourhood of $f(a)$, and so $f^{-1}(N_\epsilon(f(a)))$ is a neighbourhood of a. Thus there exists $\delta > 0$ such that $N_\delta(a) \subseteq f^{-1}(N_\epsilon(f(a)))$. Being interpreted, this says that if $d(x, a) < \delta$ then $\rho(f(x), f(a) < \epsilon$.

(b) Suppose that f is continuous on X, that U is open in Y and that $x \in f^{-1}(U)$. Then $f(x) \in U$. Since U is open, there exists $\epsilon > 0$ such that $N_\epsilon(f(x)) \subseteq U$. Since f is continuous at x, there exists $\delta > 0$ such that if $d(x', x) < \delta$ then $\rho(f(x'), f(x)) < \epsilon$. Thus $N_\delta(x) \subseteq f^{-1}(U)$, and so x is an interior point of $f^{-1}(U)$. Since this holds for all $x \in f^{-1}(U)$, $f^{-1}(U)$ is open: (i) implies (ii).

Conversely, suppose that (ii) holds. Suppose that $a \in X$ and that N is a neighbourhood of $f(a)$. Then there exists $\epsilon > 0$ such that $N_\epsilon(f(a)) \subseteq N$. Then $a \in f^{-1}(N_\epsilon(f(a))) \subseteq f^{-1}(N)$, and $f^{-1}(N_\epsilon(f(a)))$ is open, by hypothesis, so that $f^{-1}(N)$ is a neighbourhood of a. Thus f is continuous at a. Since this is true for all $a \in X$, (ii) implies (i).

Since a set is open if and only if its complement is closed, and since $f^{-1}(C(B)) = C(f^{-1}(B))$, (ii) and (iii) are equivalent.

Suppose that (iii) holds, and that $A \subseteq X$. Then $\overline{f(A)}$ is closed in Y, and so $f^{-1}(\overline{f(A)})$ is closed in X. But $A \subseteq f^{-1}(\overline{f(A)})$, and $\overline{A}$ is the smallest closed set containing A, and so $\overline{A} \subseteq f^{-1}(\overline{f(A)})$; that is, $f(\overline{A}) \subseteq \overline{f(A)}$. Thus (iii) implies (iv).

Suppose that (iv) holds, and that B is closed in Y. By hypothesis, $f(\overline{f^{-1}(B)}) \subseteq \overline{f(f^{-1}(B))}$. But $f(f^{-1}(B)) \subseteq B$, so that $\overline{f(f^{-1}(B))} \subseteq \overline{B} = B$. Thus $f(\overline{f^{-1}(B)}) \subseteq B$, and so $\overline{f^{-1}(B)} \subseteq f^{-1}(B)$. Consequently $\overline{f^{-1}(B)} = f^{-1}(B)$: $f^{-1}(B)$ is closed. Thus (iv) implies (iii). $\qquad \square$

Corollary 12.4.2 *Suppose that f is continuous on X and that A is a dense subset of X. Then $f(A)$ is dense in the metric subspace $f(X)$ of (Y, ρ). In particular, if X is separable, then so is $f(X)$.*

Proof   For $f(X) = f(\overline{A}) \subseteq \overline{f(A)}$. $\qquad \square$

Corollary 12.4.3 *Suppose that f is a bijective mapping f from a metric space (X, d) onto a metric space (Y, ρ). The following are equivalent:*

1. *f is a homeomorphism.*
2. *U is open in (X, d) if and only if $f(U)$ is open in (Y, ρ).*
3. *B is closed in (X, d) if and only if $f(B)$ is closed in (Y, ρ).*
4. *$f(\overline{A}) = \overline{f(A)}$ for every subset A of X.*

Corollary 12.4.4 *Suppose that f is a bijective mapping f from a metric space (X, d) onto a metric space (Y, ρ). Then f is a homeomorphism if and only if whenever $(x_n)_{n=1}^{\infty}$ is a sequence in X then $x_n \to x$ in (X, d) as $n \to \infty$ if and only if $f(x_n) \to f(x)$ as $n \to \infty$.*

There are two points to notice about this theorem and its proof. The first is that one needs facility at handling images and inverse images of sets. The second and more important point is that the conditions, in terms of open sets and closed sets, that we have given for a function to be continuous involve the *inverse images* of sets in Y, and *not* the images of sets in X.

Exercises

12.4.1 We have defined topological notions or properties to be those that can be defined in terms of the open sets of a metric space. Show that a notion or property is topological if it can be defined in terms of each of the following.

(a) The neighbourhoods of each point.
(b) The closed sets.
(c) The mapping which sends each subset of X to its closure.
(d) The mapping which sends each subset of X to its frontier.

12.4.2 Suppose that f and g are continuous mappings from a metric space (X, d) into a metric space (Y, ρ). Show that the set

$$\{x \in X : f(x) = g(x)\}$$

is closed in X.

12.4.3 Suppose that A is a non-empty subset of a metric space (X, d). If $x \in X$ let $d(x, A) = \inf\{d(x, a) : a \in A\}$. Show that

$$\overline{A} = \{x \in X : d(x, A) = 0\}.$$

12.4.4 Suppose that A and B are disjoint closed subsets of a metric space. If $x \in X$ let $g(x) = d(x, A) + d(x, B)$. Show that $g(x)$ is a continuous strictly positive function on X. Let $h(x) = d(x, A)/g(x)$. Show that g is a continuous function on X which satisfies

- $0 \leq h(x) \leq 1$, for $x \in X$;
- $h(x) = 0$, for $x \in A$;
- $h(x) = 1$, for $x \in B$.

This is easy. Its extension to certain topological spaces is *Urysohn's lemma*, which we shall prove later (Theorem 13.4.6); the proof is much harder.

13

Topological spaces

13.1 Topological spaces

The results of the previous section show that many important results concerning metric spaces depend only on the topology. We now generalize this, by introducing the notion of a *topological space*. This is traditionally defined in terms of open sets. A topological space is a set X, together with a collection τ of subsets of X which satisfy:

- the empty set and X are in τ;
- if $\mathcal{O} \subseteq \tau$ then $\bigcup_{O \in \mathcal{O}} O \in \tau$;
- if O_1 and O_2 are in τ then $O_1 \cap O_2 \in \tau$.

Then τ is the *topology* on X, and the sets in τ are called *open* sets. The conditions say that the empty set and X are open, that the union of an arbitrary collection of open sets is open, and that the intersection of finitely many open sets is open.

The first example of a topological space is given by taking the open sets of a metric space (X, d) for the topology on X; this is the *metric space topology* on X. A topological space (X, τ) is said to be *metrizable* if there is a metric d on X such that τ is the set of open sets of the metric space (X, d).

Why do we make this definition? First, there are many important examples of topological spaces in various areas of mathematics, including not only analysis but also logic, algebra and algebraic geometry, which are not given by a metric. In fact, we shall not need any of these, but it is as well to know that they exist. Secondly, metric spaces have a rich structure, and it is appropriate to develop topological properties of metric spaces in a purely topological way – this helps us to appreciate the nature of these properties. Thirdly, there are many examples of topological spaces with weird and wonderful properties, and it is entertaining to investigate them; we shall do this in Section 13.6.

Starting from a topology τ on a set X, we can immediately set up the machinery that has been defined for metric spaces. Suppose that $x \in X$ and that A is a subset of X. Here are the definitions.

- A subset N of X is a *neighbourhood*, or τ-*neighbourhood*, of x if there is an open set O such that $x \in O \subseteq N$. The set of neighbourhoods of x is denoted by $\mathcal{N}_x$. A subset $\mathcal{B}$ of $\mathcal{N}_x$ is called a *base of neighbourhoods* of x if whenever $N \in \mathcal{N}_x$ there exists $B \in \mathcal{B}$ with $B \subseteq N$.
- A subset $M^*(x)$ of X is a *punctured neighbourhood* of x if there is a neighbourhood M of x such that $M^*(x) = M \setminus \{x\}$.
- x is a *limit point*, or *accumulation point*, of A if $M^*(x) \cap A \neq \emptyset$ for every punctured neighbourhood $M^*(x)$ of x. The set of limit points of A is called the *derived set* of A, and is denoted by A'. A is said to be *perfect* if $A = A'$.
- x is a *closure point* of A if $N \cap A \neq \emptyset$ for every $N \in \mathcal{N}_x$. The set of closure points of A is called the *closure* of A, and is denoted by $\overline{A}$. A is said to be *closed* if $A = \overline{A}$. A is said to be *dense* in X if $\overline{A} = X$. (X, τ) is *separable* if there is a countable subset C of X which is dense in X.
- x is an *isolated point* of A if there exists $N \in \mathcal{N}_x$ such that $N \cap A = \{x\}$. If $i(A)$ is the set of isolated points of A then $A' \cap i(A) = \emptyset$ and $\overline{A} = A' \cup i(A)$.
- x is an *interior point* of A if $A \in \mathcal{N}_x$. The set of interior points of A is the *interior* of A; it is denoted by A°.
- The *frontier*, or *boundary*, ∂A is the set $\overline{A} \setminus A^\circ$.
- A subset $\mathcal{B}$ of a topology τ on a set X is a *base* for the topology if every open set is the union of subsets in $\mathcal{B}$.
- Suppose that $(x_n)_{n=1}^\infty$ is a sequence in X. Then $x_n \to x$ as $n \to \infty$ if for each $N \in \mathcal{N}_x$ there exists $n_0 \in \mathbf{N}$ such that $x_n \in N$ for all $n \geq n_0$.
- Suppose that f is a mapping from A into a topological space (Y, σ), that b is a limit point of A, and that $l \in Y$. Then $f(x) \to l$ as $x \to b$ in A (in words, $f(x)$ *tends* to, or *converges* to, l as x tends to b in A) if whenever N is a neighbourhood of l then there is a punctured neighbourhood $M^*(x)$ of b such that $f(M^*(x) \cap A) \subseteq N$.
- Suppose that f is a mapping from X into a topological space (Y, σ). Then $f(x)$ is *continuous at* x if, whenever N is a neighbourhood of $f(x)$ then $f^{-1}(N)$ is a neighbourhood of x. f is *continuous on* X (or, simply, is continuous), if it is continuous at each point of X. If τ_1 and τ_2 are topologies on X and the identity mapping $i : (X, \tau_1) \to (X, \tau_2)$ is continuous, then we say that τ_1 is *finer* or *stronger* than τ_2, and that τ_2 is *coarser* or *weaker* than τ_1. This happens if and only if $\tau_2 \subseteq \tau_1$.

- Suppose that f is a bijective mapping from X onto a topological space (Y, σ). If f and f^{-1} are both continuous, then f is called a *homeomorphism* of (X, τ) onto (Y, σ).

Before investigating the use of these definitions, let us give some examples of topological spaces. The reader should verify that in each instance the conditions for being a topology are satisfied.

1. If X is any set, let $\tau = \{\emptyset, X\}$. This is the *indiscrete topology*.
2. If X is any set, let $\tau = P(X)$, the set of all subsets of X. This is the *discrete topology*. It is the metric space topology defined by the discrete metric d, where $d(x, y) = 1$ if $x \neq y$ and $d(x, x) = 0$.
3. Suppose that Y is a subset of a topological space (X, τ). Then $\tau_Y = \{O \cap Y : O \in \tau\}$ is a topology on Y, called the *subspace* topology. (Y, τ_Y) is then a *topological subspace* of (X, τ). Topological subspaces inherit many, but not all, of the properties of the larger space.
4. Suppose that q is a mapping of a topological space (X, τ) onto a set S. The collection $\{U \subseteq S : q^{-1}(U) \in \tau\}$ of subsets of S is a topology on S, the *quotient topology*. In many cases, it is very badly behaved, and quotient topologies are a rich source of idiosyncracies and counterexamples.
5. If X is an infinite set, let τ_f be the collection of subsets of X with a finite complement, together with the empty set. This is the *cofinite topology*.
6. If X is an uncountable set, let τ_σ be the collection of subsets of X with a countable complement, together with the empty set. This is the *cocountable* topology.
7. Let $\tau_- = \{\emptyset\} \cup \{(-\infty, a) : a \in \mathbf{R}\} \cup \{\mathbf{R}\}$. Then τ is a *one-sided topology* on $\mathbf{R}$; another is $\tau_+ = \{\emptyset\} \cup \{(a, \infty) : a \in \mathbf{R}\} \cup \{\mathbf{R}\}$.
8. Let P denote the vector space of complex polynomials in two variables. If S is a subset of P, let

$$U_S = \{(z_1, z_2) \in \mathbf{C}^2 : p(z_1, z_2) \neq 0 \text{ for } p \in S\}.$$

Then it can be shown that the collection of sets $\{U_S : S \subseteq P\}$ is a topology on $\mathbf{C}^2$, the *Zariski topology*. This definition can be extended to other settings in algebraic geometry and in ring theory, where it is an important tool. (It does not have any clear use in analysis.)

We now establish some elementary results about topological spaces. In many cases, the arguments are similar to those for the real line, or for metric spaces, and the details are left to the reader.

Proposition 13.1.1 *Suppose that $\mathcal{B}$ is a collection of subsets of a set X which satisfies*

(i) If $B_1, B_2 \in \mathcal{B}$ then $B_1 \cap B_2 \in \mathcal{B}$, and

(ii) $\cup \{B : B \in \mathcal{B}\} = X$.

Then there is a unique topology τ on X for which $\mathcal{B}$ is a base.

Proof   Let τ be the collection of unions of sets in $\mathcal{B}$. Then the empty set is in τ (the union of the empty set of subsets of $\mathcal{B}$) and $X \in \tau$, by (ii). Clearly the union of sets in τ is in τ, and so it remains to show that finite intersections of sets in τ are in τ. For this, it is sufficient to show that if $U = \cup_{C \in \mathcal{C}} C$ and $V = \cup_{D \in \mathcal{D}} D$ are in τ (where $\mathcal{C}$ and $\mathcal{D}$ are subsets of $\mathcal{B}$), then $U \cap V \in \tau$. But this holds, since $U \cap V = \cup_{C \in \mathcal{C}, D \in \mathcal{D}} (C \cap D)$, which is in τ.

It follows from the construction that τ is unique. For if σ is a topology on X for which $\mathcal{B}$ is a base, then $\sigma \subseteq \tau$, by the definition of a base, and $\tau \subseteq \sigma$, since the union of open sets is open. $\square$

Let us give an example. The subsets $[a, b)$ of $\mathbf{R}$, where $a < b$, satisfy the conditions of the proposition, and so define a topology, the *right half-open interval topology* on $\mathbf{R}$. Note that

$$(a, b) = \cup\{[(1 - \lambda)a + \lambda b, b) : 0 < \lambda < 1\},$$

so that (a, b) is open in this topology; from this it follows that the usual topology on $\mathbf{R}$ is weaker than the right half-open interval topology.

Proposition 13.1.2   *Suppose that A and B are subsets of a topological space (X, d).*

(i) *If $A \subseteq B$ then $\overline{A} \subseteq \overline{B}$.*

(ii) *$\overline{A}$ is closed.*

(iii) *$\overline{A}$ is the smallest closed set containing A: if C is closed and $A \subseteq C$ then $\overline{A} \subseteq C$.*

Proof   (i) follows trivially from the definition of closure.

(ii) Suppose that $b \notin \overline{A}$. Then there exists a neighbourhood N of b such that $N \cap A = \emptyset$, and there exists an open set U such that $b \in U \subseteq N$. If $x \in U$ then $x \notin \overline{A}$, since $U \cap A = \emptyset$. Thus b is not in the closure of $\overline{A}$. Since this holds for all $b \notin \overline{A}$, $\overline{A}$ is the closure of $\overline{A}$ and $\overline{A}$ is closed.

(iii) By (i), $\overline{A} \subseteq \overline{C} = C$. $\square$

Theorem 13.1.3   *Suppose that (Y, τ_Y) is a topological subspace of a topological space (X, τ) and that $A \subseteq Y$. Let $\overline{A}^Y$ denote the closure of A in Y, and $\overline{A}^X$ the closure in X.*

(i) *$\overline{A}^Y = \overline{A}^X \cap Y$.*

(ii) *A is closed in Y if and only if there exists a closed set B in X such that $A = B \cap Y$.*

Proof (i) Certainly $\overline{A}^Y \subseteq \overline{A}^X$, so that $\overline{A}^Y \subseteq \overline{A}^X \cap Y$. On the other hand, if $y \in \overline{A}^X \cap Y$ and N is a τ-neighbourhood of y in X, then there exists an open subset U in X such that $y \in U \subseteq N$, and $U \cap A \neq \emptyset$. Now $U \cap Y$ is a τ_Y open set, $y \in U \cap Y$ and $(U \cap Y) \cap A = U \cap A$ is not empty. Thus $y \in \overline{A}^Y$.

(ii) If A is closed in Y, then $A = \overline{A}^Y = \overline{A}^X \cap Y$, so that we can take $B = \overline{A}^X$. Conversely if B is closed in X and $A = B \cap Y$, then $\overline{A}^X \subseteq B$, so that $\overline{A}^Y = \overline{A}^X \cap Y \subseteq B \cap Y = A$, and A is closed in Y. $\qquad\square$

Proposition 13.1.4 *Suppose that A and B are subsets of a topological space (X, d).*

(i) If $A \subseteq B$ then $A^\circ \subseteq B^\circ$.

(ii) $C(A^\circ) = \overline{C(A)}$.

(iii) A is open if and only if $C(A)$ is closed.

(iv) A° is open.

(v) A° is the largest open set contained in A: if U is open and $U \subseteq A$ then $U \subseteq A^\circ$.

Proof This follows by making obvious modifications to the proof of Proposition 12.3.10. $\qquad\square$

Thus, in the examples above, a subset F of (X, τ_f) is closed if and only if it is finite, or the whole space, and a subset F of (X, τ_σ) is closed if and only if it is countable, or the whole space.

Proposition 13.1.5 *Suppose that (X, τ) is a topological space, that $x \in X$ and that $\mathcal{N}_x$ is the collection of neighbourhoods of x. Then $\mathcal{N}_x$ is a filter: that is*

(i) each $N \in \mathcal{N}_x$ is non-empty;

(ii) if $N_1, N_2 \in \mathcal{N}_x$ then $N_1 \cap N_2 \in \mathcal{N}_x$;

(iii) if $N \in \mathcal{N}_x$ and $N \subseteq M$ then $M \in \mathcal{N}_x$.

A subset O of X is open if and only if it is a neighbourhood of each of its points.

Proof (i) Since $x \in N$, N is not empty.

(ii) There exist open sets O_1 and O_2 such that $x \in O_1 \subseteq N_1$ and $x \in O_2 \subseteq N_2$. Then $O_1 \cap O_2$ is open, and $x \in O_1 \cap O_2 \subseteq N_1 \cap N_2$.

(iii) This is trivial.

If O is open, then it follows from the definition of neighbourhood that O is a neighbourhood of each of its points. Conversely, if the condition is satisfied, then $O = O^\circ$, and so O is open. $\qquad\square$

The analogues of Theorem 12.2.3 and the sandwich principle also hold; simple modifications to the proofs are needed, and the details are left to the reader.

Composition also works well; the proof is easy, but the result is of fundamental importance.

Theorem 13.1.6 *Suppose that f is a mapping from a topological space (X, τ) into a topological space (Y, ρ) and that g is a mapping from Y into a topological space (Z, σ). If f is continuous at $a \in X$ and g is continuous at $f(a)$, then $g \circ f$ is continuous at a.*

Proof   If $N \in \mathcal{N}_{g(f(a))}$ then, since g is continuous at $f(a)$, $g^{-1}(N) \in \mathcal{N}_{f(a)}$, and since f is continuous at a, $f^{-1}(g^{-1}(N)) \in \mathcal{N}_a$. The result follows, since $(g \circ f)^{-1}(N) = f^{-1}(g^{-1}(N))$. $\qquad\qquad\square$

The next result corresponds to Theorem 12.4.1. We give some details of the proof, though the proof is essentially the same.

Theorem 13.1.7 *Suppose that f is a mapping from a topological space (X, τ) into a topological space (Y, σ). The following are equivalent.*
   (i) f is continuous on X.
   (ii) If U is an open subset of Y then $f^{-1}(U)$ is open in X.
   (iii) If F is a closed subset of Y then $f^{-1}(F)$ is closed in X.
   (iv) If A is a subset of X then $f(\overline{A}) \subseteq \overline{f(A)}$.

Proof   Suppose that f is continuous, that U is open in Y and that $x \in f^{-1}(U)$. Then $f(x) \in U$, and $U \in \mathcal{N}_{f(x)}$. Since f is continuous at x, $f^{-1}(U) \in \mathcal{N}_x$, and so x is an interior point of $f^{-1}(U)$. Since this holds for all $x \in f^{-1}(U)$, $f^{-1}(U)$ is open: (i) implies (ii).

Conversely, suppose that (ii) holds. Suppose that $a \in X$ and that N is a neighbourhood of $f(a)$. Then there exists an open set U in Y such that $f(a) \in U \subseteq N$. Then $a \in f^{-1}(U) \subseteq f^{-1}(N)$, and $f^{-1}(U)$ is open, by hypothesis, so that $f^{-1}(N)$ is a neighbourhood of a. Thus f is continuous at a. Since this is true for all $a \in X$, (ii) implies (i).

Since a set is open if and only if its complement is closed, and since $f^{-1}(C(B)) = C(f^{-1}(B))$, (ii) and (iii) are equivalent.

The proof of the equivalence of (iii) and (iv) is exactly the same as the proof in Theorem 12.4.1. $\qquad\qquad\square$

Corollary 13.1.8 *Suppose that f is continuous and that A is a dense subset of X. Then $f(A)$ is dense in the topological subspace $f(X)$ of (Y, ρ). In particular, if X is separable, then so is $f(X)$.*

Proof   This follows from condition (iv). □

Corollary 13.1.9   *Suppose that f is a bijective mapping f from a topological space (X, d) onto a topological space (Y, ρ). The following are equivalent:*

(i) f is a homeomorphism.

(ii) U is open in (X, d) if and only if $f(U)$ is open in (Y, ρ).

(iii) B is closed in (X, d) if and only if $f(B)$ is closed in (Y, ρ).

(iv) $f(\overline{A}) = \overline{f(A)}$ for every subset A of X.

(v) $f(A^\circ) = (f(A))^\circ$ for every subset A of X.

What about sequences? We shall see that there are some positive results, but that, in general, sequences are inadequate for the definition of topological properties.

Proposition 13.1.10   *Suppose that A is a subset of a topological space (X, τ) and that $b \in X$.*

(i) If there exists a sequence $(a_j)_{j=1}^\infty$ in $A \setminus \{b\}$ such that $a_j \to b$ as $j \to \infty$ then b is a limit point of A.

(ii) If there exists a sequence $(a_j)_{j=1}^\infty$ in A such that $a_j \to b$ as $j \to \infty$, then b is a closure point of A.

The converses of (i) and (ii) are false.

Proof   (i) If $M^*(b)$ is a punctured neighbourhood of b then there exists $j_0 \in \mathbf{N}$ such that $a_j \in M^*(b)$ for $j \geq j_0$. Thus $A \cap M^*(b) \neq \emptyset$, so that b is a limit point of A.

The proof of (ii) is exactly similar.

We shall use the same example to show that the converses do not hold. Suppose that X is an uncountable set, with the cocountable topology τ_σ described above. Suppose that A is any uncountable proper subset of X. If $x \in X$ and if $M^*(x)$ is any punctured neighbourhood of X, then $A \cap M^*(x)$ is non-empty, so that $A' = \overline{A} = X$: every point of X is a limit point, and therefore a closure point, of A. Suppose that $a_j \in A$ and that $a_j \to b$ as $j \to \infty$. Then $N = X \setminus \{a_j : a_j \neq b\}$ is a neighbourhood of b, and so there exists j_0 such that $a_j \in N$ for $j \geq j_0$. Thus $a_j = b$ for $j \geq j_0$: the sequence $(a_j)_{j=1}^\infty$ is eventually constant and $b \in A$. Thus if $c \notin A$ there is no sequence in A which converges to c. □

Proposition 13.1.11   *Suppose that (X, τ) and (Y, σ) are topological spaces, that $(a_j)_{j=1}^\infty$ is a sequence in X which converges to a as $j \to \infty$ and* *that f is a mapping from X to Y. If f is continuous at a then $f(a_j) \to f(a)$ as $j \to \infty$. The converse is false.*

Proof If $N \in \mathcal{N}(f(a))$ then $f^{-1}(N) \in \mathcal{N}_a$. There exists $j_0 \in \mathbf{N}$ such that $a_j \in f^{-1}(N)$ for $j \geq j_0$. Thus $f(a_j) \in N$ for $j \geq j_0$: $f(a_j) \to f(a)$ as $j \to \infty$.

Let X be an uncountable set, let τ_σ be the cocountable topology on X and let τ be the discrete topology. The identity mapping i from (X, τ_σ) into (X, τ) has no points of continuity, but a sequence converges in (X, τ_σ) if and only if it is eventually constant, in which case it converges in the discrete topology. $\square$

Exercises

13.1.1 Suppose that f is a mapping from a topological space (X, τ) to a topological space (Y, σ), and that A and B are two closed subsets of X whose union is X. Show that f is continuous on X if and only if its restriction to A and its restriction to B are continuous. Is the same true if 'closed' is replaced by 'open'? What if A is open and B is closed?

13.1.2 Give an example of a topology on $\mathbf{N}$ with the property that every non-empty proper subset of $\mathbf{N}$ is either open or closed, but not both.

13.1.3 Let X be an infinite set, with the cofinite topology τ_f. What are the convergent sequences? Show that a convergent sequence either converges to one point of X or to every point of X.

13.1.4 Give an example of a continuous mapping f from a metric space (X, d) to a metric space (Y, ρ) and a subset A of X for which $f(A^\circ) \not\subseteq (f(A))^\circ$, and an example for which $f(A^\circ) \not\supseteq (f(A))^\circ$.

13.1.5 Suppose that (X, τ) is a topological space and that (Y, d) is a metric space. Let $C_b(X, Y)$ be the space of bounded continuous mappings from X into Y. Show that $C_b(X, Y)$ is a closed subset of the space $(B_Y(X), d_\infty)$ of all bounded mappings from X to Y, with the uniform metric.

13.1.6 Verify that the collection of subsets $\{(-\infty, a) : a \in \mathbf{R}\}$ of $\mathbf{R}$, together with $\emptyset$ and $\mathbf{R}$, is a topology τ_- on $\mathbf{R}$.

A real-valued function on a topological space (X, τ) is said to be *upper semi-continuous* at x if, given $\epsilon > 0$ there exists a neighbourhood N of x such that $f(y) < f(x) + \epsilon$ for $y \in N$. Show that f is upper semi-continuous at x if and only if the mapping $f : (X, \tau) \to (\mathbf{R}, \tau_-)$ is continuous at x.

13.2 The product topology

Suppose that $\{(X_\alpha, \tau_\alpha)\}_{\alpha \in A}$ is a family of topological spaces. Is there a sensible way of defining a topology on $X = \prod_{\alpha \in A} X_\alpha$? In order to see how to answer this question, let us consider a simple example. The space $\mathbf{R}^d$ is the product of d copies of $\mathbf{R}$. If $x = (x_1, \ldots, x_d)$, let $\pi_j(x) = x_j$, for $1 \le j \le d$. The mapping $\pi_j : \mathbf{R}^d \to \mathbf{R}$ is the *jth coordinate projection*. If we give $\mathbf{R}^d$ and $\mathbf{R}$ their usual topologies, we notice four phenomena.

- Since $|\pi_j(x) - \pi_j(y)| \le d(x, y)$ (where d is the Euclidean metric on $\mathbf{R}^d$), each of the mappings π_j is continuous.
- For $1 \le j \le d$ let us denote the set $\{1, \ldots, d\} \setminus \{j\}$ by $d \setminus \{j\}$. Suppose that $y \in \mathbf{R}^{d \setminus \{j\}}$. If $x \in \mathbf{R}$ let $k_{y,j} : \mathbf{R} \to \mathbf{R}^d$ be defined by

$$(k_{y,j}(x))_j = x \text{ and } (k_{y,j}(x))_i = y_i \text{ for } i \in d \setminus \{j\}.$$

Let

$$C_{y,j} = \{x \in \mathbf{R}^d : x_i = y_i \text{ for } i \in d \setminus \{j\}\};$$

$C_{y,j}$ is called the *cross-section* of $\mathbf{R}^d$ at y and the mapping $k_{y,j}$ the *cross-section mapping*. Then the mapping $k_{y,j}$ is an isometry of $\mathbf{R}$ onto $C_{y,j}$.
- Suppose that $f : (X, \tau) \to \mathbf{R}^d$ is a mapping from a topological space (X, τ) into $\mathbf{R}^d$. We can then write $f(x) = (f_1(x), \ldots, f_d(x))$, where $f_j = \pi_j \circ f$. If f is continuous, then the composition $f_j = \pi_j \circ f$ is continuous. But the converse also holds. Suppose that each of the mappings f_j is continuous, that $x \in X$ and $\epsilon > 0$. Then for each j there exists a neighbourhood N_j of x for which $|f_j(y) - f_j(x)| < \epsilon/\sqrt{d}$ for $y \in N_j$. Then $N = \cap_{j=1}^d N_j$ is a

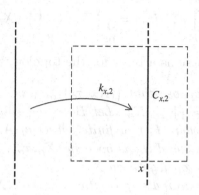

Figure 13.2. The cross-section $C_{x,2}$.

neighbourhood of x, and if $y \in N$ then

$$d(f(y), f(x)) = \left(\sum_{j=1}^{d} |f_j(y) - f_j(x)|^2 \right)^{\frac{1}{2}} < \epsilon,$$

so that f is continuous at x.

- On the other hand, suppose that $f : \mathbf{R}^d \to (Y, \sigma)$ is a mapping from $\mathbf{R}^d$ to a topological space (Y, σ), that $x \in \mathbf{R}^d$ and that $1 \leq j \leq d$. Let $\hat{x}_j(i) = x_i$ for $i \in d \setminus \{j\}$, so that $\hat{x}_j \in \mathbf{R}^{d \setminus \{j\}}$. If f is continuous, then the mapping $f \circ k_{\hat{x}_j, j} : \mathbf{R} \to (Y, \sigma)$ is continuous, for $1 \leq j \leq n$, but the converse need not be true. For example, the real-valued function f on $\mathbf{R}^2$ defined by $f(0, 0) = 0$ and $f(x, y) = xy/(x^2 + y^2)$ is continuous at every point of $\mathbf{R}^2$ except $(0, 0)$, but the mappings $x \to f(x, y') : \mathbf{R} \to \mathbf{R}$ and $y \to f(x', y)$ are continuous, for all x' and y'. We need to distinguish these phenomena carefully. We say that $f : \mathbf{R}^d \to (Y, \sigma)$ is *jointly continuous* at x if it is continuous, and say that it is *separately continuous* at x if the mapping $f \circ k_{\hat{x}_j, j}$ from $\mathbf{R}$ to (Y, σ) is continuous at x_j, for $1 \leq j \leq d$.

We use these observations to motivate the definition of the product topology on $X = \prod_{\alpha \in A} X_\alpha$. We want to define a topology τ on X for which each of the coordinate mappings $\pi_\alpha : (X, \tau) \to (X_\alpha, \tau_\alpha)$ is continuous. Thus we require that if U_α is open in X_α then

$$\pi_\alpha^{-1}(U_\alpha) = U_\alpha \times \prod_{\beta \neq \alpha} X_\beta$$

is in τ. Since finite intersections of open sets are open, we also require that if F is a finite subset of the index set A and U_α is open in X_α, for each $\alpha \in F$, then

$$\bigcap_{\alpha \in F} \pi_\alpha^{-1}(U_\alpha) = \prod_{\alpha \in F} U_\alpha \times \prod_{\beta \in A \setminus F} X_\beta$$

is in τ. We can take these as a basis for the topology we need.

Theorem 13.2.1 *Suppose that $\{(X_\alpha, \tau_\alpha)\}_{\alpha \in A}$ is a family of topological spaces, and that $X = \prod_{\alpha \in A} X_\alpha$. Let $\mathcal{B}$ be the collection of sets of the form $\bigcap_{\alpha \in F} \pi_\alpha^{-1}(U_\alpha)$, where F is a finite subset of A, π_α is the coordinate projection of X onto X_α and U_α is open in (X_α, τ_α).*

(i) $\mathcal{B}$ is a base for a topology τ on X.

(ii) Each of the coordinate projections $\pi_\alpha : (X, \tau) \to (X_\alpha, \tau_\alpha)$ is continuous.

(iii) If (Y, σ) is a topological space and $f : (Y, \sigma) \to (X, \tau)$ is a mapping, then f is continuous if and only if each of the mappings $\pi_\alpha \circ f : (Y, \sigma) \to (X_\alpha, \tau_\alpha)$ is continuous.

Further τ is the unique topology on X for which (ii) and (iii) hold.

Proof (i) Since X is the empty intersection of sets of the form $\pi_\alpha^{-1}(U_\alpha)$ $X \in \mathcal{B}$. Suppose that $U = \cap_{\alpha \in F} \pi_\alpha^{-1}(U_\alpha)$ and $V = \cap_{\beta \in G} \pi_\beta^{-1}(V_\beta)$ are in $\mathcal{B}$. Then

$$U \cap V = \left(\cap_{\alpha \in F \setminus G} \pi_\alpha^{-1}(U_\alpha)\right) \cap \left(\cap_{\alpha \in F \cap G} \pi_\alpha^{-1}(U_\alpha \cap V_\alpha)\right) \cap \left(\cap_{\beta \in G \setminus F} \pi_\beta^{-1}(V_\beta)\right),$$

which is in $\mathcal{B}$. Thus $\mathcal{B}$ is the base for a topology on X, by Proposition 13.1.1.

(ii) is a consequence of the definition of $\mathcal{B}$ (take $F = \{\alpha\}$), and shows that the condition in (iii) is necessary. On the other hand, suppose that the conditions are satisfied. If $y \in Y$ then the sets $\{B \in \mathcal{B} : f(y) \in B\}$ form a base of τ-neighbourhoods of $f(y)$. If $y \in Y$ and $N = \cap_{\alpha \in F} \pi_\alpha^{-1}(U_\alpha)$ is such a neighbourhood then

$$f^{-1}(N) = \cap_{\alpha \in F} f^{-1}(\pi_\alpha^{-1}(U_\alpha)) = \cap_{\alpha \in F}(\pi_\alpha \circ f)^{-1}(U_\alpha).$$

Since $\pi_\alpha \circ f$ is continuous, $(\pi_\alpha \circ f)^{-1}(U_\alpha)$ is a neighbourhood of y, and so therefore is $f^{-1}(N)$. Thus f is continuous.

If σ is a topology on X for which (ii) and (iii) hold, then the sets in $\mathcal{B}$ must be in σ, since the coordinate mappings are continuous, and so $\tau \subseteq \sigma$. On the other hand, consider the identity mapping $i : (X, \tau) \to (X, \sigma)$. Each of the mappings $\pi_\alpha \circ i : (X, \tau) \to (X_\alpha, \tau_\alpha)$ is continuous, and so i is continuous, by hypothesis. Thus if $U \in \sigma$ then $i^{-1}(U) = U \in \tau$, and so $\sigma \subseteq \tau$. $\qquad\square$

The topology τ of this theorem is called the *product topology* on $X = \prod_{\alpha \in A} X_\alpha$.

Similar remarks apply about the need to distinguish between joint continuity and separate continuity.

We also have a result concerning cross-sections. Suppose that $\beta \in A$ and that $y \in \prod_{A \setminus \{\beta\}} X_\alpha$. Let $k_{y,\beta} : X_\beta \to X$ be defined by

$$(k_{y,\beta}(x))_\beta = x \text{ and } (k_{y,\beta}(x))_\alpha = y_\alpha \text{ for } \alpha \in A \setminus \{\beta\}.$$

Let

$$C_{y,\beta} = \{x \in X : x_\alpha = y_\alpha \text{ for } \alpha \in A \setminus \{\beta\}\};$$

$C_{y,\beta}$ is called the *cross-section* of X at y and $k_{y,\beta}$ the *cross-section mapping*.

Corollary 13.2.2 *If $\beta \in A$ and $y \in \prod_{A \setminus \{\beta\}} X_\alpha$, the cross-section mapping $k_{y,\beta}$ is a homeomorphism of (X_β, τ_β) onto $C_{y,\beta}$, when $C_{y,\beta}$ is given the subspace topology.*

Proof   It follows from the definition of τ that $k_{y,\beta}$ is continuous. On the other hand, $k_{y,\beta}^{-1}$ is the restriction of π_β to $C_{y,\beta}$.   □

In particular, if $((X_1, \tau_1), \ldots, (X_n, \tau_n))$ is a finite product of topological spaces, then the collection of sets $U_1 \times \cdots \times U_n$, where $U_j \in \tau_j$ for $1 \leq j \leq n$, is a base for the product topology on $\prod_{j=1}^n X_j$.

Similarly, if $A = \mathbf{N}$, so that we have a product of an infinite sequence of topological spaces, then the collection of sets $U_1 \times \cdots \times U_n \times \prod_{j=n+1}^\infty X_j$, where $n \in \mathbf{N}$ and $U_j \in \tau_j$ for $1 \leq j \leq n$, is a base for the product topology on $\prod_{j=1}^\infty X_j$.

Suppose that there is a topological space (X, τ) such that $(X_\alpha, \tau_\alpha) = (X, \tau)$ for each α in A. In this case, we can identify the product $\prod_{\alpha \in A} X_\alpha$ with the space X^A of all functions from A into X. In this case, the product topology is referred to as the *topology of pointwise convergence*. Suppose that b is a limit point of a subset C of a topological space (Y, σ), that f is a mapping of C into X^A, and that $l \in X^A$. If $c \in C$, then $f(c)$ is a function on A taking values in X, which we denote by f_c, and l is also a function on A taking values in X. Then $f_c \to l$, in the topology of pointwise convergence, as $c \to b$ if and only if $f_c(\alpha) \to l(\alpha)$ in X as $c \to b$, for each $\alpha \in A$.

The product topology is a very weak topology. Consider the vector space $\mathbf{R}^{[0,1]}$ of all real-valued functions on $[0, 1]$ with the topology of pointwise convergence. This is a big set – it is bigger than $\mathbf{R}$. Nevertheless, it is separable: let us show that the countable set of all polynomials with rational coefficients is a dense subset. Suppose that $f \in \mathbf{R}^{[0,1]}$ and that N is a neighbourhood of f in the product topology. Then there exists a finite subset F of $[0, 1]$ and $\epsilon > 0$ such that

$$\{g \in \mathbf{R}^{[0,1]} : |g(t) - f(t)| < \epsilon \text{ for } t \in F\} \subseteq N.$$

There exists a real polynomial p of degree $|F| - 1$ such that $p(t) = f(t)$ for $t \in F$, and there exists a polynomial q with rational coefficients, of the same degree, such that $|q(t) - p(t)| < \epsilon$ for $t \in F$. Thus $g \in N$, so that the countable set of all polynomials with rational coefficients is a dense subset of $\mathbf{R}^{[0,1]}$.

Here is another example. Give the two-point set $\{0, 1\}$ the discrete topology. Recall that a mapping f from a set X into $\{0, 1\}$ is called an *indicator function*. We write $f = I_A$, where $A = \{x : f(x) = 1\}$, and call I_A the indicator function of A. The space $\{0, 1\}^X$ of all indicator functions on X, with the product topology, is called the *Bernoulli space* of X, and is denoted by $\Omega(X)$. The space $\Omega(\mathbf{N})$, whose elements are sequences, taking values 0 or 1, is called the *Bernoulli sequence space*. The mapping $I : A \to I_A$ is a bijection of the power set $P(X)$ onto $\Omega(X)$. We can therefore define a topology on $P(X)$ by taking the collection $\{I^{-1}(U) : U \text{ open in } \Omega(X)\}$ as the topology on $P(X)$; we call this topology the *Bernoulli topology*. If A is a subset of X, and F is a finite subset of X, let $N_F(A) = \{B \subseteq X : B \cap F = A \cap F\}$. Then the collection $\{N_F(A) : F \text{ a finite subset of } X\}$ forms a base of open neighbourhoods for A in the Bernoulli topology. Some properties of the Bernoulli space $\Omega([0,1])$ are investigated in the exercises.

Exercises

13.2.1 Suppose that (X_1, τ_1) and (X_2, τ_2) are topological spaces and that $A_1 \subseteq X_1$ and $A_2 \subseteq X_2$. Show that if $X_1 \times X_2$ is given the product topology then

$$\partial(A_1 \times A_2) = (\partial A_1 \times A_2) \cup (A_1 \times \partial A_2).$$

13.2.2 Suppose that $\{(X_\alpha, \tau_\alpha)\}_{\alpha \in A}$ is a family of topological spaces, and that A is the disjoint union of non-empty sets A_1 and A_2. Show that the natural mapping from $X = \prod_{\alpha \in A} X_\alpha$ onto $\prod_{\alpha \in A_1} X_\alpha \times \prod_{\alpha \in A_2} X_\alpha$ is a homeomorphism, when each of the spaces is given its product topology.

13.2.3 Define a partial order on the topologies on a set X by saying that $\tau_1 \leq \tau_2$ if $\tau_1 \subseteq \tau_2$. Suppose that τ_1 and τ_2 are topologies on X. Let $\delta : X \to X \times X$ be the diagonal mapping defined as $\delta(x) = (x, x)$. Let $\tau_1 \times \tau_2$ be the product topology on $X \times X$, and let $\sigma = \{\delta^{-1}(U) : U \in \tau_1 \times \tau_2\}$. Show that σ is a topology on X, and that it is the least upper bound of τ_1 and τ_2. Show that τ_1 and τ_2 have a greatest lower bound, and determine its elements. Thus the topologies on X form a lattice.

13.2.4 Give $P([0,1])$ the Bernoulli topology.
   (i) Show that $P([0,1])$ is separable. (Hint: consider step functions.)
   (ii) Let $A_n([0,1])$ be the collection of subsets of $[0,1]$ with exactly n elements and let $B_n([0,1])$ be the collection of subsets of $[0,1]$ with at most n elements. Show that $A_n([0,1])$ is a discrete subspace of $P([0,1])$.
   (iii) Show that $B_n([0,1])$ is the closure of $A_n([0,1])$.
   (iv) Show that $B_n([0,1])$ is not separable.

13.3 Product metrics

What can we say about products of metric spaces? First we consider finite products.

Proposition 13.3.1 *Suppose that $((X_j, d_j))_{j=1}^n$ is a finite sequence of metric spaces. Then there is a metric d on $X = \prod_{j=1}^n X_j$ such that the metric space topology defined by d is the same as the product topology. Further, we can choose d so that each of the cross-section mappings $k_{y,j}$ is an isometry, and so that the coordinate mappings π_j are Lipschitz mappings, with constant 1.*

Proof   There are many ways of doing this. The easiest is to consider a norm $\|.\|$ on $\mathbf{R}^n$ with the properties that if $0 \leq a_j \leq b_j$ for $1 \leq j \leq n$ then $\|(a_1, \ldots, a_n)\| \leq \|(b_1, \ldots, b_n)\|$, and that $\|e_j\| = 1$ for each unit vector e_j in $\mathbf{R}^n$ - we could, for example, take

$$\|x\| = \begin{cases} \|x\|_1 = \sum_{j=1}^n |x_j|, \text{ or} \\ \|x\|_2 = (\sum_{j=1}^n |x_j|^2)^{1/2}, \text{ or} \\ \|x\|_\infty = \max\{|x_j| : 1 \leq j \leq n\}. \end{cases}$$

We then set $d(x, y) = \|(d_1(x_1, y_1), \ldots, d_n(x_n, y_n))\|$. Then d is a metric on X. Clearly, $d(x, y) = d(y, x)$ and $d(x, y) = 0$ if and only if $x = y$; since

$$\begin{aligned} d(x, z) &= \|(d_1(x_1, z_1), \ldots, d_n(x_n, z_n))\| \\ &\leq \|(d_1(x_1, y_1) + d_1(y_1, z_1), \ldots, d_n(x_n, y_n) + d_n(y_n, z_n))\| \\ &\leq \|(d_1(x_1, y_1), \ldots, d_n(x_n, y_n))\| + \|(d_1(y_1, z_1), \ldots, d_n(y_n, z_n))\| \\ &= d(x, y) + d(y, z), \end{aligned}$$

the triangle inequality holds. If $y \in \prod_{n \setminus \{j\}} X_i$ and $x, x' \in X_j$, then $d(k_{y,j}(x), k_{y,j}(x')) = \|d_j(x, x')e_j\| = d_j(x, x')$, so that the cross-section mapping $k_{y,j}$ is an isometry.

If $1 \leq j \leq n$ and $x, y \in X$ then

$$d_j(\pi_j(x), \pi_j(y)) = \|d_j(\pi_j(x), \pi_j(y))e_j\| \leq d(x, y),$$

so that the coordinate mappings are Lipschitz mappings with constant 1. Finally, suppose that f is a mapping from a topological space (Y, σ) into X for which each of the mappings $f_j = \pi_j \circ f$ is continuous, that $y \in Y$ and that $\epsilon > 0$. For each j there exists a neighbourhood N_j of y such that if $z \in N_j$ then $d_j(f_j(z), f_j(y)) < \epsilon/n$. Let $N = \cap_{j=1}^n N_j$; N is a neighbourhood of y in Y. If $z \in N$ then

$$d(f(z), f(y)) = \|(d_1(f_1(z), f_1(y)), \ldots, d_n(f_n(z), f_n(y)))\|$$
$$\leq \|d_1(f_1(z), f_1(y))e_1\| + \cdots + \|d_n(f_n(z), f_n(y))e_n\| < \epsilon,$$

so that f is continuous. □

When (X_j, d_j) is $\mathbf{R}$, with its usual metric, the metric d is simply the metric given by the norm $\|.\|$.

A metric which satisfies the conclusions of this proposition is called a *product metric*. Suppose that each of the spaces is a normed space, and that the product metric is given by a norm. Then the norm is called a *product norm*. For example, the Euclidean norm is a product norm on $\mathbf{R}^d = \prod_{j=1}^{d}(X_j, d_j)$, where (X_j, d_j) is $\mathbf{R}$, with its usual metric, for $1 \leq j \leq d$.

Here is another example.

Proposition 13.3.2 *Suppose that (X, d) is a metric space. Then the real-valued function $(x, y) \to d(x, y)$ on $(X, d) \times (X, d)$ is a Lipschitz mapping with constant 2 when $(X, d) \times (X, d)$ is given a product metric ρ.*

Proof By Lemma 11.1.14, $|d(x, y) - d(x', y')| \leq d(x, x') + d(y, y')$; the result follows since

$$d(x, x') \leq \rho((x, y), (x', y')) \text{ and } d(y, y') \leq \rho((x, y), (x', y')).$$

□

What is more interesting is that a result similar to Proposition 13.3.1 holds for countable infinite products of metric spaces.

Theorem 13.3.3 *Suppose that $((X_j, d_j))_{j=1}^{\infty}$ is a countably infinite sequence of metric spaces. Then there is a metric ρ on $X = \prod_{j=1}^{\infty} X_j$ such that the metric space topology defined by ρ is the same as the product topology. Further, we can choose ρ so that the cross-section mappings $k_{y,j}$ are Lipschitz mappings, with constant 1.*

Proof We need a preliminary result, of interest in its own right.

Lemma 13.3.4 *Suppose that ϕ is a continuous increasing real-valued function on $[0, \infty)$ for which $\phi(t) = 0$ if and only if $t = 0$, and for which the function ψ on $(0, \infty)$ defined by $\psi(t) = \phi(t)/t$ is decreasing. If (X, d) is a metric space, then the function ρ defined by $\rho(x, y) = \phi(d(x, y))$ is a metric on X equivalent to d. If $\psi(t) \leq K$ for all $t \in (0, \infty)$ then the identity mapping $(X, d) \to (X, \rho)$ is a Lipschitz mapping with constant K.*

Proof   First we show that ρ is a metric. Certainly, $\rho(x,y) = 0$ if and only if $x = y$, and $\rho(x,y) = \rho(y,x)$. Note that if $a > 0$ and $b > 0$ then

$$\phi(a+b) = (a+b)\psi(a+b) \leq a\psi(a) + b\psi(b) = \phi(a) + \phi(b).$$

Now suppose that $x, y, z \in X$. Then

$$\rho(x,z) = \phi(d(x,z)) \leq \phi(d(x,y) + d(y,z))$$
$$\leq \phi(d(x,y)) + \phi(d(y,z)) = \rho(x,y) + \rho(y,z).$$

Thus the triangle inequality holds.

Suppose that $\epsilon > 0$. Since ϕ is continuous at 0, there exists $\delta > 0$ such that if $0 \leq t \leq \delta$ then $0 \leq \phi(t) < \epsilon$. Thus if $d(x,y) \leq \delta$ then $\rho(x,y) \leq \epsilon$, and the identity mapping $i : (X,d) \to (X,\rho)$ is continuous. Conversely, if $\eta > 0$ and $0 \leq \phi(t) < \phi(\eta)$ then $0 \leq t < \eta$. Thus if $\rho(x,y) = \phi(d(x,y)) < \phi(\eta)$ then $d(x,y) < \eta$ and the identity mapping $i : (X,\rho) \to (X,d)$ is continuous.

If ψ is bounded by K, then $\phi(t) \leq Kt$ and so $\rho(x,y) \leq Kd(x,y)$.   □

In particular, if ϕ is bounded then ρ is a bounded metric on X. Popular functions with this property are $\phi_c(t) = t/(1+ct)$, and $\phi_c(t) = \min(c,t)$, where $c > 0$; $0 \leq \phi_c(t) \leq c$. In each case the corresponding function ψ_c is bounded by 1, so that the Lipschitz constant can be taken to be 1.

We now return to the proof of Theorem 13.3.3. Again, there are many ways of defining a suitable metric. For example, let $(c_j)_{j=1}^\infty$ be a sequence of positive numbers for which $\sum_{j=1}^\infty c_j < \infty$ (for example, take $c_j = 1/2^j$). For each j let ρ_j be a metric on X_j which is equivalent to d_j, which is bounded by c_j, and for which the identity mapping $(X_j, d_j) \to (X_j, \rho_j)$ is a Lipschitz mapping, with Lipschitz constant 1. Define a real-valued function ρ on $X \times X$ by setting

$$\rho(x,y) = \sum_{j=1}^\infty \rho_j(x_j, y_j).$$

The conditions that we have imposed show that this sum is finite. First, we show that ρ is a metric on X. Clearly, $d(x,y) = d(y,x)$, and $d(x,y) = 0$ if and only if $x = y$. If $x, y, z \in X$ then

$$\rho(x,z) = \sum_{j=1}^\infty \rho_j(x_j, z_j) \leq \sum_{j=1}^\infty (\rho_j(x_j,y_j) + \rho_j(y_j,z_j))$$

$$= \sum_{j=1}^\infty \rho_j(x_j,y_j) + \sum_{j=1}^\infty \rho_j(y_j,z_j) = \rho(x,y) + \rho(y,z),$$

so that the triangle inequality holds.

Next, we show that the metric topology on X defined by ρ is the product topology; we show that it satisfies the conditions of Theorem 13.2.1. If $j \in \mathbf{N}$ and $x, y \in X$ then $\rho_j(\pi_j(x), \pi_j(y)) \leq \rho(x, y)$, so that π_j is continuous, and (ii) is satisfied. Suppose that f is a mapping from a topological space (Y, σ) into (X, ρ), and let $f_j = \pi_j \circ f$. If f is continuous, then, since each mapping π_j is continuous, each of the mappings f_j is continuous. On the other hand, suppose that each of the mappings f_j is continuous. Suppose that $y \in Y$ and that $\epsilon > 0$. We must show that $f^{-1}(N_\epsilon(f(y)))$ is a neighbourhood of y. There exists j_0 such that $\sum_{j=j_0+1}^{\infty} c_j < \epsilon/2$. For $1 \leq j \leq j_0$, let $U_j = \{x \in X_j : \rho_j(x, f_j(y)) < \epsilon/2j_0\}$; U_j is an open neighbourhood of $f_j(y)$ in X_j. Let $U = \cap_{1 \leq j \leq j_0} f_j^{-1}(U_j)$. Since each of the mappings π_j is continuous, U is an open neighbourhood of y in (Y, σ). If $z \in U$ then

$$\rho(f(z), f(y)) = \sum_{j=1}^{j_0} \rho_j(f_j(z), f_j(y)) + \sum_{j=j_0+1}^{\infty} \rho_j(f_j(z), f_j(y))$$

$$\leq \sum_{j=1}^{j_0} \epsilon/2j_0 + \sum_{j=j_0+1}^{\infty} c_j < \epsilon,$$

so that $U \subseteq f^{-1}(N_\epsilon(f(y)))$, and $f^{-1}(N_\epsilon(f(y)))$ is a neighbourhood of y.

Finally, it follows from the construction that each of the cross-section mappings $k_{y,j}$ is a Lipschitz mapping, with constant 1. $\qquad \square$

A metric which satisfies the conditions of this theorem is called an *infinite product metric*, or, simply, a *product metric*.

As examples, the metrics

$$\rho(x, y) = \sum_{j=1}^{\infty} \frac{1}{2^j} \frac{d_j(x, y)}{1 + d_j(x, y)} \quad \text{and} \quad \sigma(x, y) = \sum_{j=1}^{\infty} \min(d_j(x_j, y_j), 1/2^j)$$

are frequently used.

Let us give three examples of countable products of metric spaces. We can give the vector space $\mathbf{R}^{\mathbf{N}}$ of all real sequences the topology of pointwise convergence. This is metrizable, and a suitable product metric is

$$d(x, y) = \sum_{j=1}^{\infty} \frac{|x_j - y_j|}{2^j(1 + |x_j - y_j|)}.$$

This is separable, since the countable set of sequences with rational terms, all but finitely many of which are zero, is dense in $\mathbf{R}^{\mathbf{N}}$.

Next, let $\mathcal{H} = [0, 1]^{\mathbf{N}} = \prod_{j=1}^{\infty} I_j$, where $I_j = [0, 1]$ for $j \in \mathbf{N}$, with the product topology. Then $\mathcal{H}$ is a closed subset of $\mathbf{R}^{\mathbf{N}}$. There are many product metrics which define the product topology on $\mathcal{H}$. One such metric is $\rho(x,y) = (\sum_{j=1}^{\infty} |x_j - y_j|^2/j^2)^{1/2}$. The mapping $m : \mathcal{H} \to l_2$ defined by $m(x) = (x_j/j)_{j=1}^{\infty}$ is then an isometry of $\mathcal{H}$ onto the subset $\mathcal{H}_2$ of the inner product space l_2 defined as

$$\mathcal{H}_2 = \{x \in l_2 : 0 \le x_j \le 1/j \text{ for } 1 \le j \le \infty\}.$$

$\mathcal{H}_2$ is called the *Hilbert cube;*.

Thirdly, the *Bernoulli sequence space* $\Omega(\mathbf{N})$ can be considered as a closed subset of $\mathcal{H}$. One product metric which defines the topology on $\Omega(\mathbf{N})$ is given by setting $\delta(x,y) = 2\sum_{j=1}^{\infty} |x_j - y_j|/3^j$. With this metric, if $x \in \Omega(\mathbf{N})$ then

$$N_{2/3^j} = \{y \in \Omega(\mathbf{N}) : y_i = x_i \text{ for } 1 \le i \le j\}.$$

If $x \in \Omega(\mathbf{N})$, let $s(x) = 2\sum_{j=1}^{\infty} x_j/3^j$. Then s is an isometry of $(\Omega(\mathbf{N}), \delta)$ onto Cantor's ternary set. As we have seen, $\Omega(\mathbf{N})$ can be identified with $P(\mathbf{N})$, with the product topology, and so $P(\mathbf{N})$, with the product topology, is homeomorphic to Cantor's ternary set.

Exercises

13.3.1 Suppose that $(X_i, d_i)_{i=1}^j$ is a finite sequence of metric spaces. If $x = (x_i)_{i=1}^j$ and $y = (y_i)_{i=1}^j$ are in $X = \prod_{i=1}^j X_j$ let

$$\rho_1(x,y) = \sum_{i=1}^j d_i(x_i, y_i) \text{ and } \rho_\infty(x,y) = \max_{1 \le i \le j} d_i(x_i, y_i).$$

Show that ρ_1 and ρ_∞ are product metrics on X and that if ρ is any product metric on X, then $\rho_\infty \le \rho \le \rho_1$. Deduce that any two product metrics on X are Lipschitz equivalent.

13.3.2 Let $\omega = \mathbf{R}^\mathbf{N}$ be the vector space of all real sequences. Show that there is no norm on ω which defines the product topology on ω.

13.4 Separation properties

If τ is the metric space topology of a metric space (X,d) then τ has certain properties which other topologies do not possess. In this section and the next, we shall introduce some of these properties.

If X is a set with the indiscrete topology, it is not possible to distinguish between points topologically. In order to be able to do so, it is necessary to introduce separation properties. We shall introduce five of them.

- A topological space (X, τ) is a T_1 *space* if each singleton set is closed.
- A topological space (X, τ) is a *Hausdorff space* if whenever x and y are distinct points of X there exist disjoint open sets U and V such that $x \in U$, $y \in V$.
- A topological space (X, τ) is *regular* if it is Hausdorff and whenever A is a closed subset of X and x is an element of X which is not in A, then there exist disjoint open sets U and V such that $x \in U$, $A \subseteq V$.
- A topological space (X, τ) is *completely regular* if it is Hausdorff and whenever A is a closed subset of X and x is an element of X which is not in A then there exists a continuous function $f : X \to [0, 1]$ such that $f(x) = 0$ and $f(a) = 1$ for $a \in A$.
- A topological space (X, τ) is *normal* if it is Hausdorff and whenever A and B are disjoint closed subsets of X then there exist disjoint open sets U and V such that $A \subseteq U$, $B \subseteq V$.

Unfortunately, terminology varies from author to author; the issue is whether or not the Hausdorff condition should be included in the last three definitions. It is therefore sensible to be cautious, and, for example, to refer to a 'regular Hausdorff space'. As we shall see, the conditions are listed in increasing order of restrictiveness.

It follows from the definition that a topological space (X, τ) is a T_1 *space* if whenever x and y are distinct points of X there exists an open set U such that $y \in U$ and $x \notin U$, so that a Hausdorff space is a T_1 space.

Proposition 13.4.1 *In a Hausdorff space, limits are unique. Suppose that f is a continuous mapping from a topological space (Y, σ) into a Hausdorff topological space (X, τ), that b is a limit point of X and that $f(x) \to l$ as $x \to b$. Then l is unique.*

Proof   If $f(x) \to m$ as $x \to b$, and $l \neq m$ then there exist disjoint open sets U and V in X such that $l \in U$ and $m \in V$. But then there exist punctured neighbourhoods $N_U^*(b)$ and $N_V^*(b)$ of b such that $f(N_U^*(b)) \subseteq U$ and $f(N_V^*(b)) \subseteq V$. But this implies that $N_U^* \cap N_V^* = \emptyset$, contradicting the fact that b is a limit point of Y. $\qquad\square$

In fact, the condition is also necessary (Exercise 13.4.1).

Proposition 13.4.2 *A topological space (X, τ) is regular if and only if it is Hausdorff, and each point has a base of neighbourhoods consisting of closed sets.*

Proof   Suppose that (X, τ) is regular, that $x \in X$, and that $N \in \mathcal{N}_x$. There exists an open set O such that $x \in O \subseteq N$. Then $X \setminus O$ is closed, and

$x \notin X \setminus O$, and so there exist disjoint open sets U and V with $x \in U$ and $X \setminus O \subseteq V$. Then

$$x \in U \subseteq X \setminus V \subseteq O \subseteq N,$$

so that $X \setminus V$ is a closed neighbourhood of x contained in N. Thus the closed neighbourhoods of x form a base of neighbourhoods of x. The proof of the converse is left as an exercise (Exercise 13.4.2). □

Proposition 13.4.3 *A completely regular topological space (X, τ) is regular.*

Proof Suppose that A is a closed subset of X and that $x \notin A$. Then there exists a continuous mapping of X into $[0, 1]$ with $f(x) = 0$ and $f(a) = 1$ for $a \in A$. Let

$$U = \{y \in X : f(y) < 1/2\}, \ V = \{y \in X : f(y) > 1/2\}.$$

Then U and V are disjoint open subsets of X, and $x \in U$, $A \subseteq V$. □

Complete regularity has a different character to the other separation conditions, since it involves real-valued functions; but a great deal of analysis is concerned with continuous real-valued functions on a topological space.

Proposition 13.4.4 *A topological space (X, τ) is completely regular if and only if whenever $x \in X$, $\mathcal{B}_x$ is a base of neighbourhoods of x and $B \in \mathcal{B}_x$ then there exists a continuous function $f : X \to [0, 1]$ such that $f(x) = 0$ and $f(y) = 1$ for $y \notin B$.*

Proof Suppose that (X, τ) is completely regular and that B is a basic neighbourhood of a point $x \in X$. Then $x \in B^\circ \subseteq B$, and $X \setminus B^\circ$ is closed, and so there exists a continuous function $f : X \to [0, 1]$ such that $f(x) = 0$ and $f(y) = 1$ for $y \notin B^\circ$. Thus $f(y) = 1$ for $y \notin B$.

Conversely, suppose that the conditions are satisfied, that $x \in X$, that A is a closed subset of X and that $x \notin A$. Then there exists $B \in \mathcal{B}_x$ such that $B \cap A = \emptyset$. There exists a continuous function $f : X \to [0, 1]$ such that $f(x) = 0$ and $f(y) = 1$ for $y \notin B$. Then $f(a) = 1$ for $a \in A$. □

It is easy to verify that a topological subspace of a T_1 space (Hausdorff space, regular space, completely regular space) is a T_1 space (Hausdorff space, regular space, completely regular space), and that a *closed* subspace of a normal space is normal. As we shall see (Example 13.6.9), not every subspace of a normal space is normal.

Proposition 13.4.5 *If $\{(X_\alpha, \tau_\alpha) : \alpha \in A\}$ is a family of T_1 space (Hausdorff spaces, regular spaces, completely regular spaces) then the product*

$X = \prod_{\alpha \in A} X_\alpha$ is a T_1 space (Hausdorff space, regular space, completely regular space) when it is given the product topology τ.

Proof Suppose x and y are distinct elements of X. There exists $\alpha \in A$ such that $x_\alpha \neq y_\alpha$. If (X_α, τ_α) is T_1 then there exists $U_\alpha \in \tau_\alpha$ such that $x_\alpha \in U_\alpha$ and $y_\alpha \notin U_\alpha$. Then $U = \pi_\alpha^{-1}(U_\alpha)$ is open in X and $x \in U$, $y \notin U$, so that X is T_1. An exactly similar argument shows that X is Hausdorff if each of the spaces (X_α, τ_α) is Hausdorff.

Suppose next that each space (X_α, τ_α) is regular. If $x \in X$ then each x_α has a base $\mathcal{B}_{x,\alpha}$ of neighbourhoods consisting of closed sets. Then the collection

$$\{\cap_{\alpha \in F} \pi_\alpha^{-1}(N_\alpha(x_\alpha)) : F \text{ finite}, \ N_\alpha(x_\alpha) \in \mathcal{B}_{x,\alpha}\}$$

of subsets of X is a base of neighbourhoods of x consisting of closed sets. Thus (X, τ) is regular.

Finally, suppose that each (X_α, τ_α) is completely regular, and that $\cap_{\alpha \in F} \pi_\alpha^{-1}(N_\alpha(x))$ is a basic neighbourhood of x. For each $\alpha \in F$ there exists a continuous mapping $f_\alpha : X_\alpha \to [0,1]$ such that $f_\alpha(x_\alpha) = 0$ and $f_\alpha(y_\alpha) = 1$ for $y_\alpha \notin N_\alpha(x_\alpha)$. Set $f(x) = \max_{\alpha \in F} f_\alpha(\pi_\alpha(x))$. Since each mapping $f_\alpha \circ \pi_\alpha$ is continuous on (X, τ), the function f is a continuous function on (X, τ). Further, $f(x) = 0$, and $f(y) = 1$ if $y \notin \cap_{\alpha \in F} \pi_\alpha^{-1}(N_\alpha(x))$. Thus (X, τ) is completely regular. $\square$

We shall see (Example 13.6.11) that the product of two normal spaces need not be normal.

A normal space is completely regular; this follows immediately from the principal result of this section.

Theorem 13.4.6 (Urysohn's lemma) *If A and B are disjoint closed subsets of a normal topological space (X, τ) then there exists a continuous mapping $f : X \to [0,1]$ such that $f(a) = 0$ for $a \in A$ and $f(b) = 1$ for $b \in B$.*

Proof Let D be the set of dyadic rational numbers in $[0,1]$ - numbers of the form $p/2^n$ with p and n in $\mathbf{Z}^+$, and $p \leq 2^n$. Using the axiom of dependent choice, we define a family $\{U_d : d \in D\}$ of open subsets of X with the properties that

- if $d_1 < d_2$ then $\overline{U}_{d_1} \subseteq U_{d_2}$, and
- $A \subseteq U_0$ and $U_1 = X \setminus B$.

We begin by setting $U_1 = X \setminus B$. Since A and B are disjoint closed sets, there exist disjoint open subsets V and W such that $A \subseteq V$ and $B \subseteq W$.

Then, since $X \setminus W$ is closed,

$$A \subseteq V \subseteq \overline{V} \subseteq X \setminus W \subseteq X \setminus B = U_1,$$

and so we can take $U_0 = V$.

We now re-iterate this argument. Suppose that we have defined U_d for all d which can be written in the form $d = p/2^m$, with $m \leq n$. Suppose that $d = (2k + 1)/2^{n+1}$. Let $l = k/2^n$ and let $r = (k + 1)/2^n$. Then $l < d < r$, and U_l and U_r have been defined. Since $\overline{U}_l$ and $X \setminus U_r$ are disjoint closed sets, there exist disjoint open subsets V and W such that $\overline{U}_l \subseteq V$ and $X \setminus U_r \subseteq W$. Then, since $X \setminus W$ is closed,

$$\overline{U}_l \subseteq V \subseteq \overline{V} \subseteq X \setminus W \subseteq U_r.$$

We can therefore take $U_d = V$.

We now use this family to define the function f. If $x \in B$, we set $f(x) = 1$ and if $x \in X \setminus B$ we set $f(x) = \inf\{d \in D : x \in U_d\}$. Then $0 \leq f(x) \leq 1$, $f(b) = 1$ for $b \in B$, and, since $A \subseteq U_0$, $f(a) = 0$ for $a \in A$. It remains to show that f is continuous. For this we use the fact that D is dense in $[0,1]$.

First, suppose that $0 < \alpha \leq 1$. Let $V_\alpha = \cup\{U_d : d < \alpha\}$. V_α is open; we shall show that $V_\alpha = \{x \in X : f(x) < \alpha\}$. If $f(x) < \alpha$ there exists $d \in D$ with $f(x) < d < \alpha$, and so $x \in U_d \subseteq V_\alpha$. Thus $\{x \in X : f(x) < \alpha\} \subseteq V_\alpha$. On the other hand, if $x \in V_\alpha$ then $x \in U_d$ for some $d < \alpha$, and so $f(x) < \alpha$. Thus $V_\alpha \subseteq \{x \in X : f(x) < \alpha\}$. Consequently, $V_\alpha = \{x \in X : f(x) < \alpha\}$.

Next, suppose that $0 \leq \beta < 1$. Let $W_\beta = \cup\{X \setminus \overline{U}_d : d > \beta\}$. W_β is open; we shall show that $W_\beta = \{x \in X : f(x) > \beta\}$. If $f(x) > \beta$ there exists $d \in D$ with $\beta < d < f(x)$, and so $x \notin U_d$. Thus $x \in W_\beta$, and so $\{x \in X : f(x) > \beta\} \subseteq W_\beta$. On the other hand, if $x \in W_\beta$ then $x \in X \setminus \overline{U}_d$ for some $d > \beta$. There exists $e \in D$ with $d < e < \beta$. Then $x \in X \setminus U_e$, so that $f(x) \geq e > d > \beta$. Thus $W_\beta \subseteq \{x \in X : f(x) > \beta\}$. Consequently, $W_\beta = \{x \in X : f(x) > \beta\}$

Thus if $0 \leq \beta < \alpha \leq 1$ then

$$\{x \in X : \beta < f(x) < \alpha\} = W_\beta \cap V_\alpha$$

is an open set; from this it follows that f is continuous.   □

Note that Exercise 12.4.4 shows that this theorem holds for metric spaces. If A and B are closed disjoint subsets of a metric space (X, d), and f is a continuous mapping of X into $[0, 1]$ for which $f(a) = 0$ for $a \in A$ and $f(b) = 1$ for $b \in B$, then $U = \{x \in X : f(x) < 1/2\}$ and $V = \{x \in X : f(x) > 1/2\}$ are disjoint open sets which separate A and B. Consequently, a metric space is normal.

Exercises

13.4.1 Suppose that (X, τ) is a topological space with the property that whenever f is a continuous mapping from a topological space (Y, σ) into (X, τ), b is a limit point of Y, and $f(x) \to l$ as $x \to b$, then l is unique. Show that (X, τ) is Hausdorff.

13.4.2 Suppose that every point x in a Hausdorff topological space has a base of neighbourhoods consisting of closed sets. Show that the space is regular.

13.4.3 Suppose that f is a continuous mapping from a topological space (X, τ) into a topological space (Y, σ). The *graph* G_f of f is the set $\{(x, f(x)) : x \in X\}$. Show that if (Y, σ) is a T_1 space then G_f is closed in $X \times Y$, when $X \times Y$ is given the product topology. Give an example to show that the T_1 condition cannot be dropped.

13.4.4 Suppose that (X, τ) is a topological space. Show that the following are equivalent:
 (a) (X, τ) is Hausdorff;
 (b) the diagonal$\{(x, x) : x \in X\}$ is closed in $(X, \tau) \times (X, \tau)$;
 (c) whenever f and g are continuous mappings from a topological space (Y, σ) into (X, τ), the set $\{y \in Y : f(y) = g(y)\}$ is closed in Y.

13.4.5 Suppose that (X, τ) is not a Hausdorff space, and that y and z cannot be separated by open sets. Let $i : X \to X$ be the identity mapping. Show that $i(x) \to y$ and $i(x) \to z$ as $x \to y$.

13.5 Countability properties

There are several countability properties that a topological space (X, τ) might possess. We list the three most important of these:

- (X, τ) is *first countable* if each point has a countable base of neighbourhoods;
- (X, τ) is *second countable* if there is a countable base for the topology;
- (X, τ) is *separable* if there is a countable subset of X which is dense in X.

A metric space is first countable (the sequence $(N_{1/n}(x))_{n=1}^{\infty}$ is a countable base of neighbourhoods of x) but need not be second countable (consider an uncountable set with the discrete metric).

Here are some elementary consequences of the definitions.

Proposition 13.5.1 *(i) A subspace of a first countable topological space is first countable.*

(ii) A countable product of first countable topological spaces is first countable.

(iii) A countable product of second countable spaces is second countable.

(iv) A countable product of separable topological spaces is separable.

(v) If f is a continuous mapping of a separable topological space (X, τ) into a topological space (Y, σ) then $f(X)$, with the subspace topology, is separable. In particular, the quotient of a separable topological space is separable.

Proof   (i), (ii) (iii) and (v) are easy consequences of the definitions, and the details are left as exercises for the reader.

(iv) We give the proof for a countably infinite product; the proof for a finite product is easier. Suppose that $((X_j, \tau_j))_{j=1}^{\infty}$ is a sequence of separable topological spaces, and that $(X, \tau) = \prod_{j=1}^{\infty}(X_j, \tau_j)$. Let C_j be a countable dense subset of X_j, for $1 \leq j < \infty$. If any X_j is empty, then the product is empty, and therefore separable. Otherwise, choose $y_j \in C_j$ for $1 \leq j < \infty$. We consider $y = (y_j)_{j=1}^{\infty}$ as a base point in the product. For $1 \leq j < \infty$, let

$$A_j = \{x \in X : x_i \in C_i \text{ for } 1 \leq i \leq j, \ x_i = y_i \text{ for } i > j\}.$$

Then $(A_j)_{j=1}^{\infty}$ is an increasing sequence of countable subsets of X. Let $A = \cup_{j=1}^{\infty} A_j$. A is countable; we show that it is dense in X. Suppose that $x \in X$ and that $N \in \mathcal{N}_x$. Then there exists $j_0 \in \mathbf{N}$ and neighbourhoods $N_j \in \mathcal{N}_{x_j}$ for $1 \leq j \leq j_0$ such that $N \supseteq \cap_{j=1}^{j_0} \pi_j^{-1}(N_j)$. Since there exists $a \in A_{j_0}$ such that $a_j \in N_j$ for $1 \leq j \leq j_0$, $N \cap A$ is not empty.   □

Here are some results concerning first countability. The last three show that in the presence of first countability, certain topological properties can be expressed in terms of convergent sequences.

Proposition 13.5.2   *Suppose that (X, τ) is a first countable topological space, that A is a subset of X, and that $x \in X$.*

(i) There is a decreasing sequence of neighbourhoods of x which is a base of neighbourhoods of x.

(ii) The element x is a limit point of A if and only there is a sequence in $A \setminus \{x\}$ such that $x_n \to x$ as $n \to \infty$.

(iii) The element x is a closure point of A if and only if there is a sequence in A such that $x_n \to x$ as $n \to \infty$.

(iv) If f is a mapping from X into a topological space (Y, σ) then f is continuous at x if and only if whenever $(x_n)_{n=1}^{\infty}$ is a sequence in X for which $x_n \to x$ as $n \to \infty$ then $f(x_n) \to f(x)$.

Proof (i) Suppose that $(N_j)_{j=1}^{\infty}$ is a countable base of neighbourhoods of x. Let $M_j = \cap_{i=1}^{j} N_i$. Then $(M_j)_{j=1}^{\infty}$ satisfies the requirements; it is a *decreasing base of neighbourhoods* of x.

(ii) The condition is certainly sufficient. It is also necessary. Suppose that x is a limit point of A. Let $\{M_k : k \in \mathbf{N}\}$ be a decreasing countable base of neighbourhoods of x. For each $k \in \mathbf{N}$ there exists $x_k \in (M_k \setminus \{x\}) \cap A$. If $N \in \mathcal{N}_x$, there exists k such that $M_k \subseteq N$. Then $x_n \in N \setminus \{x\}$ for $n \geq k$, and so $x_n \to x$ as $n \to \infty$.

(iii) is proved in exactly the same way.

(iv) If f is continuous, then the condition is satisfied. Suppose that the condition is satisfied, and that f is not continuous at x. Then there exists a neighbourhood N of $f(x)$ such that $f^{-1}(N)$ is not a neighbourhood of x. Thus for each $k \in \mathbf{N}$ there exists $x_k \in M_k \setminus f^{-1}(N)$. Then $x_k \to x$ as $k \to \infty$, but $f(x_k) \notin N$ for any $k \in \mathbf{N}$, and so $f(x_k) \not\to f(x)$ as $k \to \infty$. $\square$

Here are some results concerning second countability.

Proposition 13.5.3 *(i) A second countable topological space (X, τ) is first countable and separable.*

(ii) A metric space (X, d) is second countable if and only if it is separable.

Proof (i) Suppose that $(U_j)_{j=1}^{\infty}$ is a basis for τ. If $x \in X$ then $\{U_j : x \in U_j\}$ is a base of neighbourhoods for x, and so X is first countable. If $U_j \neq \emptyset$, choose $x_j \in U_j$. Then $A = \{x_j : U_j \neq \emptyset\}$ is a countable subset of X; we show that A is dense in X. If $x \in X$ and $N \in \mathcal{N}_x$, there exists a basic open set U_j such that $x \in U_j \subseteq N$. Then $x_j \in A \cap U_j \subseteq A \cap N$, so that $N \cap A \neq \emptyset$.

(ii) By (i), we need only prove that a separable metric space (X, d) is second countable. Let C be a countable dense subset of X. We shall show that the countable set $\mathcal{B} = \{N_{1/n}(x) : n \in \mathbf{N} : x \in C\}$ is a base for the topology. Suppose that U is open. Let $V = \cup\{B \in \mathcal{B} : B \subseteq U\}$; V is an open set contained in U. Suppose that $x \in U$; then there exists $\epsilon > 0$ such that $N_\epsilon(x) \in U$. Choose n such that $1/n < \epsilon/2$. There exists $c \in C$ such that $d(c, x) < 1/n$. If $y \in N_{1/n}(c)$ then

$$d(y, x) \leq d(y, c) + d(c, x) < 1/n + 1/n < \epsilon,$$

so that $y \in U$. Thus $x \in N_{1/n}(c) \subseteq V$. Thus $U \subseteq V$, and so $U = V$. Consequently $\mathcal{B}$ is a base for the topology. $\square$

Corollary 13.5.4　*A subspace of a separable metric space is separable.*

Proof　For it is a subspace of a second countable space, and so is second countable. But a second countable space is separable.　　　　　□

This result does not extend to topological spaces (Example 13.6.11).

We now prove two substantial theorems concerning second countable topological spaces.

Theorem 13.5.5　*A regular second countable topological space (X, τ) is normal.*

Proof　Let $\mathcal{B}$ be a countable base for the topology. Suppose that C and D are disjoint closed subsets of X. Let $\mathcal{C} = \{U \in \mathcal{B} : \overline{U} \cap C = \emptyset\}$ and let $\mathcal{D} = \{U \in \mathcal{B} : \overline{U} \cap D = \emptyset\}$; let $V_1, V_2, \ldots$ be an enumeration of $\mathcal{C}$ and let $W_1, W_2, \ldots$ be an enumeration of $\mathcal{D}$. For $1 \le j, k < \infty$, let

$$P_j = W_j \setminus \left(\bigcup_{i=1}^{j} \overline{V}_i \right) \text{ and } Q_k = V_k \setminus \left(\bigcup_{i=1}^{k} \overline{W}_i \right);$$

P_j and Q_k are open. If $j \ge k$ then $P_j \cap V_k = \emptyset$, and so $P_j \cap Q_k = \emptyset$. Similarly, $P_j \cap Q_k = \emptyset$ if $k > j$. Let $P = \cup_{j=1}^{\infty} P_j$, $Q = \cup_{k=1}^{\infty} Q_k$. Then P and Q are open and,

$$P \cap Q = \cup_{j,k \in \mathbf{N}} (P_j \cap Q_k) = \emptyset.$$

If $x \in C$ then, since (X, τ) is regular, there exists $W_j \in \mathcal{D}$ such that $x \in W_j$. But $x \notin \overline{V}_i$ for $1 \le i \le j$, and so $x \in P_j \subseteq P$. Since this holds for all $x \in C$, $C \subseteq P$. Similarly, $D \subseteq Q$. Thus (X, τ) is normal.　　　　□

Recall that $\mathcal{H}$ is the set $[0, 1]^{\mathbf{N}} = \prod_{j=1}^{\infty} I_j$, where $I_j = [0, 1]$ for $j \in \mathbf{N}$, with the product topology.

Theorem 13.5.6　(Urysohn's metrization theorem)　*A regular second countable topological space (X, τ) is metrizable. There exists a homeomorphism f of (X, τ) onto a subspace $f(X)$ of $\mathcal{H}$.*

Proof　Since $\mathcal{H}$ is metrizable, and a subspace of a metrizable space is metrizable, it is sufficient to prove the second statement. Let $\mathcal{B}$ be a countable base for the topology. Let

$$S = \{(U, V) \in \mathcal{B} \times \mathcal{B} : \overline{U} \subseteq V\}.$$

S is a countable set; let us enumerate it as $(s_i)_{i=1}^{\infty}$.

If $s_i = (U_i, V_i) \in S$ then by Urysohn's lemma there exists a continuous mapping $f_i : (X, \tau) \to [0, 1]$ such that $f_i(x) = 0$ if $x \in \overline{U}_i$ and $f_i(x) = 1$ if $x \notin V_i$. We can therefore define a mapping $f : (X, \tau) \to \mathcal{H}$ by setting $(f(x))_i = f_i(x)$. Since each of the mappings f_i is continuous, f is continuous.

We now use two very similar arguments to show first that f is injective and secondly that $f^{-1} : f(X) \to X$ is continuous. Suppose that $x \neq z$. Since (X, τ) is a T_1 space, there exists $V \in \mathcal{B}$ such that $x \in V$ and $z \notin V$ and, since (X, τ) is regular, there exists $U \in \mathcal{B}$ such that $x \in U$ and $\overline{U} \subseteq V$. Then $(U, V) \in S$; let $(U, V) = s_i$. Then $f_i(x) = 0$ and $f_i(z) = 1$. Consequently $f(x) \neq f(z)$.

It remains to show that f^{-1} is continuous. Suppose that $x \in X$ and that $N \in \mathcal{N}_x$. There exists $V \in \mathcal{B}$ such that $x \in V \subseteq N$, and, since (X, τ) is regular, there exists $U \in \mathcal{B}$ such that $x \in U \subseteq \overline{U} \subseteq V$. Then $(U, V) \in S$; let $(U, V) = s_i$. Then $f_i(x) = 0$ and $f_i(z) = 1$ if $z \notin N$. Let $M = \{y \in \mathcal{H} : y_i < 1\}$. Then M is an open neighbourhood of $f(x)$, and if $y \in M \cap f(X)$ then $f^{-1}(y) \in N$. Thus $f^{-1} : f(X) \to X$ is continuous. $\qquad\square$

13.6 *Examples and counterexamples*

(This section can be omitted on a first reading.)

We now describe a collection of examples of topological spaces which illustrate the connections between the various ideas that we have introduced. The descriptions frequently include statements that need checking: the reader should do so. First, quotients can behave badly.

Example 13.6.1 An equivalence relation $\sim$ on $\mathbf{R}$ for which the quotient space is uncountable, and for which the quotient topology is the indiscrete topology.

Define a relation $\sim$ on $\mathbf{R}$ by setting $x \sim y$ if $x - y \in \mathbf{Q}$. This is clearly an equivalence relation; let $q : \mathbf{R} \to \mathbf{R}/\sim$ be the quotient mapping. Each equivalence class is countable, and so there must be uncountably many equivalence classes. Suppose that U is a non-empty open set in $\mathbf{R}/\sim$, so that $q^{-1}(U)$ is a non-empty open subset of $\mathbf{R}$, and so contains an open interval (a, b). If $x \in \mathbf{R}$ there exists $r \in \mathbf{Q}$ such that $x - r \in (a, b)$, and so $q(x) \in U$. Thus $U = q(\mathbf{R}) = \mathbf{R}/\sim$.

Next we consider the relations between the various separation properties.

Example 13.6.2 A T_1 space which is not Hausdorff.

Let X be an infinite set, and let τ_f be the cofinite topology on X. The finite subsets are closed, so that (X, τ_f) is a T_1 space. If $U = X \setminus F$ and $V = X \setminus G$ are non-empty open sets then $U \cap V = X \setminus (F \cup G)$ is non-empty. Thus (X, τ_f) is not Hausdorff.

Example 13.6.3 A quotient of a closed interval which is a T_1 space but not Hausdorff.

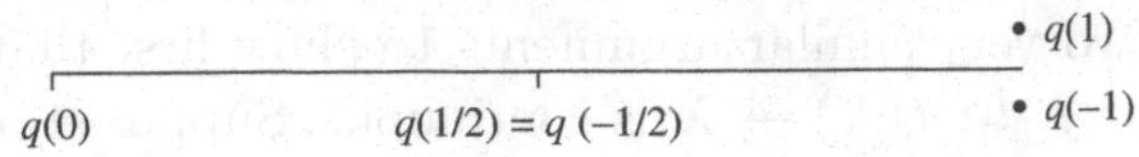

Figure 13.6a. A T_1 space that is not Hausdorff.

Define a partition of $[-1, 1]$ by taking $\{-1\}$, $\{0\}$, $\{1\}$ and $\{-t, t\}$, for $0 < t < 1$, as the sets of the partition. (Fold the interval $[-1, 1]$ over, and stick corresponding points together, except for the points -1 and 1.) Let X be the corresponding quotient space and let $q : [-1, 1] \to X$ be the quotient mapping. Give $[-1, 1]$ its usual topology, and give X the quotient topology τ_q. Since the equivalence classes are closed in $[-1, 1]$, X is a T_1 space. On the other hand, if U and V are open sets in X containing $q(-1)$ and $q(1)$ respectively, then there exists $\epsilon > 0$ such that

$$\{(\eta, -\eta) : 1 - \epsilon < \eta < 1\} \subseteq U \cap V,$$

and so X is not Hausdorff.

Example 13.6.4 A separable first countable Hausdorff space which is not regular.

Let $X = \mathbf{R}^2$, let $L = \{(x, y) \in \mathbf{R}^2 : x > 0, y = 0\}$ and let $P = (0, 0)$. We define a topology τ' on X by saying that U is open if $U \setminus \{P\}$ is open in $\mathbf{R}^2$ in the usual topology and if $P \in U$ then there exists $\epsilon > 0$ such that $N_\epsilon(P) \setminus L \subseteq U$. The reader should verify that this is indeed a topology. (X, τ') is separable, since the countable set $\{(r, s) \in X : r, s \in \mathbf{Q}\}$ is dense in X, and it is clearly first countable. Then τ' is a topology which is finer than the usual topology, and so (X, τ') is Hausdorff. Since $P \notin \bar{L}$, L is closed. Suppose that U and V are τ'-open subsets of X with $P \in U$ and $L \subseteq V$. Then there exists $\epsilon > 0$ such that $N_\epsilon(P) \setminus L \subseteq U$. Then $(\epsilon/2, 0) \in L \subseteq V$, and so there exists $\delta > 0$ such that $N_\delta((\epsilon/2, 0)) \subseteq V$. Since $(N_\epsilon(P) \setminus L) \cap N_\delta((\epsilon/2, 0))$ is not empty, $U \cap V$ is not empty. Thus (X, τ') is not regular.

Example 13.6.5 A quotient of the first countable space $\mathbf{R}$ which is separable and normal, but not first countable.

We define an equivalence relation on $\mathbf{R}$ by setting $x \sim y$ if $x = y$ or if x and y are both integers; in other words, we identify all the integers. We consider the quotient space $\mathbf{R}/\sim$, with the quotient topology. Let $q : \mathbf{R} \to \mathbf{R}/\sim$ be the quotient mapping. Then $\mathbf{R}/\sim$ is separable. It is also normal. Suppose that A and B are disjoint closed subsets of $\mathbf{R}/\sim$. Suppose first that $q(0) \notin A \cup B$. Then $q^{-1}(A)$ and $q^{-1}(B)$ are disjoint closed sets in $\mathbf{R} \setminus \mathbf{Z}$, and so there exist disjoint open subsets U and V in $\mathbf{R} \setminus \mathbf{Z}$, such that $q^{-1}(A) \subseteq U$ and $q^{-1}(B) \subseteq V$. Then $q(U)$ and $q(V)$ are open and disjoint in $\mathbf{R}/\sim$, and $A \subseteq q(U)$ and $B \subseteq q(V)$. Suppose secondly that $q(0) \in A \cup B$, and suppose, without loss of generality, that $q(0) \in A$. Then $q^{-1}(A)$ and $q^{-1}(B)$ are disjoint closed sets in $\mathbf{R}$, and $\mathbf{Z} \subseteq q^{-1}(A)$. Then there exist disjoint open subsets U and V in $\mathbf{R}$, such that $q^{-1}(A) \subseteq U$ and $q^{-1}(B) \subseteq V$. Then $q(U)$ and $q(V)$ are open and disjoint in $\mathbf{R}/\sim$, and $A \subseteq q(U)$ and $B \subseteq q(V)$. Thus $\mathbf{R}/\sim$ is normal.

Now suppose that $(N_j)_{j=1}^{\infty}$ is a sequence of neighbourhoods of $q(0)$ in $\mathbf{R}/\sim$. Then for each $j \in \mathbf{Z}$ there exists $0 < \epsilon_j < 1$ such that $(j - \epsilon_j, j + \epsilon_j) \subseteq q^{-1}(N_j)$. Let $M = (-\infty, 1) \cup (\cup_{j=1}^{\infty}(j - \epsilon_j/2, j + \epsilon_j/2))$. Then $q(M)$ is a neighbourhood of $q(0)$ in $\mathbf{R}/\sim$, and $N_j \not\subseteq q(M)$ for $j \in \mathbf{Z}$. Thus $(N_j)_{j=1}^{\infty}$ is not a base of neighbourhoods of $q(0)$, and so $\mathbf{R}/\sim$ is not first countable.

There exist topological spaces which are regular, but not completely regular, but these are too complicated to describe here.[1]

Before describing the next few examples, we need to prove an easy but important result about the usual topology on $\mathbf{R}$. This is a special case of Baire's category theorem, which is proved in Section 14.7.

Theorem 13.6.6 (Osgood's theorem) *Suppose that $(U_n)_{n=1}^{\infty}$ is a sequence of dense open subsets of $\mathbf{R}$. Then $\cap_{n=1}^{\infty} U_n$ is dense in $\mathbf{R}$.*

Proof Suppose that (a_0, b_0) is an open interval in $\mathbf{R}$. We must show that $(a_0, b_0) \cap (\cap_{n=1}^{\infty} U_n)$ is not empty. Since U_1 is dense in $\mathbf{R}$, the set $(a_0, b_0) \cap U_1$ is not empty. Since $(a_0, b_0) \cap U_1$ is open, there exists a non-empty open interval (a_1, b_1) such that

$$(a_1, b_1) \subseteq [a_1, b_1] \subseteq (a_0, b_0) \cap U_1.$$

We now iterate the argument. Suppose we have defined non-empty open intervals (a_j, b_j) such that

$$(a_j, b_j) \subseteq [a_j, b_j] \subseteq (a_{j-1}, b_{j-1}) \cap U_j,$$

for $1 \leq j < n$. Then $(a_{n-1}, b_{n-1}) \cap U_n$ is a non-empty open set, and so there exists a non-empty open interval (a_n, b_n) such that

$$(a_n, b_n) \subseteq [a_n, b_n] \subseteq (a_{n-1}, b_{n-1}) \cap U_n.$$

The sequence $(a_n)_{n=0}^{\infty}$ is increasing, and is bounded above by b_m, for each $m \in \mathbf{N}$, and so converges to a limit a. If $n \in \mathbf{N}$ then $a_n < a_{n+1} \leq a \leq b_{n+1} < b_n$, so that $a \in (a_n, b_n) \subseteq U_n$. Thus $(a_0, b_0) \cap (\cap_{n=1}^{\infty} U_n) \neq \emptyset$. $\square$

---

[1] See Example 90 in Lynn Arthur Steen and J. Arthur Seebach, Jr., *Counterexamples in Topology*, Dover, 1995. This is a wonderful comprehensive collection of counterexamples.

Corollary 13.6.7 *Suppose that $(C_n)_{n=1}^{\infty}$ is a sequence of subsets of $\mathbf{R}$ whose union is $\mathbf{R}$. Then there exist $n \in \mathbf{N}$ and a non-empty interval (c, d) such that $(c, d) \subset \overline{C}_n$.*

Proof   If not, each of the open sets $U_n = \mathbf{R} \setminus \overline{C}_n$ is dense in $\mathbf{R}$, and $\cap_{n=1}^{\infty} U_n = \emptyset$.   □

In other words, $\overline{C}_n$ has a non-empty interior.

Example 13.6.8 (The Niemytzki space)   A separable first countable completely regular space (H, τ') which is not normal, and which has a non-separable subspace.

Let H be the closed upper half-space $H = \{(x, y) \in \mathbf{R}^2 : y \geq 0\}$, let L be the real axis $L = \{(x, y) \in \mathbf{R}^2 : y = 0\}$ and let $U = H \setminus L$ be the open upper half-space. Let τ be the usual topology on H. If $(x, 0) \in L$ and $\epsilon > 0$ let

$$D_\epsilon(x) = \{(u, v) \in U : (u - x)^2 + (v - \epsilon)^2 < \epsilon^2\}.$$

$D_\epsilon(x)$ is the open disc with centre (x, ϵ) and radius ϵ, and L is the tangent to $D_\epsilon(x)$ at $(x, 0)$. Let $M_\epsilon(x) = \{(x, 0)\} \cup D_\epsilon(x)$, and let $T_\epsilon(x)$ be the boundary of $D_\epsilon(x)$ in U:

$$T_\epsilon(x) = \{(u, v) \in U : (u - x)^2 + (v - \epsilon)^2 = \epsilon^2\}.$$

If $(x, 0) \in L$ let $\mathcal{M}_x = \{M_\epsilon(x) : \epsilon > 0\}$, and let $\mathcal{U} = \cup_{(x,0) \in L} \mathcal{M}_x$. Let σ be the collection of all unions of sets in $\mathcal{U}$ and let

$$\tau' = \{V \cup S : V \in \tau, S \in \sigma\}.$$

The reader should verify that τ' is a topology; it is a topology on H finer than the usual topology. The two subspace topologies on U are the same, but if $(x, 0) \in L$ then the sets in $\mathcal{M}_x$ form a base of τ'-neighbourhoods of $(x, 0)$.

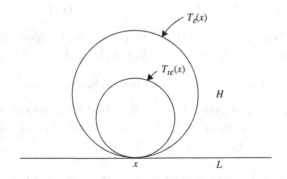

Figure 13.6b. The Niemytzki space.

(H, τ') is separable, since the countable set $\{(r, s) \in U : r, s \in \mathbf{Q}\}$ is dense in H. It is first countable: for example, if $(x, 0) \in L$, the sequence $(M_{1/n}(x))_{n=1}^{\infty}$ is a base of neighbourhoods of $(x, 0)$.

(H, τ') is completely regular. Suppose that C is closed in H and that $z \notin C$. We consider two cases. First, suppose that $z \in U$. Then there exists $N_{\epsilon}(z) \subseteq U \setminus C$. Let $f(y) = \|y - z\| / \epsilon$ for $y \in N_{\epsilon}(z)$, and $f(y) = 1$ otherwise. Then f is continuous, $f(z) = 0$ and $f(c) = 1$ for $c \in C$. Secondly, suppose that $z = (x, 0) \in L$. Then there exists $\epsilon > 0$ such that $M_{\epsilon}(x) \cap C = \emptyset$. If $w \in D_{\epsilon}(x)$, there exists a unique $0 < t < 1$ such that $w \in T_{t\epsilon}(x)$. Set $f(w) = t$, set $f(z) = 0$, and set $f(w) = 1$ if $w \notin M_{\epsilon}(x)$. Let us show that f is a continuous function on H. It is clearly continuous at every point of H other than z; since $\{w : f(w) < t\} = M_{t\epsilon}(x)$ for $0 < t < 1$, f is also continuous at z. Further, $f(z) = 0$ and $f(c) = 1$ for $c \in C$.

The subspace L has the discrete topology, since $M_{\epsilon}(x) \cap L = \{(x, 0)\}$, for $\epsilon > 0$. Since L is uncountable, it is not separable. Further, it is closed in (H, τ'). Thus any subset of L is closed in H. In particular, the sets

$$A = \{(x, 0) : x \text{ irrational}\} \text{ and } B = \{(q, 0) : q \text{ rational}\}$$

are disjoint closed subsets of H. We shall show that if V and W are open sets in H such that $A \subseteq V$ and $B \subseteq W$ then $V \cap W \neq \emptyset$, so that (H, τ') is not normal.

For $n \in \mathbf{N}$, let $A_n = \{x \in \mathbf{R} : (x, 0) \in A \text{ and } M_{1/n}(x) \subseteq V\}$. Then $\cup_{n=1}^{\infty} A_n = \{x \in \mathbf{R} : x \text{ irrational}\}$. Hence

$$\mathbf{R} = (\cup_{n=1}^{\infty} A_n) \cup \left(\bigcup_{q \in \mathbf{Q}} \{q\} \right).$$

This is a countable union of sets, and so by Corollary 13.6.7, the closure (in $\mathbf{R}$, with its usual topology) of one of the sets contains an open interval (c, d). This clearly cannot be one of the singleton sets $\{q\}$, and so there exists $n \in \mathbf{N}$ such that $\overline{A}_n \supseteq (c, d)$. Suppose now that $q \in \mathbf{Q} \cap (c, d)$. Then there exists $\epsilon > 0$ such that $M_{\epsilon}(q) \subseteq W$; we can suppose that $\epsilon < 1/n$. There exists $x \in A_n$ such that $|x - q| < \epsilon$. Then $(x, \epsilon) \in M_{\epsilon}(q) \cap M_{1/n}(x)$. Since $M_{\epsilon}(q) \subseteq W$ and $M_{1/n}(x) \subseteq V$, $V \cap W \neq \emptyset$.

Example 13.6.9   A normal topological space with a subspace which is not normal.

Add an extra point P to the Niemytzki space (H, τ') to obtain a larger set H^+, and define a topology τ^+ in H^+ by taking as open sets those subsets V of H^+ for which

- $V \cap H \in \tau'$, and
- if $P \in V$ then $L \setminus V$ is finite.

Then (H, τ') is a topological subspace of (H^+, τ^+) which is not normal. The τ^+-closed neighbourhoods of P form a base of neighbourhoods of P (why?), and so (H^+, τ^+) is regular.

Suppose that A and B are disjoint closed subsets of (H^+, τ^+). If C is a closed subset of H^+ for which $C \cap L$ is infinite, then $P \in C$. Thus either $A \cap L$ is finite, or $B \cap L$ is finite, or both; without loss of generality, suppose that $A \cap L = F$ is finite. Since (H^+, τ^+) is regular, there exist disjoint open subsets V_1 and W_1 such that $F \subseteq V_1$ and $B \subseteq W_1$. Now $(B \cup L) \setminus \{P\}$ and $A \setminus V_1$ are disjoint, and are closed in H in the usual topology, and so there exist subsets V_2 and W_2 which are open in H in the usual topology such that $A \setminus V_1 \subseteq V_2$ and $(B \cup L) \setminus \{P\} \subseteq W_2$. Then $V = V_1 \cup V_2$ and $W = W_1 \cap (W_2 \cup \{P\})$ are disjoint open subsets of (H^+, τ^+), and $A \subseteq V$, $B \subseteq W$. Thus (H^+, τ^+) is normal.

In fact, every completely regular space is homeomorphic to a subspace of a normal space; this is too difficult to prove here.

Example 13.6.10   A first countable separable normal topological space which is not second countable.

Let $\mathcal{B}$ be the collection of half-open half-closed intervals $[a, b)$ in $\mathbf{R}$. This clearly satisfies the conditions for it to be the base for a topology τ' on $\mathbf{R}$. $(\mathbf{R}, \tau')$ is separable, since the rationals are dense; it is first countable, since the sets $\{[x, x + 1/n) : n \in \mathbf{N}\}$ form a base of neighbourhoods of x. Suppose that A and B are disjoint closed subsets of $(\mathbf{R}, \tau')$. If $a \in A$ there exists a largest l_a in $(a, a + 1]$ such that $[a, l_a) \cap B = \emptyset$; let $U = \cup_{a \in A}[a, l_a)$. Then U is an open set containing A and disjoint from B. Similarly, if $b \in B$ there exists a largest m_b in $(b, b + 1]$ such that $[b, m_b) \cap A = \emptyset$; let $V = \cup_{b \in B}[b, m_b)$. Then V is an open set containing B and disjoint from A. But if $a \in A$ and $b \in B$ then $[a, l_a) \cap [b, m_b) = \emptyset$, and so $U \cap V = \emptyset$; thus $(\mathbf{R}, \tau')$ is normal.

Suppose that $\mathcal{B}'$ is a base for the topology. For each $x \in \mathbf{R}$ there exists $B_x \in \mathcal{B}'$ such that $x \in B_x \subseteq [x, x + 1)$. But if $x \neq y$ then $B_x \neq B_y$, so that $\mathcal{B}'$ cannot be countable: (X, τ') is not second countable.

Example 13.6.11   The product of two first countable separable normal topological spaces which is not normal, and which has a non-separable subspace.

Consider $(X, \tau) = (\mathbf{R}, \tau') \times (\mathbf{R}, \tau')$, where $(\mathbf{R}, \tau')$ is the space of the previous example. Then the line $L = \{(x, -x) : x \in \mathbf{R}\}$ is closed, and its subspace topology is the discrete topology, so that L is not separable. All of the subsets of L are closed in X; an argument just like the one for the Niemytzki space shows that (X, τ) is not normal.

We can sum up some of our conclusions in a table. In the context of topological spaces, this shows that the choice of the word 'normal' is clearly not at all appropriate!

	subspace	quotient	countable product	uncountable product
T_1	Yes	No	Yes	Yes
Hausdorff	Yes	No	Yes	Yes
Regular	Yes	No	Yes	Yes
Completely regular	Yes	No	Yes	Yes
Normal	No	No	No	No
First Countable	Yes	No	Yes	No
Second Countable	No	No	Yes	No
Separable	No	Yes	Yes	No

Exercises

13.6.1 Suppose that $(x_n)_{n=1}^{\infty}$ is a convergent sequence in the space of Example 13.6.3. Show that it converges to one or two points.

13.6.2 In Example 13.6.4, characterize the sequences which converge to P.

13.6.3 In Example 13.6.5, characterize the sequences which converge to $q(0)$.

13.6.4 Show that an example similar to the Niemytzki space can be obtained by replacing the open sets $D_\epsilon(x)$ by triangular regions, of a fixed shape. Let $0 < \alpha < 1$. Let $R_\epsilon(x) = \{(u, v) \in U : |u - x| < \epsilon, \alpha|u - x| < v < \alpha\epsilon\}$, and replace the sets $D_\epsilon(x)$ by the sets $R_\epsilon(x)$.

13.6.5 Suppose that f is a bounded real-valued function on $[0, 1]$ for which $f(x) > 0$ for all $x \in [0, 1]$. Use Osgood's theorem to show that the upper integral $\overline{\int_0^1} f(x)\, dx$ is strictly positive.

13.6.6 Show that the topology of Example 13.6.10 is finer than the usual topology on $\mathbf{R}$. Characterize the convergent sequences.

14

Completeness

14.1 Completeness

The general principle of convergence played an essential role in the analysis on $\mathbf{R}$; similar ideas are just as important in analysis on a metric space. As we shall see, these are not topological ideas.

The definitions are straightforward. Suppose that (X, d) is a metric space. A sequence $(x_n)_{n=1}^{\infty}$ taking values in X is a *Cauchy sequence* if whenever $\epsilon > 0$ there exists n_0 such that $d(x_m, x_n) < \epsilon$ for $m, n \geq n_0$.

Proposition 14.1.1 *(i) A sequence $(x_n)_{n=1}^{\infty}$ in a metric space which converges to a limit l is a Cauchy sequence.*

(ii) If a Cauchy sequence $(x_n)_{n=1}^{\infty}$ in a metric space has a subsequence $(x_{n_k})_{k=1}^{\infty}$ which converges to l, then $x_n \to l$ as $n \to \infty$.

Proof (i) Given $\epsilon > 0$ there exists n_0 such that $d(x_n, l) < \epsilon/2$ for $n \geq n_0$. If $m, n \geq n_0$ then $d(x_m, x_n) \leq d(x_m, l) + d(l, x_n) < \epsilon$, by the triangle inequality.

(ii) Given $\epsilon > 0$ there exists N such that $d(x_m, x_n) < \epsilon/2$ for $m, n \geq N$, and there exists K with $n_K \geq N$ such that $d(x_{n_k}, l) < \epsilon/2$ for $k \geq K$. If $n \geq n_K$ then $d(x_n, l) \leq d(x_n, x_{n_K}) + d(x_{n_K}, l) < \epsilon$, again by the triangle inequality, so that $x_n \to l$ as $n \to \infty$. $\qquad\square$

A metric space (X, d) is *complete* if every Cauchy sequence in X is convergent to a point in X. Thus the general principle of convergence says that $\mathbf{R}$, with the usual metric, is complete. Let us give some examples.

Theorem 14.1.2 *A sequence $(x^{(n)})_{n=1}^{\infty}$ in $\mathbf{R}^d$ or $\mathbf{C}^d$, with its usual metric, is a Cauchy sequence if and only if each of the coordinate sequences $(x_j^{(n)})_{n=1}^{\infty}$, for $1 \leq j \leq d$, is a Cauchy sequence in $\mathbf{R}$.*

$\mathbf{R}^d$ and $\mathbf{C}^d$, with their usual metrics, are complete.

Proof If $(x^{(n)})_{n=1}^{\infty}$ is a Cauchy sequence in $\mathbf{R}^d$ (or $\mathbf{C}^d$) and $1 \le j \le d$, then, since $|x_j^{(m)} - x_j^{(n)}| \le d_2(x^{(m)}, x^{(n)})$, the sequence $(x^{(n)})_{n=1}^{\infty}$ is a Cauchy sequence in $\mathbf{R}$ (or $\mathbf{C}$). Conversely, if each of the sequences $(x_j^{(n)})_{n=1}^{\infty}$, for $1 \le j \le d$, is a Cauchy sequence, then there exists n_0 such that

$$|x_j^{(m)} - x_j^{(n)}| < \epsilon/d^{1/2} \text{ for } m, n \ge n_0 \text{ and } 1 \le j \le d.$$

Thus $d_2(x^{(m)}, x^{(n)}) < \epsilon$ for $m, n \ge n_0$, and $(x^{(n)})_{n=1}^{\infty}$ is a Cauchy sequence.

Thus if $(x^{(n)})_{n=1}^{\infty}$ is a Cauchy sequence then each of the sequences $(x_j^{(n)})_{n=1}^{\infty}$ is a Cauchy sequence, and so converges to a limit x_j, by the general principle of convergence. Hence $x^{(n)} \to x = (x_1, \ldots x_d)$ as $n \to \infty$: thus $\mathbf{R}^d$ and $\mathbf{C}^d$, with their usual metrics, are complete. $\square$

Proposition 14.1.3 *(i) A closed metric subspace A of a complete metric space (X, d) is complete.*

(ii) A complete metric subspace B of a metric space (X, d) is closed in X.

Proof (i) Suppose that $(x_n)_{n=1}^{\infty}$ is a Cauchy sequence in A. Then $(x_n)_{n=1}^{\infty}$ is a Cauchy sequence in X. Since (X, d) is complete, x_n converges to an element l in X as $n \to \infty$. But A is closed, and so $l \in A$. Thus x_n converges to an element of A, and A is complete.

(ii) Let b be a closure point of B in X. Then there exists a sequence $(b_n)_{n=1}^{\infty}$ in B which converges to b. Thus $(b_n)_{n=1}^{\infty}$ is a Cauchy sequence in B. But B is complete, and so b_n converges to a point l of B as $n \to \infty$. By the uniqueness of limits, $b = l$. Thus any closure point of B belongs to B, and so B is closed. $\square$

Many of the metric spaces that we shall consider are spaces of functions. The next result, and its corollary, lie behind a great number of results concerning such spaces.

Theorem 14.1.4 *If S is a non-empty set and (Y, d) is a complete metric space then the space $B_Y(S)$ of bounded mappings of S into Y is complete under the uniform metric d_∞.*

Proof The proof follows a pattern common to many proofs of completeness. There are three steps. We start with a Cauchy sequence $(f_n)_{n=1}^{\infty}$. First we identify what the limit f should be, secondly we verify that it is an element of $B_Y(S)$, and thirdly we prove that $f_n \to f$ as $n \to \infty$.

Suppose then that $(f_n)_{n=1}^{\infty}$ is a Cauchy sequence in $(B_Y(S), d_\infty)$. If $s \in S$ then $d(f_m(s), f_n(s)) \le d_\infty(f_m, f_n)$, and so $(f_n(s))_{n=1}^{\infty}$ is a Cauchy sequence in Y. Since (Y, d) is complete, there exists $f(s) \in Y$ such that $f_n(s) \to f(s)$

as $n \to \infty$. We claim that the function $f : s \to f(s)$ is in $B_Y(X)$, and that $f_n \to f$ in the uniform metric.

Take $\epsilon = 1$. There exists n_0 such that $d_\infty(f_m, f_n) < 1$ for $m, n \geq n_0$, and so if $s \in S$ then $d(f_m(s), f_{n_0}(s)) < 1$ for $m \geq n_0$. By Proposition 13.3.2 (which we shall use repeatedly), $d(f_m(s), f_{n_0}(s)) \to d(f(s), f_{n_0}(s))$ as $m \to \infty$, and so $d(f(s), f_{n_0}(s)) \leq 1$. Thus if $s, t \in S$ then

$$d(f(s), f(t)) \leq d(f(s), f_{n_0}(s)) + d(f_{n_0}(s), f_{n_0}(t)) + d(f_{n_0}(t), f(t))$$

$$\leq \mathrm{diam}\,(f_{n_0}(S)) + 2,$$

so that $f \in B_Y(S)$.

Finally we show that $f_n \to f$ as $n \to \infty$. Suppose that $\epsilon > 0$. There exists n_1 such that $d_\infty(f_m, f_n) < \epsilon/2$ for $m, n \geq n_1$, and so if $s \in S$ then $d(f_m(s), f_n(s)) < \epsilon/2$ for $m, n \geq n_1$. Suppose that $n \geq n_1$. Since $d(f_m(s), f_n(s)) \to d(f(s), f_n(s))$ as $m \to \infty$, $d(f(s), f_n(s)) \leq \epsilon/2$. Since this holds for all $s \in S$, $d_\infty(f, f_n) \leq \epsilon/2 < \epsilon$. But this holds for all $n \geq n_1$, and so $f_n \to f$ as $n \to \infty$. $\qquad\square$

A Cauchy sequence in $(B_Y(S), d_\infty)$ is called a *uniform Cauchy sequence*.

Corollary 14.1.5 (The general principle of uniform convergence)   *If (X, τ) is a topological space and (Y, ρ) is a complete metric space, then the space $C_b(X, Y)$ of bounded continuous mappings of X into Y is complete under the uniform metric d_∞; a uniformly Cauchy sequence $(f_n)_{n=1}^\infty$ of bounded continuous functions converges uniformly to a bounded continuous function.*

Proof   For $C_b(X, Y)$ is closed in $B_Y(X)$, by Theorem 12.3.7. $\qquad\square$

The first important theoretical point to make is that completeness is *not* a topological property. To see this, consider $\mathbf{R}$ and $(-\pi/2, \pi/2)$, each with the usual metric d. $(\mathbf{R}, d)$ is complete, by the general principle of convergence, but $((-\pi/2, \pi/2), d)$ is not, since it is not closed in $\mathbf{R}$. Now consider the mapping $j = \tan^{-1}$ from $\mathbf{R}$ onto $(-\pi/2, \pi/2)$. This is a homeomorphism; we use it to define a new metric ρ on $\mathbf{R}$, setting $\rho(x, y) = |j(x) - j(y)|$. Then j is an isometry of $(\mathbf{R}, \rho)$ onto $(-\pi/2, \pi/2), d)$, and so $(\mathbf{R}, \rho)$ is not complete. But d and ρ are equivalent metrics, since j is a homeomorphism. Thus the two metrics d and ρ define the same topology on $\mathbf{R}$: but $(\mathbf{R}, d)$ is complete, and $(\mathbf{R}, \rho)$ is not.

We need a stronger equivalence to preserve completeness. A mapping f from a metric space (X, d) to a metric space (Y, ρ) is said to be *uniformly continuous* if for each $\epsilon > 0$ there exists $\delta > 0$ such that if $x, y \in X$ and $d(x, y) < \delta$ then $\rho(f(x), f(y)) < \epsilon$. The important feature of this definition is that while δ may depend upon ϵ (and usually does), it does not depend upon x or y. A Lipschitz mapping is uniformly continuous, and a uniformly continuous function is continuous. The real-valued function $f(x) = x^2$ on $\mathbf{R}$ is an example of a continuous function which is not uniformly continuous.

Proposition 14.1.6 *Suppose that f is a uniformly continuous mapping from a metric space (X, d) into a metric space (Y, ρ).*

(i) If $(x_n)_{n=1}^{\infty}$ is a Cauchy sequence in X then $(f(x_n))_{n=1}^{\infty}$ is a Cauchy sequence in Y.

(ii) Suppose that (g_n) is a sequence of functions from a set S to X which converges uniformly on S to a function g. Then $f \circ g_n$ converges uniformly on S to $f \circ g$.

Proof The proof of (i), which follows almost immediately from the definitions, is left as an exercise for the reader. The proof of (ii) is as easy. Given $\epsilon > 0$ there exists $\delta > 0$ such that if $d(x, x') < \delta$ then $\rho(f(x), f(x')) < \epsilon$. There exists $n_0 \in \mathbf{N}$ such that if $n \geq n_0$ then $d(g_n(s), g(s)) < \delta$ for all $s \in S$. Thus $\rho(f(g_n(s)), f(g(s))) < \epsilon$ for $n \geq n_0$ and $s \in S$. □

A bijective mapping f from a metric space (X, d) onto a metric space (Y, ρ) is a *uniform homeomorphism* if f and f^{-1} are both uniformly continuous. Two metrics d and d' on a set X are *uniformly equivalent* if the identity mapping $i : (X, d) \to (X, d')$ is a uniform homeomorphism.

Corollary 14.1.7 *If f is a uniform homeomorphism from a metric space (X, d) onto a metric space (Y, ρ) then (X, d) is complete if and only if (Y, ρ) is complete. If two metrics d and d' on a set X are uniformly equivalent then (X, d) is complete if and only if (X, d') is complete.*

As an important example, let us consider a product $X = \prod_{i=1}^{\infty}(X_i, d_i)$ of an infinite sequence of metric spaces. In Theorem 13.3.3 we constructed a metric ρ on X for which the ρ-metric topology is the product topology, and for which the cross-section mappings $k_{y,j}$ are Lipschitz mappings. Inspection of the construction shows that each of the coordinate mappings π_j is uniformly continuous. A metric d on X with all of these properties is called an *uniform product metric*.

Corollary 14.1.8 *Suppose that d is a uniform product metric on the product $X = \prod_i(X_i, d_i)$ of non-empty metric spaces (X_i, d_i). Then (X, d) is complete if and only if each metric space (X_i, d_i) is complete.*

Proof This result holds for finite and infinite products. Suppose first that each metric space (X_i, d_i) is complete. Suppose that $(x^{(n)})_{n=1}^{\infty}$ is a

Cauchy sequence in (X, d). For each i, the sequence $(\pi_i(x^{(n)}))_{n=1}^{\infty}$ is a Cauchy sequence in (X_i, d_i), since the coordinate mapping π_i is uniformly continuous, and therefore converges to an element l_i of X_i. But this implies that $x^{(n)} \to l$ as $n \to \infty$, where $l = (l_i)$. Thus (X, d) is complete.

Conversely suppose that (X, d) is complete. Let $(x_i^{(n)})_{n=1}^{\infty}$ be a Cauchy sequence in (X_i, d_i). Choose $y \in \prod_{j \in \mathbf{N}, i \neq j} X_j$. Since the cross-section mapping $k_{y,i}$ is a Lipschitz mapping, it is uniformly continuous, and so $(k_{y,i}(x_i^{(n)}))_{n=1}^{\infty}$ is a Cauchy sequence in (X, d). Since (X, d) is complete, $k_{y,i}(x_i^{(n)})$ converges to an element l of X as $n \to \infty$, Then $x_i^{(n)} \to \pi_i(l)$ as $n \to \infty$, since π_i is continuous and $x_i^{(n)} = \pi_i(k_{y,i}(x_i^{(n)}))$. Thus (X_i, d_i) is complete. $\qquad \square$

The next result provides a powerful test for completeness.

Proposition 14.1.9 *Suppose that X is a subset of a complete metric space (Y, ρ) and that d is a metric on X for which*
  (i) the inclusion mapping $j : (X, d) \to (Y, \rho)$ is uniformly continuous, and
  (ii) for each $x \in X$ and each $\epsilon > 0$ the closed ϵ-neighbourhood $M_\epsilon(x) = \{x' \in X : d(x', x) \leq \epsilon\}$ of x in X is ρ-closed in Y.
  Then (X, d) is complete.

Proof   Suppose that $(x_n)_{n=1}^{\infty}$ is a Cauchy sequence in (X, d), and that $\epsilon > 0$. There exists n_0 such that $d(x_m, x_n) < \epsilon/2$ for $m, n \geq n_0$. Since j is uniformly continuous, $(x_n)_{n=1}^{\infty}$ is a ρ-Cauchy sequence, and since (Y, ρ) is complete, there exists $y \in Y$ such that $\rho(x_n, y) \to 0$ as $n \to \infty$. If $m > n \geq n_0$ then $x_m \in M_{\epsilon/2}(x_n)$. Since $M_{\epsilon/2}(x_n)$ is ρ-closed, it follows that $y \in M_{\epsilon/2}(x_n)$. Thus $y \in X$ and $d(x_n, y) \leq \epsilon/2 < \epsilon$. Since this holds for all $n \geq n_0$, $x_n \to y$ as $n \to \infty$. $\qquad \square$

We shall give applications of this in Proposition 14.2.3 and Corollary 14.2.4.

Next we prove a fundamental extension result. This depends in an essential way on the relation between uniform continuity and completeness.

Theorem 14.1.10 *Suppose that A is a dense subset of a metric space (X, d), and that f is a uniformly continuous mapping from A to a complete metric space (Y, ρ). Then there is a unique continuous mapping $\widetilde{f}$ from X to Y which extends f: $\widetilde{f}(a) = f(a)$ for $a \in A$. Further, $\widetilde{f}$ is uniformly continuous. If f is a Lipschitz mapping with constant K then so is $\widetilde{f}$, and if f is an isometry then so is $\widetilde{f}$.*

Proof   Suppose that $x \in X$. There exists a sequence $(a_n)_{n=1}^{\infty}$ in A such that $a_n \to x$ as $n \to \infty$. Then $(a_n)_{n=1}^{\infty}$ is a Cauchy sequence in A, and so $(f(a_n))_{n=1}^{\infty}$ is a Cauchy sequence in (Y, d). Since Y is complete $f(a_n)$ converges to some element y of Y. We show that y does not depend on the choice of the approximating sequence $(a_n)_{n=1}^{\infty}$. Suppose that $a'_n \to x$ as $n \to \infty$. Then, as before, there exists $y' \in Y$ such that $f(a'_n) \to y'$ as $n \to \infty$. But then the sequence $(a_1, a'_1, a_2, a'_2, \ldots)$ converges to x, and the sequence $(f(a_1), f(a'_1), f(a_2), f(a'_2), \ldots)$ converges in Y. The subsequences $(f(a_n))_{n=1}^{\infty}$ and $(f(a'_n))_{n=1}^{\infty}$ must therefore converge to the same limit; thus $y = y'$. We set $\tilde{f}(x)$ to be the common limit. We have thus defined the mapping $\tilde{f}$ from X to Y; $\tilde{f}$ clearly extends f (consider constant sequences).

Next we show that $\tilde{f}$ is uniformly continuous. Suppose that $\epsilon > 0$. There exists $\delta > 0$ such that if $a, b \in A$ and $d(a, b) < \delta$ then $\rho(f(a), f(b)) < \epsilon$. Suppose that $x, y \in X$ and that $d(x, y) < \delta$. Let $\eta = \delta - d(x, y)$. There exist sequences $(a_n)_{n=1}^{\infty}$ and $(b_n)_{n=1}^{\infty}$ in A such that $a_n \to x$ and $b_n \to y$ as $n \to \infty$, and so there exists n_0 such that $d(x, a_n) < \eta/2$ and $d(y, b_n) < \eta/2$ for $n \geq n_0$. By the triangle inequality,

$$d(a_n, b_n) \leq d(a_n, x) + d(x, y) + d(y, b_n) < \delta$$

for $n \geq n_0$, and so $\rho(f(a_n), f(b_n)) < \epsilon$ for $n \geq n_0$. Since $f(a_n) \to \tilde{f}(x)$ and $f(b_n) \to \tilde{f}(y)$, it follows from Proposition 13.3.2 that $\rho(\tilde{f}(x), \tilde{f}(y)) \leq \epsilon$.

There is only one continuous extension. For if $\tilde{f}$ and $\overline{f}$ are two continuous extensions then $\{x \in X : \tilde{f}(x) = \overline{f}(x)\}$ is a closed subset of X (Exercise 10.4.2) which contains A, and so contains $\overline{A} = X$.

If $x, y \in X$ there exist sequences $(a_n)_{n=1}^{\infty}$ and $(b_n)_{n=1}^{\infty}$ in A such that $a_n \to x$ and $b_n \to y$ as $n \to \infty$. Since $\tilde{f}$ is continuous,

$$f(a_n) = \tilde{f}(a_n) \to \tilde{f}(x) \text{ and } f(b_n) = \tilde{f}(b_n) \to \tilde{f}(y) \text{ as } n \to \infty.$$

Thus $\rho(\tilde{f}(x), \tilde{f}(y)) = \lim_{n \to \infty} \rho(f(a_n), f(b_n))$, by Proposition 13.3.2. Thus if f is a Lipschitz mapping with constant K, then

$$\rho(\tilde{f}(x), \tilde{f}(y)) \leq K \lim_{n \to \infty} d(a_n, b_n) = K d(x, y),$$

so that $\tilde{f}$ is also a Lipschitz mapping with constant K. Similarly, if f is an isometry then $\rho(\tilde{f}(x), \tilde{f}(y)) = \lim_{n \to \infty} d(a_n, b_n) = d(x, y)$, so that $\tilde{f}$ is an isometry.   □

We can characterize completeness in terms of sequences of sets.

Theorem 14.1.11 *Suppose that (X, d) is a metric space. The following are equivalent.*

  (i) (X, d) is complete.

  (ii) If $(A_n)_{n=1}^{\infty}$ is a decreasing sequence of non-empty closed subsets of X for which $\operatorname{diam}(A_n) \to 0$ *as $n \to \infty$ then $\cap_{n=1}^{\infty} A_n$ is non-empty.*

  If so, then $\cap_{n=1}^{\infty} A_n$ is a singleton set $\{a\}$ and if $a_n \in A_n$ for each n then $a_n \to a$ as $n \to \infty$.

Proof Suppose first that (X, d) is complete, and that $(A_n)_{n=1}^{\infty}$ is a decreasing sequence of non-empty subsets of X for which $\operatorname{diam} A_n \to 0$ as $n \to \infty$. Pick $a_n \in A_n$ for each $n \in \mathbf{N}$. We shall show that $(a_n)_{n=1}^{\infty}$ is a Cauchy sequence. Suppose that $\epsilon > 0$. There exists n_0 such that $\operatorname{diam}(A_n) < \epsilon$ for $n \geq n_0$. If $m > n \geq n_0$ then $a_m \in A_n$, so that $d(a_m, a_n) < \epsilon$. Thus $(a_n)_{n=1}^{\infty}$ is a Cauchy sequence; since (X, d) is complete, it converges to an element a of X. Suppose that $n \in \mathbf{N}$. Since $a_m \in A_n$ for $m \geq n$, and since A_n is closed, $a \in A_n$. Thus $a \in \cap_{n=1}^{\infty} A_n$. Further, $\operatorname{diam}(\cap_{n=1}^{\infty} A_n) \leq \operatorname{diam} A_m$ for each $m \in \mathbf{N}$, so that $\operatorname{diam}(\cap_{n=1}^{\infty} A_n) = 0$. Thus $A = \{a\}$.

  Conversely, suppose that (ii) holds and that (x_n) is a Cauchy sequence in (X, d). Let $T_n = \{x_n, x_{n+1}, \ldots\}$: $(T_n)_{n=1}^{\infty}$ is the *tail sequence*. Since (x_n) is a Cauchy sequence, $\operatorname{diam}(T_n) \to 0$ as $n \to \infty$. Let $F_n = \overline{T}_n$. $(F_n)_{n=1}^{\infty}$ is a decreasing sequence of non-empty closed sets, and $\operatorname{diam}(F_n) = \operatorname{diam}(T_n)$ (Proposition 12.3.3), so that $\operatorname{diam}(F_n) \to 0$ as $n \to \infty$. Thus $\cap_{n=1}^{\infty} F_n$ is non-empty. Suppose that $x \in \cap_{n=1}^{\infty} F_n$. Since $x \in F_n$ for each n, $d(x, x_n) \leq \operatorname{diam} F_n$, and so $d(x, x_n) \to 0$ as $n \to \infty$. Thus $x_n \to x$ as $n \to \infty$, and (X, d) is complete. $\square$

Corollary 14.1.12 *Suppose that $(A_\epsilon)_{0 < \epsilon \leq \epsilon_0}$ is a family of non-empty closed sets of a complete metric space (X, d), with $A_\epsilon \subseteq A_{\epsilon'}$ for $0 < \epsilon < \epsilon' \leq \epsilon_0$. If $\operatorname{diam} A_\epsilon \to 0$ as $\epsilon \to 0$ then $\cap_{0 < \epsilon \leq \epsilon_0} A_\epsilon$ is non-empty, and is a singleton set.*

Proof $\cap_{0 < \epsilon \leq \epsilon_0} A_\epsilon = \cap_{n=1}^{\infty} B_n$, where $B_n = A_{\epsilon_0/n}$. $\square$

  Although completeness is not a topological property, it has topological consequences. It is therefore of interest to know when there is an equivalent complete metric on a metric space. A metric space (X, d) is said to be *topologically complete* if there is an equivalent metric ρ on X for which (X, ρ) is complete. We use uniform product metrics to provide information about this.

  A subset of a topological space (X, τ) is a G_δ *set* if it is the intersection of a sequence of open sets (and is an F_σ *set* if it is the union of a sequence of closed sets). If (X, τ) is T_1 space, then the complement of a countable set is a G_δ set. Thus the set of irrational numbers in $\mathbf{R}$ is a G_δ subset of $\mathbf{R}$ and the set of transcendental numbers in $\mathbf{C}$ is a G_δ subset of $\mathbf{C}$. The following theorem is therefore quite surprising.

Theorem 14.1.13 *Suppose that (X, d) is a complete metric space, and that A is a G_δ subset of X. Then the metric subspace A is topologically complete.*

Proof First suppose that A is open. If $A = X$, we can take $\rho = d$. Otherwise, let $X \times \mathbf{R}$ have a uniform product metric σ. Then $(X \times \mathbf{R}, \sigma)$ is complete. Consider the injective mapping f from A to $X \times \mathbf{R}$ defined by $f(a) = (a, 1/d(a, C(A)))$. Since $d(a, C(A)) > 0$ for $a \in A$, this is well-defined. Since the mapping $a \to 1/d(a, C(A))$ is continuous on A, f is continuous. We show that $f(A)$ is closed in $X \times \mathbf{R}$. Suppose that

$$f(a_n) \to y = (x, \lambda) \in X \times \mathbf{R} \text{ as } n \to \infty,$$

so that $a_n \to x$ and $1/d(a_n, C(A)) \to \lambda$ as $n \to \infty$. Since

$$|d(a_m, C(A)) - d(a_n, C(A))| \le d(a_m, a_n),$$

the sequence $(d(a_n, C(A)))_{n=1}^\infty$ is bounded, so that $1/d(a_n, C(A))$ does not tend to 0 as $n \to \infty$. Thus $\lambda \ne 0$, and $d(a_n, C(A)) \to 1/\lambda$. Since $d(a_n, C(A)) \to d(x, C(A))$, it follows that $d(x, C(A)) = 1/\lambda$, and so $x \in A$. Consequently $f(a_n) \to f(x)$ as $n \to \infty$, so that $f(A)$ is closed in $X \times \mathbf{R}$, and $(f(A), \sigma)$ is complete. The mapping $f : (A, d) \to (X \times \mathbf{R}, \sigma)$ is continuous, and the inverse mapping $f^{-1} : f(A) \to A$ is a Lipschitz mapping with constant 1, since $d(a, a') \le \sigma(f(a), f(a'))$, so that f is a homeomorphism of (A, d) onto $(f(A), \sigma)$. Thus if we define $\rho(a, a') = \sigma(f(a), f(a'))$ then ρ is a complete metric on A equivalent to d.

Next, suppose that $A = \cap_{j=1}^\infty U_j$ is a G_δ set. For each j there is a metric σ_j on U_j, equivalent to the restriction of d to U_j, under which U_j is complete. Let $U = \prod_{j=1}^\infty (U_j, \sigma_j)$, and let σ be a uniform product metric on U. Then (U, σ) is complete. If $a \in A$, let $i_j : A \to U_j$ be the inclusion map and let $i : A \to U$ be defined as $i(a) = (i_j(a))_{j=1}^\infty$. Then i is a continuous injective map of (A, d) into (U, σ). We show that $i(A)$ is closed in U. Suppose that $i(a_n) \to u$ as $n \to \infty$. Then, for each $j \in \mathbf{N}$, $i_j(a_n) \to u_j$ as $n \to \infty$. Since the metric σ_j on U_j is equivalent to the metric d, $d(a_n, u_j) \to 0$ as $n \to \infty$. Since this is true for each j, there exists l in X such that $u_j = l$ for each $j \in \mathbf{N}$. Thus $l \in \cap_{n=1}^\infty U_j = A$, and $u = i(l)$. Thus $i(A)$ is closed in (U, σ), and so $(i(A), \sigma)$ is complete. If $i(a_n) \to i(a)$ in $(i(A), \sigma)$, then $i_1(a_n) \to i_1(a)$ in (U_1, σ_1) as $n \to \infty$. But d and σ_1 are equivalent on U_1, and so $a_n \to a$ in

(A, d). Thus $i^{-1} : (i(A), \sigma) \to (A, d)$ is continuous, and i is a homeomorphism of (A, d) onto $(i(A), \sigma)$. Thus if we define $\rho(a, a') = \sigma(i(a), i(a'))$ then ρ is a complete metric on A equivalent to d. $\qquad\qquad\qquad\qquad\qquad\qquad\qquad\square$

Exercises

14.1.1 Give an example of a bijection j from a metric space (X, d) onto a metric space (Y, ρ) which is uniformly continuous, while j^{-1} is not continuous.

14.1.2 Give an example of a bijection j from a metric space (X, d) onto a metric space (Y, ρ) which is uniformly continuous, while j^{-1} is continuous, but not uniformly continuous.

14.1.3 Suppose that f is a differentiable function on $\mathbf{R}$. Show that f is uniformly continuous if the derivative f' is a bounded function on $\mathbf{R}$. Does the converse hold?

14.1.4 Suppose that Y is a dense subset of a metric space (X, d) and that any Cauchy sequence in Y converges to an element of X. Show that (X, d) is complete.

14.1.5 Define a metric ρ on $\mathbf{N}$ by setting $\rho(m, n) = |1/m - 1/n|$. Suppose that $(x_n)_{n=1}^{\infty}$ is a sequence in a metric space (X, d). Set $f(n) = x_n$. Show that $(x_n)_{n=1}^{\infty}$ is a Cauchy sequence if and only if $f : (\mathbf{N}, \rho) \to (X, d)$ is uniformly continuous.

14.1.6 Suppose that f is a continuous bijection from a complete metric space (X, d) onto a metric space (Y, ρ) and that f^{-1} is uniformly continuous. Show that (Y, ρ) is complete.

14.1.7 If $x \in X = \prod_{j=1}^{\infty}\{0, 1\}_j$, let $f(x) = 2\sum_{j=1}^{\infty} x_j/3^j$. Show that f is a uniform homeomorphism of X onto Cantor's ternary set.

14.1.8 The set $\{0, 1\}$ becomes an abelian group when we define $0 + 0 = 1 + 1 = 0$, $0 + 0 = 0 + 1 = 1$. Use this to define an abelian group structure on $X = \prod_{j=1}^{\infty}\{0, 1\}_j$. Show that the mapping $(x, y) \to x + y : X \times X \to X$ is jointly uniformly continuous. What about the mapping $x \to -x$?

14.1.9 Suppose that (X, d) is a metric space which is not complete. Show that there is an unbounded continuous real-valued function on X, and a bounded continuous real-valued function on X which is not uniformly continuous.

14.1.10 Suppose that $(x_n)_{n=1}^{\infty}$ is a sequence in an ultrametric space (X, d) for which $d(x_n, x_{n+1}) \to 0$ as $n \to \infty$. Show that $(x_n)_{n=1}^{\infty}$ is a Cauchy sequence.

14.2 Banach spaces

A normed space $(E, \|.\|)$ which is complete under the metric defined by the norm is called a *Banach space*, after the Polish mathematician, Stefan Banach. Similarly, an inner-product space which is complete under the metric defined by the inner-product norm is called a *Hilbert space*, after the German mathematician, David Hilbert.

Proposition 14.2.1 *If $(E, \|.\|)$ is a Banach space then the normed space $(B_E(S), \|.\|_\infty)$ of bounded functions on S taking values in E is a Banach space.*

Proof This is an immediate corollary of Theorem 14.1.4. □

In particular, the spaces $(B_{\mathbf{R}}(S), \|.\|_\infty)$ and $(B_{\mathbf{C}}(S), \|.\|_\infty)$ of bounded real-valued functions and bounded complex-valued functions on S are Banach spaces.

Proposition 14.2.2 *If (X, τ) is a topological space and $(E, \|.\|)$ is a Banach space then $(C_b(X, E), \|.\|_\infty)$ is a Banach space.*

Proof This is a special case of the general principle of uniform convergence. □

Proposition 14.2.3 *Suppose that $(F, \|.\|_F)$ is a Banach space, and that E is a linear subspace of F equipped with a norm $\|.\|_E$ for which the inclusion mapping $(E, \|.\|_E) \to (F, \|.\|_F)$ is continuous. If the closed unit ball $B_E = \{x \in E : \|x\|_E \le 1\}$ is closed in $(F, \|.\|_F)$ then $(E, \|.\|_E)$ is a Banach space.*

Proof The inclusion mapping is uniformly continuous, and $M_\epsilon(x) = x + \epsilon B_E$ is closed in $(F, \|.\|_F)$, since translation and multiplication by non-zero scalars are homeomorphisms. The result therefore follows from Proposition 14.1.9. □

Corollary 14.2.4 *The space $(l_1, \|.\|_1)$ is a Banach space and the space $(l_2, \|.\|_2)$ is a Hilbert space.*

Proof We give the proof for l_2: the proof for l_1 is exactly similar. l_2 is a linear subspace of the Banach space $(l_\infty, \|.\|_\infty)$, and the inclusion mapping $(l_2, \|.\|_2) \to (l_\infty, \|.\|_\infty)$ is continuous, since $\|x\|_\infty \le \|x\|_2$ for $x \in l_2$. It is therefore sufficient to show that B_{l_2} is closed in $(l_\infty, \|.\|_\infty)$. Suppose that $(x^{(n)})_{n=1}^\infty$ is a sequence in B_{l_2} which converges uniformly to $x \in l_\infty$. Then

$x_j^{(n)} \to x_j$ as $n \to \infty$, for each $j \in \mathbf{N}$, and so if $k \in \mathbf{N}$ then

$$\sum_{j=1}^{k} |x_j|^2 = \lim_{n \to \infty} \sum_{j=1}^{k} |x_j^{(n)}|^2 \leq 1.$$

Since this holds for all $k \in \mathbf{N}$, $\sum_{j=1}^{\infty} |x_j|^2 \leq 1$, so that $x \in B_{l_2}$. $\square$

As we shall see, there are many other applications of Proposition 14.2.3.

We can consider infinite series in a normed space, and these can be used as a test for completeness. A series $\sum_{j=0}^{\infty} a_j$ in a normed space $(E, \|.\|)$ is said to *converge absolutely* if $\sum_{j=0}^{\infty} \|a_j\|$ converges.

Proposition 14.2.5 *Suppose that $(a_j)_{j=1}^{\infty}$ is a sequence in a Banach space $(E, \|.\|)$ for which $\sum_{j=0}^{\infty} a_j$ converges absolutely. Then $\sum_{j=0}^{\infty} a_j$ converges, and $\left\| \sum_{j=0}^{\infty} a_j \right\| \leq \sum_{j=0}^{\infty} \|a_j\|$.*

Conversely, if every absolutely convergent series in a normed space $(E, \|.\|)$ converges then $(E, \|.\|)$ is complete.

Proof Suppose that $\epsilon > 0$. There exists n_0 such that if $n > m \geq n_0$ then $\sum_{j=m+1}^{n} \|a_j\| < \epsilon$. Let $s_n = \sum_{j=1}^{n} a_j$. By the triangle inequality, if $n > m \geq n_0$ then

$$\|s_n - s_m\| = \|a_{m+1} + \cdots + a_n\| \leq \sum_{j=m+1}^{n} \|a_j\| < \epsilon,$$

so that (s_n) is a Cauchy sequence in $(E, \|.\|)$. Since $(E, \|.\|)$ is complete, there exists $s \in E$ such that $s_n \to s$ as $n \to \infty$; that is, $\sum_{j=1}^{\infty} a_j = s$. Since the function $x \to \|x\|$ is continuous on E,

$$\left\| \sum_{j=0}^{\infty} a_j \right\| = \lim_{n \to \infty} \|s_n\| \leq \lim_{n \to \infty} \sum_{j=1}^{n} \|a_j\| = \sum_{j=1}^{\infty} \|a_j\|.$$

Conversely, suppose that every absolutely convergent series in $(E, \|.\|)$ converges. Let $(x_n)_{n=1}^{\infty}$ be a Cauchy sequence in $(E, \|.\|)$. There exists a strictly increasing sequence $(n_j)_{j=1}^{\infty}$ in $\mathbf{N}$ such that if $n > m \geq n_j$ then $\|x_n - x_m\| < 1/2^j$. Let $a_1 = x_{n_1}$, and let $a_j = x_{n_j} - x_{n_{j-1}}$ for $j > 1$. Then $\|a_j\| < 1/2^{j-1}$ for $j > 1$ and so $\sum_{j=1}^{\infty} \|a_j\| < \infty$. Thus $\sum_{j=1}^{\infty} a_j$ converges, to s say. But $\sum_{j=1}^{k} a_j = x_{n_k}$, and so $x_{n_k} \to s$ as $n \to \infty$. Thus $x_n \to s$ as $n \to \infty$, by Proposition 14.1.1, and so $(E, \|.\|)$ is complete. $\square$

As special cases, we have the following corollaries.

Corollary 14.2.6 (Cauchy's test for Banach spaces) *The series* $\sum_{j=1}^{\infty} a_j$ *converges if* $\limsup_{j\to\infty} \|a_j\|^{1/j} < 1$ *and it does not converge if* $\limsup_{j\to\infty} \|a_j\|^{1/j} > 1$.

Corollary 14.2.7 (D'Alembert's ratio test for Banach spaces) *Suppose that* $a_j \neq 0$ *for all* j. *If* $\limsup_{j\to\infty} \|a_{j+1}\|/\|a_j\| < 1$ *then* $\sum_{j=1}^{\infty} a_j$ *converges. If* $\limsup_{j\to\infty} \|a_{j+1}\|/\|a_j\| > 1$ *then* $\sum_{j=1}^{\infty} a_j$ *does not converge.*

Corollary 14.2.8 (Weierstrass' uniform M test) *Suppose that* (X, τ) *is a topological space, that* $(E, \|.\|)$ *is a Banach space and that* $(f_j)_{j=1}^{\infty}$ *is a sequence in* $(C_b(X, E), \|.\|_\infty)$. *If* $\|f_j\|_\infty \leq M_j$ *for each* $j \in \mathbf{N}$, *and* $\sum_{j=1}^{\infty} M_j < \infty$, *then the series* $\sum_{j=1}^{\infty} f_j(x)$ *converges uniformly to a bounded continuous function on* X.

When the conditions of this corollary are met, we say that the series $\sum_{j=1}^{\infty} f_j(x)$ converges *absolutely uniformly* on X.

The following special case is particularly useful.

Corollary 14.2.9 *Suppose that, for each* $j \in \mathbf{N}$, $(f_j^{(n)})_{n=1}^{\infty}$ *is a sequence in a Banach space* $(E, \|.\|)$ *which converges to an element* l_j *of* E, *that* $\left\|f_j^{(n)}\right\| \leq M_j$ *for each* $n \in \mathbf{N}$ *and each* $j \in \mathbf{N}$, *and that* $\sum_{j=1}^{\infty} M_j < \infty$. *Then the series* $\sum_{j=1}^{\infty} f_j^{(n)}$ *converges uniformly on* $\mathbf{N}$ *to a sequence* $(s_n)_{n=1}^{\infty}$ *in* E, *the series* $\sum_{j=1}^{\infty} l_j$ *converges to an element* s_∞ *of* E, *and* $s_n \to s_\infty$ *as* $n \to \infty$.

Proof Let $\bar{\mathbf{N}} = \mathbf{N} \cup \{\infty\}$ with the metric $d(m, n) = |1/m - 1/n|$, $d(n, +\infty) = 1/n$. Let $g_j(n) = f_j^{(n)}$, for $n \in \mathbf{N}$, and let $g_j(\infty) = l_j$. Then $g_j \in C_b(\bar{\mathbf{N}}, E)$, and $\|g_j\|_\infty \leq M_j$. Thus $\sum_{j=1}^{\infty} g_j$ converges uniformly to an element s of $C_b(\bar{\mathbf{N}}, E)$; this gives the result. (It is as easy to prove the result directly.) □

There is also a test for products.

Corollary 14.2.10 (Weierstrass' uniform M-test for products) *Suppose that* (X, τ) *is a topological space and that* $(f_j)_{j=1}^{\infty}$ *is a sequence in* $(C_b(X, \mathbf{R}), \|.\|_\infty)$. *If* $\|f_j\|_\infty \leq M_j$ *for each* $j \in \mathbf{N}$, *and* $\sum_{j=1}^{\infty} M_j < \infty$, *then the infinite product* $\prod_{j=1}^{\infty}(1 + f_j(x))$ *converges uniformly to a bounded continuous function on* X.

Proof   Since $M_j \to 0$ as $j \to \infty$ there exists $N \in \mathbf{N}$ such that $M_j \leq 1/2$ for $j \geq N$. We use the mean-value theorem. If $|t| \leq 1/2$, then

$$0 < \frac{d}{dt} \log(1+t) \leq 2, \text{ so that } |\log(1+f_j(x))| \leq 2|f_j(x)| \leq 2M_j$$

for $j \geq N$ and $x \in X$. Thus $\sum_{j=N+1}^{\infty} \log(1+f_j(x))$ converges absolutely and uniformly to a bounded continuous function $g(x)$. The exponential function exp has a bounded derivative on $[-\|g\|_\infty, \|g\|_\infty]$, and so is uniformly continuous on $[-\|g\|_\infty, \|g\|_\infty]$. It therefore follows that $\prod_{j=N+1}^{\infty}(1+f_j(x))$ converges uniformly to $e^{g(x)}$. Finally, $\prod_{j=1}^{\infty}(1+f_j(x))$ converges uniformly to $(\prod_{j=1}^{N}(1+f_j(x)))e^{g(x)}$.   □

Exercises

14.2.1 Suppose that $(E, \|.\|)$ is a real normed space. Show that the mapping $(x,y) \to x+y : E \times E \to E$ is uniformly continuous, but that the mapping $(\lambda, x) \to \lambda x : \mathbf{R} \times E \to E$ is continuous, but not uniformly continuous.

14.2.2 *(Hardy's test)* (a) Suppose that $(a_j)_{j=0}^{\infty}$ is a null sequence of real or complex numbers for which $\sum_{j=0}^{\infty} |a_j - a_{j+1}| < \infty$ and that $(b_j)_{j=0}^{\infty}$ is a sequence in a Banach space $(E, \|.\|)$ for which the sequence of partial sums $(\sum_{j=0}^{n} b_j)_{n=0}^{\infty}$ is bounded. Show that $\sum_{j=0}^{\infty} a_j b_j$ converges.
(b) Suppose that $(a_j)_{j=0}^{\infty}$ is a null sequence of real or complex numbers for which the sequence of partial sums $(\sum_{j=0}^{n} a_j)_{n=0}^{\infty}$ is bounded and that $(b_j)_{j=0}^{\infty}$ is a sequence in a Banach space $(E, \|.\|)$ for which $\sum_{j=0}^{\infty} \|b_j - b_{j+1}\| < \infty$. Show that $\sum_{j=0}^{\infty} a_j b_j$ converges.

14.2.3 *(Hardy's uniform test)* Suppose that $(a_j)_{j=0}^{\infty}$ is a sequence in the real or complex Banach space $(B(S), \|.\|_\infty)$ for which

$$\sum_{j=0}^{\infty} \|a_j - a_{j+1}\|_\infty < \infty$$

and that $(b_j)_{j=0}^{\infty}$ is a null sequence in $B(S)$ for which the sequence of partial sums $(\sum_{j=0}^{n} b_j)_{n=0}^{\infty}$ is bounded. Show that $\sum_{j=0}^{\infty} a_j b_j$ converges uniformly.

14.2.4 *(Dirichlet's test)* Suppose that $(a_j)_{j=0}^{\infty}$ is a decreasing null sequence of positive numbers and that $(b_j)_{j=0}^{\infty}$ is a sequence in a Banach space $(E, \|.\|)$ for which the sequence of partial sums $(\sum_{j=0}^{n} b_j)_{n=0}^{\infty}$ is bounded. Show that $\sum_{j=0}^{\infty} a_j b_j$ converges, to s say. Show further that if

$$s_m = \sum_{j=0}^{m} a_j b_j \text{ and } M = \sup_n \left\| \sum_{j=0}^{n} b_j \right\|$$

then $\|s - s_m\| \leq 2a_{m+1}M$.

14.2.5 *(Abel's test)* Suppose that $(a_j)_{j=0}^{\infty}$ is a decreasing sequence of positive numbers and that $(b_j)_{j=0}^{\infty}$ is a sequence in a Banach space $(E, \|.\|)$ for which $\sum_{j=0}^{\infty} b_j$ converges. Show that $\sum_{j=0}^{\infty} a_j b_j$ converges.

14.2.6 This exercise and the next one give two classical applications of Weierstrass' uniform M-test. Formulae such as these go back to Euler. Let $P_k(0) = 1$ and

$$P_k(j) = \left(1 - \frac{1}{2k+1}\right)\left(1 - \frac{2}{2k+1}\right)\cdots\left(1 - \frac{2j}{2k+1}\right),$$

for $1 \leq j \leq k$. Show that

$$\frac{1}{2i}\left(\left(1 + \frac{i\theta}{2k+1}\right)^{2k+1} - \left(1 - \frac{i\theta}{2k+1}\right)^{2k+1}\right)$$

$$= \sum_{j=0}^{k} \frac{(-1)^j}{(2j+1)!} P_k(j)\theta^{2j+1}.$$

Let $X = \{0\} \cup \{1/(2k+1) : k \in \mathbf{N}\} \subseteq \mathbf{R}$. Let

$$f_j(0) = \frac{(-1)^j}{(2j+1)!}\theta^{2j+1},$$

$$f_j\left(\frac{1}{2k+1}\right) = \frac{(-1)^j}{(2j+1)!}P_k(j)\theta^{2j+1} \text{ for } 0 \leq j \leq k,$$

$$= 0 \text{ for } j > k.$$

(a) Show that f_j is a continuous function on X.
(b) Show that $\|f_j\|_{\infty} = |\theta|^{2j+1}/(2j+1)!$.
(c) Use Weierstrass' uniform M-test to show that

$$\frac{1}{2i}\left(\left(1 + \frac{i\theta}{2k+1}\right)^{2k+1} - \left(1 - \frac{i\theta}{2k+1}\right)^{2k+1}\right) \to \sin\theta$$

as $k \to \infty$.

14.2.7 Let

$$
f_j(y) = \begin{cases} \dfrac{1}{\pi^2 j^2} & \text{for } y = 0, \\[2mm] y^2 \left(\dfrac{1 + \cos 2jy\pi}{1 - \cos 2jy\pi} \right) = y^2 \left(\dfrac{1 + \cos 2jy\pi}{2 \sin^2 jy\pi} \right) & \text{for } 0 < y \le 1/2j, \\[2mm] 0 & \text{for } 1/2j < y \le 1. \end{cases}
$$

(a) Show that $f_j \in C[0,1]$.
(b) Show that $\|f_j\|_\infty \le 1/2\pi j^2$. (Use the inequality $\sin\theta \ge 2\theta/\pi$ for $0 \le \theta \le \pi/2$.)
(c) Use Weierstrass' uniform M-test to show that

$$
\sum_{j=1}^{\lfloor \pi/y \rfloor} y^2 \left(\frac{1 + \cos 2jy\pi}{1 - \cos 2jy\pi} \right) \to \sum_{j=1}^{\infty} \frac{1}{j^2 \pi^2}
$$

as $y \searrow 0$.

14.2.8 Give a direct proof of Corollary 14.2.9.

14.3 Linear operators

When we consider linear mappings between normed spaces, then continuity and uniform continuity are the same. Indeed, we can say more.

Theorem 14.3.1 *Suppose that* $(E_1, \|.\|_1)$ *and* $(E_2, \|.\|_2)$ *are normed spaces and that* T *is a linear mapping from* E_1 *to* E_2. *The following are equivalent:*

(i) $K = \sup\{\|Tx\|_2 : \|x\|_1 \le 1\} < \infty$;
(ii) there exists $C \in \mathbf{R}$ *such that* $\|Tx\|_2 \le C\|x\|_1$, *for all* x *in* E_1;
(iii) T *is Lipschitz;*
(iv) T *is uniformly continuous on* E_1;
(v) T *is continuous on* E_1;
(vi) T *is continuous at* 0.

Proof　(i) implies (ii): (ii) is trivially satisfied if $x = 0$. Otherwise, let $x_1 = x/\|x\|_1$. Then

$$
\|T(x)\|_2 = \|T(\|x\|_1 \, x_1)\|_2 = \|\|x\|_1 \, T(x_1)\|_2 = \|x\|_1 \, \|T(x_1)\|_2 \le K\|x\|_1 \,.
$$

(ii) implies (iii): $\|T(x_1) - T(x_2)\|_2 = \|T(x_1 - x_2)\|_2 \le C\|x_1 - x_2\|_1 \,.$
Obviously (iii) implies (iv), (iv) implies (v) and (v) implies (vi).

(vi) implies (i): There exists $\delta > 0$ such that if $\|x\|_1 \leq \delta$ then $\|T(x)\|_2 \leq 1$. If $\|x\|_1 \leq 1$ then $\|\delta x\|_1 \leq \delta$, so that

$$\|T(x)\|_2 = \left\|\delta^{-1} T(\delta x)\right\|_2 = \delta^{-1} \|T(\delta x)\|_2 \leq \delta^{-1}.$$

$\square$

We denote the set of continuous linear mappings from V_1 to V_2 by $L(V_1, V_2)$. We write $L(V)$ for $L(V, V)$. A continuous linear mapping from V_1 to V_2 is also called a *bounded linear mapping*, or a *linear operator*; a continuous linear mapping from V to itself is called an *operator on V*.

Two norms $\|.\|_1$ and $\|.\|_2$ on a vector space E are *equivalent* if the corresponding metrics are equivalent.

Corollary 14.3.2   *If $\|.\|_1$ and $\|.\|_2$ are equivalent norms on a vector space E then $(E, \|.\|_1)$ is a Banach space if and only if $(E, \|.\|_2)$ is.*

Proof   For they are uniformly equivalent, and so the result follows from Corollary 14.1.7. $\square$

We have the following extension theorem.

Theorem 14.3.3   *Suppose that F is a dense linear subspace of a normed space $(E, \|.\|_E)$, and that T is a continuous linear mapping from F to a Banach space $(G, \|.\|_G)$. Then there is a unique continuous linear mapping $\widetilde{T}$ from E to G which extends T: $\widetilde{T}(y) = T(y)$ for $y \in F$. If T is an isometry then so is $\widetilde{T}$.*

Proof   By Theorem 14.3.1, T is uniformly continuous, and so by Theorem 14.1.10 there is a unique continuous extension $\widetilde{T}$, which is an isometry if T is. We must show that $\widetilde{T}$ is linear. Suppose that $x, y \in E$ and that α, β are scalars. There exist sequences $(x_n)_{n=1}^{\infty}$ and $(y_n)_{n=1}^{\infty}$ in F such that $x_n \to x$ and $y_n \to y$ as $n \to \infty$. Then $\alpha x_n + \beta y_n \to \alpha x + \beta y$ as $n \to \infty$, and so

$$\widetilde{T}(\alpha x + \beta y) = \lim_{n \to \infty} \widetilde{T}(\alpha x_n + \beta y_n) = \lim_{n \to \infty} T(\alpha x_n + \beta y_n)$$

$$= \lim_{n \to \infty} (\alpha T(x_n) + \beta T(y_n)) = \alpha \lim_{n \to \infty} T(x_n) + \beta \lim_{n \to \infty} T(y_n)$$

$$= \alpha \lim_{n \to \infty} \widetilde{T}(x_n) + \beta \lim_{n \to \infty} \widetilde{T}(y_n) = \alpha \widetilde{T}(x) + \beta \widetilde{T}(y).$$

$\square$

Theorem 14.3.4   *(i) $L(V_1, V_2)$ is a linear subspace of the vector space of all linear mappings from V_1 to V_2.*

(ii) If $T \in L(V_1, V_2)$, set $\|T\| = \sup\{\|T(x)\|_2 : \|x\|_1 \leq 1\}$. Then $\|T\|$ is a norm on $L(V_1, V_2)$, the operator norm.

(iii) If $T \in L(V_1, V_2)$, and $x \in V_1$ then $\|T(x)\|_2 \leq \|T\| \cdot \|x\|_1$.

Proof. (i): We use condition (i) of Theorem 14.3.1. Suppose that $S, T \in L(V_1, V_2)$ and that α is a scalar. Then

$$\sup\{\|(\alpha T)(x)\|_2 : \|x\|_1 \leq 1\} = |\alpha| \sup\{\|T(x)\|_2 : \|x\|_1 \leq 1\},$$

so that $\alpha T \in L(V_1, V_2)$ and

$$\sup\{\|(S + T)(x)\|_2 : \|x\|_1 \leq 1\}$$
$$\leq \sup\{\|S(x)\|_2 : \|x\|_1 \leq 1\} + \sup\{\|T(x)\|_2 : \|x\|_1 \leq 1\},$$

so that $S + T \in L(V_1, V_2)$.

(ii): If $\|T\| = 0$, then $T(x) = 0$ for x with $\|x\| \leq 1$, and so $T(x) = 0$ for all x: thus $T = 0$. $\|\alpha T\| = |\alpha| \|T\|$ and $\|S + T\| \leq \|S\| + \|T\|$, by the equation and inequality that we have established to prove (i).

(iii): This is true if $x = 0$. Otherwise, let $y = x / \|x\|_1$. Then $\|y\|_1 = 1$, so that

$$\|T(x)\|_2 = \|T(\|x\|_1 \, y)\|_2 = \|x\|_1 \|T(y)\|_2 \leq \|T\| \|x\|_1.$$

Theorem 14.3.5 *If $(E_1, \|.\|_1)$ is a normed space and $(E_2, \|.\|_2)$ is a Banach space then $L(E_1, E_2)$ is a Banach space under the operator norm.*

Proof The proof is like the proof of Theorem 14.1.4. Let (T_n) be a Cauchy sequence in $L(E_1, E_2)$. First we identify what the limit must be. Since, for each $x \in E_1$, $\|T_n(x) - T_m(x)\|_2 \leq \|T_n - T_m\| \|x\|_1$, $(T_n(x))$ is a Cauchy sequence in E_2, which converges, by the completeness of E_2, to $T(x)$, say. Secondly, we show that T is a linear mapping from E_1 to E_2. This follows, since

$$T(\alpha x + \beta y) - \alpha T(x) - \beta T(y) = \lim_{n \to \infty} (T_n(\alpha x + \beta y) - \alpha T_n(x) - \beta T_n(y)) = 0,$$

for all $x, y \in E_1$ and all scalars α, β. Thirdly we show that T is continuous. There exists N such that $\|T_n - T_m\| \leq 1$, for $m, n \geq N$. Then

$$\|(T - T_N)(x)\|_2 = \lim_{n \to \infty} \|(T_n - T_N)(x)\|_2 \leq \|x\|_1,$$

for each $x \in E_1$, so that $T - T_N \in L(E_1, E_2)$. Since $L(E_1, E_2)$ is a vector space, $T = (T - T_N) + T_N \in L(E_1, E_2)$. Finally we show that $T_n \to T$. Given $\epsilon > 0$ there exists M such that $\|T_n - T_m\| \leq \epsilon$, for $m, n \geq M$. Then if $m \geq M$, and $x \in E_1$,

$$\|(T - T_m)(x)\|_2 = \lim_{n \to \infty} \|(T_n - T_m)(x)\|_2 \leq \epsilon \|x\|_1 \,,$$

so that $\|T - T_m\| \leq \epsilon$. $\qquad\qquad\qquad\qquad\qquad\qquad\qquad\qquad\qquad\qquad$ □

A *linear functional* on a vector space V is a linear mapping from V into the underlying scalar field. The vector space of continuous linear functionals on a normed space $(E, \|.\|)$ is called the *dual space* E'; it is given the *dual norm* $\|\phi\|' = \{\sup |\phi(x)| : \|x\| \leq 1\}$. This is simply the operator norm from $(E, \|.\|)$ into the scalars.

Corollary 14.3.6 *The dual space $(E', \|.\|')$ of a normed space $(E, \|.\|)$ is a Banach space.*

Let us consider one important example. We need a definition. A mapping T from a complex vector space E into a complex vector space F is *conjugate linear* if

$$T(x + y) = T(x) + T(y) \text{ and } T(\alpha x) = \overline{\alpha} T(x) \text{ for } x, y \in E, \alpha \in \mathbf{C}.$$

Theorem 14.3.7 (The Fréchet–Riesz representation theorem) *Suppose that H is a real or complex Hilbert space. If $x, y \in H$, let $l_y(x) = \langle x, y \rangle$. Then $l_y \in H'$, and the mapping $l : H \to H'$ is an isometry of H onto H'. It is linear if H is real and is conjugate linear if H is complex.*

Proof We consider the complex case: the real case is easier. The function l_y is a linear mapping of H into $\mathbf{C}$. Since

$$|l_y(x)| = |\langle x, y \rangle| \leq \|x\| \cdot \|y\| \,,$$

$l_y \in H'$ and $\|l_y\|' \leq \|y\|$. If $y \neq 0$ then $l_y(y/\|y\|) = \|y\|$, so that $\|l_y\|' \geq \|y\|$. Thus $\|l_y\|' = \|y\|$. Clearly $l_{y_1 + y_2} = l_{y_1} + l_{y_2}$, so that $\|l_{y_2} - l_{y_1}\|' = \|y_1 - y_2\|$, for $y_1, y_2 \in H$: l is an isometry of H into H'. Since

$$l_{\alpha y}(x) = \langle x, \alpha y \rangle = \overline{\alpha} \langle x, y \rangle = \overline{\alpha} l_y(x),$$

l is conjugate linear.

The important part of the proof is the proof that l is surjective: every continuous linear functional ϕ on H is represented in terms of the inner product; there exists $y \in H$ such that $\phi(x) = \langle x, y \rangle$ for all $x \in H$.

If $\phi = 0$, then $\phi = l_0$. Otherwise, by scaling, we can suppose that $\|\phi\|' = 1$. First we show that there is a unique y in the closed unit ball B of H such that $\phi(y) = 1$. We use Theorem 14.1.11. Let $A_n = \{x \in B : \Re\phi(x) \geq 1 - 1/n\}$. Since $\|\phi\|' = 1 = \sup_{x \in B} |\phi(x)|$, A_n is non-empty, and clearly $(A_n)_{n=1}^{\infty}$ is a decreasing sequence of closed sets. Suppose that $x_1, x_2 \in A_n$. By the parallelogram law,

$$\|x_1 + x_2\|^2 + \|x_1 - x_2\|^2 = 2(\|x_1\|^2 + \|x_2\|^2) \le 4.$$

Since $\Re\phi(x_1 + x_2) = \Re\phi(x_1) + \Re\phi(x_2) \ge 2(1 - 1/n)$, $\|x_1 + x_2\| \ge 2(1 - 1/n)$. It follows from this that

$$\|x_1 - x_2\|^2 \le 4 - 4(1 - 1/n)^2 = 8/n - 4/n^2 \le 8/n.$$

Thus $\operatorname{diam}(A_n) \to 0$ as $n \to \infty$, so that $\cap_{n \in \mathbf{N}} A_n$ is a singleton y. Then y is the unique element of B for which $\phi(y) = 1$.

We now show that $\phi(x) = \langle x, y \rangle$ for all $x \in H$. Let $w = x - \langle x, y \rangle y$. Then $\langle w, y \rangle = 0$, so that w is orthogonal to y. Suppose that $\phi(w) = re^{i\theta} \ne 0$. If $t > 0$ then

$$1 + 2rt + r^2 t^2 = (\phi(y + e^{-i\theta}tw))^2 \le \left\|y + e^{-i\theta}tw\right\|^2 = 1 + t^2 \|w\|^2,$$

so that $2r \le t(\|w\|^2 - r^2)$. Since this holds for all positive t, $\|w\|^2 > r^2$. Set $t = r/(\|w\|^2 - r^2)$: then $2r \le r$, giving a contradiction. Thus $\phi(w) = 0$: that is,

$$\phi(x) = \phi(w) + \langle x, y \rangle \phi(y) = \langle x, y \rangle.$$

$\square$

This has the following consequence.

Theorem 14.3.8 *Suppose that H and K are Hilbert spaces and that $T \in L(H, K)$. Then there exists a unique $S \in L(K, H)$ such that*

$$\langle T(x), y \rangle = \langle x, S(y) \rangle \quad \text{for all } x \in H, y \in K.$$

Further, $\|S\| = \|T\|$.

Proof   Suppose that $y \in K$. Then the mapping $x \to \langle T(x), y \rangle$ is a continuous linear functional, and so there exists a unique element, $S(y)$ say, in H such that $\langle T(x), y \rangle = \langle x, S(y) \rangle$. Then

$$\langle x, S(y_1 + y_2) \rangle = \langle T(x), y_1 + y_2 \rangle = \langle T(x), y_1 \rangle + \langle T(x), y_2 \rangle$$
$$= \langle x, S(y_1) \rangle + \langle x, S(y_2) \rangle,$$

so that $S(y_1 + y_2) = S(y_1) + S(y_2)$, and

$$\langle x, S(\alpha y) \rangle = \langle T(x), \alpha y \rangle = \overline{\alpha} \langle T(x), y \rangle = \overline{\alpha} \langle x, S(y) \rangle = \langle x, \alpha S(y) \rangle,$$

so that $S(\alpha y) = \alpha S(y)$; S is a linear mapping.

If $y \in K$ then

$$\|S(y)\|^2 = \langle S(y), S(y) \rangle = \langle TS(y), y \rangle \le \|T\| \cdot \|S(y)\| \cdot \|y\|,$$

so that $\|S(y)\| \le \|T\| \cdot \|y\|$. Thus S is continuous, and $\|S\| \le \|T\|$. Similarly, if $x \in H$ then

$$\|T(x)\|^2 = \langle T(x), T(x) \rangle = \langle x, ST(x) \rangle \le \|S\| \cdot \|T(x)\| \cdot \|x\|,$$

so that $\|T(x)\| \le \|S\| \, \|x\|$, and $\|T\| \le \|S\|$. Hence $\|S\| = \|T\|$. $\qquad \square$

S is the *adjoint* of T. If H and K are real, then S is denoted by T', and if H and K are complex, then S is denoted by T^*.

Exercises

14.3.1 Suppose that $(E, \|.\|_E)$, $(F, \|.\|_F)$ and $(G, \|.\|_G)$ are normed spaces and that B is a bilinear mapping from $E \times F$ into G. Show that B is continuous if and only if there exists $M \ge 0$ such that $\|B(x, y)\|_G \le M \|x\|_E \|y\|_F$ for all $(x, y) \in E \times F$.

14.3.2 Suppose that $(E, \|.\|_E)$, $(F, \|.\|_F)$ and $(G, \|.\|_G)$ are normed spaces. If $T \in L(E, L(F, G))$, and $x \in E$, $y \in F$, let $j(T)(x, y) = (T(x))(y)$. Show that $j(T)$ is a continuous bilinear mapping from $E \times F$ into G. Show that j is a bijective linear mapping of $L(E, L(F, G))$ onto the vector space $B(E, F; G)$ of continuous bilinear mappings from $E \times F$ into G. If $b \in B(E, F; G)$, let

$$\|b\| = \sup\{\|b(x, y)\|_G : \|x\|_E \le 1, \|y\|_F \le 1\}.$$

Show that this is a norm on $B(E, F; G)$, and that with this norm the mapping j is an isometry. Deduce that if G is a Banach space, then so is $B(E, F; G)$.

14.3.3 Suppose that E and F are Euclidean spaces, with orthonormal bases $(e_1, \ldots, e_m)$ and $(f_1, \ldots, f_n)$ and that $T \in L(E, F)$ is represented by a matrix (t_{ij}) with respect to these bases. What matrix represents the adjoint T'?

14.3.4 If $x \in l_2$, let $R(x)_1 = 0$ and $(R(x))_n = x_{n-1}$ for $n > 1$: R is the *right shift* on l_2. What is R'? Show that $R'R$ is the identity on l_2. What is RR'?

14.3.5 If H is a Hilbert space and $T \in L(H)$, then T is *unitary* (in the complex case) or *orthogonal* (in the real case), if it is an isometry of H onto itself. Show that T is unitary (orthogonal) if and only if $TT^* = T^*T = I$ $(TT' = T'T = I)$.

14.3.6 Suppose that K is a closed linear subspace of a Hilbert space H.

   (i) Show that if $x \in H$ then there is a unique point $P(x)$ in K such that $\|x - P(x)\| = \inf\{\|y - x\| : y \in K\}$.

   (ii) Show that $P(x)$ is the unique point in K with $x - P(x) \in K^\perp$.

   (iii) Show that the mapping $x \to P(x)$ is linear and that P is continuous, with $\|P\| \leq 1$. When is $\|P\|$ less than 1?

     P is the *orthogonal projection* of H onto K.

   (iv) Show that $P = P^*$ $(P = P')$ and that $H = K \oplus K^\perp$.

     [Compare this with the construction in Proposition 11.4.3, when H is finite-dimensional.]

14.3.7 Suppose that ϕ is a non-zero continuous linear functional on a Hilbert space H, and that $\phi(x_0) = 1$. Let N be the null-space of ϕ, and let P be the orthogonal projection of H onto N. Let $z_0 = x_0 - P(x_0)$. Show that $\phi(z_0) = 1$ and that $\phi(x) = \langle x, z_0 \rangle / \|z_0\|^2$, for $x \in H$.

    (This gives another proof of the Fréchet–Riesz representation theorem.)

14.4 *Tietze's extension theorem*

(This section can be omitted on a first reading.)

As an application of the results of the two previous sections, we prove Tietze's extension theorem. We need a preliminary result, of interest in its own right.

Theorem 14.4.1 *Suppose that $(E_1, \|.\|_1)$ is a Banach space and that $(E_2, \|.\|_2)$ is a normed space; let their closed unit balls be B_1 and B_2, respectively. Suppose that $T \in L(E_1, E_2)$ and that there exist $0 < t < 1$ and $\epsilon > 0$ such that $\epsilon B_2 \subseteq (1-t)T(B_1) + t\epsilon B_2$ - that is, if $y \in \epsilon B_2$ there exist $x \in B_1$ and $z \in \epsilon B_2$ such that $y = (1-t)T(x) + tz$. Then the following hold:*

  (i) $\epsilon B_2 \subseteq T(B_1)$;

  (ii) T is surjective;

  (iii) If U is open in E_1 then $T(U)$ is open in E_2;

  (iv) $(E_2, \|.\|_2)$ is a Banach space.

Proof (i) Suppose that $z_0 \in \epsilon B_2$. Then there exist $x_1 \in B_1$ and $z_1 \in \epsilon B_2$ such that $z_0 = (1-t)T(x_1) + tz_1$; iterating this, there exist sequences $(x_n)_{n=1}^\infty$

in B_1 and $(z_n)_{n=1}^{\infty}$ in ϵB_2 such that $z_n = (1-t)T(x_{n+1}) + tz_{n+1}$, for $n \in \mathbf{N}$. Thus

$$z_0 = (1-t)T(x_1) + (1-t)T(tx_2) + \cdots + (1-t)T(t^n x_{n+1}) + t^{n+1} z_{n+1}$$
$$= (1-t)T(x_1 + tx_2 + \cdots + t^n x_{n+1}) + t^{n+1} z_{n+1}.$$

Since

$$\sum_{n=1}^{\infty} \|t^{n-1} x_n\| = \sum_{n=1}^{\infty} t^{n-1} \|x_n\| \leq \sum_{n=1}^{\infty} t^{n-1} = 1/(1-t),$$

$\sum_{n=1}^{\infty} t^{n-1} x_n$ converges absolutely to an element x of E_1, and $\|x\| \leq 1/(1-t)$. Since $t^{n+1} z_{n+1} \to 0$ as $n \to \infty$, it follows that $z_0 = (1-t)T(x) = T((1-t)x)$. Since $\|(1-t)x\| = (1-t)\|x\| \leq 1$, $z_0 \in T(B_1)$.

(ii) $T(E_1) = T(\cup_{n=1}^{\infty} nB_1) = \cup_{n=1}^{\infty} nT(B_1) \supseteq \cup_{n=1}^{\infty} n\epsilon B_2 = E_2$.

(iii) Suppose that $y = T(x) \in T(U)$. There exists $\delta > 0$ such that $M_\delta(x) = x + \delta B_1 \subseteq U$. We show that $y + \epsilon\delta B_2 \subseteq T(U)$, so that $T(U)$ is open. If $z = y + w \in y + \epsilon\delta B_2$ then there exists $v \in \delta B_1$ such that $w = T(v)$. Thus

$$z = T(x) + T(v) = T(x+v) \in T(x + \delta B_1) \subseteq T(U).$$

Hence $y + \epsilon\delta B_2 \subseteq T(U)$.

(iv) In order to show that $(E_2, \|.\|_2)$ is a Banach space, we use Proposition 14.2.5. By homogeneity, if $y \in E_2$ there exists $x \in E_1$ with $\|x\|_1 \leq \|y\|_2/\epsilon$ for which $T(x) = y$. Suppose that $(y_n)_{n=1}^{\infty}$ is a sequence in E_2 with $\sum_{n=1}^{\infty} \|y_n\|_2 < \infty$. For each $n \in \mathbf{N}$ there exists $x_n \in E_1$ with $\|x_n\|_1 \leq \|y_n\|_2/\epsilon$ such that $T(x_n) = y_n$. Thus $\sum_{n=1}^{\infty} \|x_n\|_1 < \infty$. Since $(E_1, \|.\|_1)$ is a Banach space, $\sum_{n=1}^{\infty} x_n$ converges in E_1, to s, say. Since T is continuous, $\sum_{n=1}^{\infty} y_n = \sum_{n=1}^{\infty} T(x_n) = T(s)$. Thus $(E_2, \|,\|_2)$ is a Banach space, by Proposition 14.2.5. $\quad\square$

Corollary 14.4.2 *Suppose that* $(E_1, \|.\|_1)$ *is a Banach space, that* $(E_2, \|.\|_2)$ *is a normed space and that* $T \in L(E_1, E_2)$. *Suppose that there exists* $\eta > 0$ *such that* $\overline{T(B_1)} \supset \eta B_2$. *If* U *is open in* E_1 *then* $T(U)$ *is open in* E_2.

Proof   Since $\overline{T(B_1)} \subset T(B_1) + (\eta/2)B_2$, it follows that $\eta B_2 \subseteq T(B_1) + (\eta/2)B_2$. Set $\epsilon = \eta/2$ and $t = 1/2$; then $\epsilon B_2 \subseteq (1-t)T(B_1) + t\epsilon B_2$, and the result follows from the theorem. $\quad\square$

Theorem 14.4.3 (Tietze's extension theorem) *Suppose that* f *is a bounded continuous real-valued function on a closed subset* A *of a normal topological space* (X, τ). *Let*

$$M = \sup\{f(a) : a \in A\}, \quad m = \inf\{f(a) : a \in A\}.$$

Then there exists a continuous real-valued function g on X such that $g(a) = f(a)$ for $a \in A$, and $m \leq g(x) \leq M$ for $x \in X$.

Proof   The result is obviously true if f is constant. Otherwise, by considering $f - (M + m)/2$, we can suppose that $m = -M$, and by considering $f/\|f\|$, we can suppose that $M = 1$ and $m = -1$. Let R be the restriction mapping from $C_b(X) \to C_b(A)$; $R(f) = f_{|A}$. We show that R satisfies the conditions of Theorem 14.4.1, with $t = 2/3$ and $\epsilon = 1$. Let

$$B = \{a \in A : f(a) \geq 1/3\} \text{ and } C = \{a \in A : f(a) \leq -1/3\}.$$

Then B and C are disjoint closed subsets of X, and so by Urysohn's lemma there exists $g \in C_b(X)$ with $\|g\|_\infty \leq 1/3$ such that $g(b) = 1/3$ for $b \in B$ and $g(c) = -1/3$ for $c \in C$. Then

$$\begin{aligned}
0 \leq f(x) - g(x) \leq 2/3, &\quad \text{for } x \in B, \\
0 > f(x) - g(x) \geq -2/3, &\quad \text{for } x \in C, \\
|f(x) - g(x)| \leq |f(x)| + |g(x)| \leq 2/3 &\quad \text{for } x \in A \setminus (B \cup C).
\end{aligned}$$

Thus $\|f - R(g)\|_\infty \leq 2/3$. Let us set $p = 3g$ and $q = (3/2)(f - R(g))$. Then $\|p\|_\infty \leq 1$, $\|q\|_\infty \leq 1$ and $f = (1/3)R(p) + (2/3)q$. Consequently, if $f \in C_b(A)$ there exists $g \in C_b(X)$ with $R(g) = f$ and $\|g\|_\infty = \|f\|_\infty$.   □

We can drop the requirement that f is bounded.

Corollary 14.4.4   *(i) If $m < f(x) < M$ there exists a continuous function g on X such that $R(g) = f$ and such that $m < g(x) < M$ for $x \in X$.*

(ii) If F is a continuous function on A then there exists a continuous function G on X such that $G(a) = F(a)$ for $a \in A$.

Proof   (i) Again, we can suppose that $M = -m = 1$. There exists $h \in C_b(X)$ such that $R(h) = f$ and $\|h\|_\infty \leq 1$. Let $D = \{x \in X : |h(x)| = 1\}$. Then D is a closed subset of X disjoint from A. By Urysohn's Lemma, there exists $k \in C_b(X)$ with $0 \leq k \leq 1$ for which $k(a) = 1$ for $a \in A$ and $k(d) = 0$ for $d \in D$. Then $g = h.k$ has the required properties.

(ii) Let $f = \tan^{-1} \circ F$, Then $f \in C_b(A)$ and $-\pi/2 < f(a) < \pi/2$ for $a \in A$. By (i), there exists $g \in C_b(X)$ with $R(g) = f$ and $-\pi/2 < g(x) < \pi/2$ for $x \in X$. Let $G = \tan \circ g$.   □

14.5 The completion of metric and normed spaces

Starting with the field $\mathbf{Q}$ of rational numbers, we constructed the field $\mathbf{R}$ of real numbers. This fills up the gaps in the rationals – any Cauchy sequence converges – but does so in an efficient way, since any real number is the limit of a sequence of rational numbers.

We can do the same for any metric space. A *completion* of a metric space (X, d) is a complete metric space $(\hat{X}, \hat{d})$, together with an isometric mapping j of X onto a dense subset $j(X)$ of $\hat{X}$. We have the following fundamental theorem.

Theorem 14.5.1 *Any metric space (X, d) has a completion. The completion is essentially unique: if $((\hat{X}, \hat{d}), j)$ and $((\bar{X}, \bar{d}), j')$ are completions of (X, d) then there is a unique isometry k of $(\hat{X}, \hat{d})$ onto $(\bar{X}, \bar{d})$ such that $j' = k \circ j$.*

Proof   We give two proofs of the existence of a completion. The first is short, but quite artificial. We have shown in Example 11.5.12 that there is an isometry j of (X, d) into $(B(X), \|.\|_\infty)$, and have shown that $(B(X), \|.\|_\infty)$ is a Banach space. We therefore take $\hat{X}$ to be the closure $\overline{j(X)}$ of $j(X)$ in $B(X)$, and take $\hat{d}$ to be the subspace metric. Then $(\hat{X}, \hat{d})$ is complete (Proposition 14.1.3) and $j(X)$ is dense in $(\hat{X}, \hat{d})$.

The second proof is longer but more natural, and is useful when we consider normed spaces. If $(a_n)_{n=1}^\infty$ is a Cauchy sequence in X then $j(a_n)$ must converge to a unique element of the completion, so that $(a_n)_{n=1}^\infty$ determines an element of the completion. In general, however, there are many Cauchy sequences which determine this element. We therefore define the elements of the completion of (X, d) to be equivalence classes of Cauchy sequences in X.

Let Y be the set of all Cauchy sequences in (X, d). Suppose that $a = (a_n)_{n=1}^\infty$ and $b = (b_n)_{n=1}^\infty$ are in Y. If $\epsilon > 0$ then there exists $n_0 \in \mathbf{N}$ such that $d(a_m, a_n) < \epsilon/2$ and $d(b_m, b_n) < \epsilon/2$ for $m, n \geq n_0$. It follows from Proposition 13.3.2 that

$$|d(a_m, b_m) - d(a_n, b_n)| \leq d(a_m, a_n) + d(b_m, b_n) < \epsilon, \text{ for } m, n \geq n_0.$$

Thus $(d(a_n, b_n))_{n=1}^\infty$ is a Cauchy sequence of real numbers, which converges, by the general principle of convergence, to a limit $p(a, b)$. Clearly $p(a, b) = p(b, a)$, and

$$p(a, c) = \lim_{n \to \infty} d(a_n, c_n)$$

$$\leq \lim_{n \to \infty} d(a_n, b_n) + \lim_{n \to \infty} d(b_n, c_n) = p(a, b) + p(b, c),$$

so that p is a pseudometric on Y. We now apply Proposition 11.1.13: there exists an equivalence relation $\sim$ on Y and a metric $\hat{d}$ on the quotient space

$Y/\sim$ (which we denote by $\hat{X}$) such that the quotient mapping $q : Y \to \hat{X}$ satisfies $\hat{d}(q(a), q(b)) = p(a, b)$, for $a, b \in Y$.

Next we define the mapping $j : X \to \hat{X}$. If $x \in X$, let $x_n = x$ for all $n \in \mathbf{N}$; the constant sequence $c(x) = (x_n)_{n=1}^{\infty}$ is certainly a Cauchy sequence. We set $j(x) = q(c(x))$. If $x, x' \in X$ then

$$p(c(x), c(x')) = \lim_{n \to \infty} d(x_n, y_n) = d(x, y),$$

so that $\hat{d}(j(x), j(y)) = d(x, y)$, and so j is an isometry of (X, d) into $(\hat{X}, \hat{d})$.

We now show that $j(X)$ is dense in $(\hat{X}, \hat{d})$. Suppose that $\hat{x} \in \hat{X}$ and that $\hat{x} = q(a)$ where $a = (a_n)_{n=1}^{\infty} \in Y$. If $N \in \mathbf{N}$ then $p(a, c(a_N)) = \lim_{n \to \infty} d(a_n, a_N) \to 0$ as $N \to \infty$. Thus $\hat{d}(\hat{x}, j(a_N)) \to 0$ as $N \to \infty$, so that $j(X)$ is dense in $(\hat{X}, \hat{d})$.

The metric space $(\hat{X}, \hat{d})$, together with the isometry j, will be the completion of (X, d).

We now come to the crux of the proof, and show that $(\hat{X}, \hat{d})$ is complete. Suppose that $(\hat{x}^{(k)})_{k=1}^{\infty}$ is a Cauchy sequence in $(\hat{X}, \hat{d})$. Since $j(X)$ is dense in $(\hat{X}, \hat{d})$, there exist $x_k \in X$ with $\hat{d}(\hat{x}^{(k)}, j(x_k)) < 1/k$, for $k \in \mathbf{N}$. Then

$$|\hat{d}(j(x_k), j(x_l)) - \hat{d}(\hat{x}^{(k)}, \hat{x}^{(l)})| \leq 1/k + 1/l,$$

so that $(j(x_k))_{k=1}^{\infty}$ is a Cauchy sequence in $(\hat{X}, \hat{d})$. Since j is an isometry, $(x_k)_{k=1}^{\infty}$ is a Cauchy sequence in (X, d). Let $\hat{x} = q((x_k)_{k=1}^{\infty})$. Then $\hat{d}(j(x_l), \hat{x}) = \lim_{k \to \infty} d(x_l, x_k)$ and so $\hat{d}(j(x_l), \hat{x}) \to 0$ as $l \to \infty$, since $(x_k)_{k=1}^{\infty}$ is a Cauchy sequence. Thus $j(x_l) \to \hat{x}$ as $l \to \infty$. Consequently

$$\hat{d}(\hat{x}^{(l)}, \hat{x}) \leq \hat{d}(\hat{x}^{(l)}, j(x_l)) + \hat{d}(j(x_l), \hat{x}) \leq 1/l + \hat{d}(j(x_l), \hat{x}) \to 0$$

as $l \to \infty$. Thus $\hat{x}_l \to \hat{x}$ as $l \to \infty$.

Finally, we show that the completion is essentially unique. The mapping $j' \circ j^{-1}$ is an isometry from $j(X)$ into $(\bar{X}, \bar{d})$, and so is uniformly continuous. By Theorem 14.1.10, there is a unique continuous extension $k : (\hat{X}, \hat{d}) \to (\bar{X}, \bar{d})$, and k is an isometry. $k(\hat{X})$ is therefore complete, and so is closed in $\bar{X}$. But $j'(X) \subseteq k(\hat{X})$, and $j'(X)$ is dense in $\bar{X}$, and so $k(\hat{X}) = \bar{X}$: k is surjective. $\square$

Because of the essential uniqueness of completion, we usually talk about *the* completion of a metric space, and consider X as a dense subset of its completion (just as we consider the field $\mathbf{Q}$ of rational numbers as a subfield of the field $\mathbf{R}$ of real numbers).

There is a corresponding result for normed spaces.

Theorem 14.5.2 *Suppose that $(E, \|.\|)$ is a normed space. There exists a Banach space $(\hat{E}, \|.\|\,\hat{})$, and an isometric linear mapping $j : E \to \hat{E}$ such that $j(E)$ is a dense linear subspace of $\hat{E}$. $(\hat{E}, \|.\|\,\hat{})$ is the completion of E. It is essentially unique: if $((\bar{E}, \|.\|\,\bar{}), j')$ is another completion then there is a unique linear isometry k of $(\hat{E}, \|.\|\,\hat{})$ onto $(\bar{E}, \|.\|\,\bar{})$ such that $j' = k \circ j$.*

Proof Consider the second construction of Theorem 14.5.1, using Cauchy sequences. The space Y of Cauchy sequences has a natural vector space structure: define

$$(a_n)_{n=1}^\infty + (b_n)_{n=1}^\infty = (a_n + b_n)_{n=1}^\infty, \quad \lambda(a_n)_{n=1}^\infty = (\lambda a_n)_{n=1}^\infty,$$

verifying that the sum and scalar product are in Y. The pseudometric p is given by the seminorm π, where $\pi(a) = p(a, 0) = \lim_{n \to \infty} \|a_n\|$. The equivalence class $q(0)$ to which 0 belongs is

$$N = \{(a_n)_{n=1}^\infty : a_n \to 0 \text{ as } n \to \infty\},$$

which is a linear subspace of Y. Further, $q(a) = a + N$, so that $\hat{E}$ is the quotient vector space E/N, and j is a linear mapping of E into $\hat{E}$. If we set $\|\hat{x}\|\,\hat{} = \hat{d}(\hat{x}, \hat{0})$, then $\|.\|\,\hat{}$ is a norm on $\hat{E}$ which defines $\hat{d}$, and under which $\hat{E}$ is a Banach space. The facts that j is an isometry and that $j(E)$ is dense on $\hat{E}$ come from Theorem 14.5.1, as does the existence of an isometry $k : \hat{E} \to \bar{E}$. It remains to show that k is linear. If $\hat{x}, \hat{y} \in \hat{E}$, there exist sequences $(x_n)_{n=1}^\infty$ and $(y_n)_{n=1}^\infty$ in E such that $j(x_n) \to \hat{x}$ and $j(y_n) \to \hat{y}$ as $n \to \infty$. Then $j(x_n + y_n) \to \hat{x} + \hat{y}$, and so, using the continuity of j, j' and k,

$$k(\hat{x}) + k(\hat{y}) = k(\lim_{n \to \infty} j(x_n)) + k(\lim_{n \to \infty} j(y_n)) = \lim_{n \to \infty} j'(x_n) + \lim_{n \to \infty} j'(y_n)$$

$$= \lim_{n \to \infty} (j'(x_n) + j'(y_n)) = \lim_{n \to \infty} (j'(x_n + y_n))$$

$$= k(\lim_{n \to \infty} (j(x_n + y_n))) = k(\hat{x} + \hat{y}).$$

Scalar multiplication is treated in a similar way. $\square$

Again, we consider $(E, \|.\|)$ as a dense linear subspace of its completion $(\hat{E}, \|.\|\,\hat{})$.

Exercise

14.5.1 Define the notion of a convergent sequence in **Q** and a Cauchy sequence in **Q**, using rational numbers, rather than real numbers.

Show how the second proof of Theorem 14.5.1 can be used to construct the completion $\hat{Q}$ of Q. Show that $\hat{Q}$ is an ordered field, and that every non-empty subset of $\hat{Q}$ which is bounded above has a least upper bound.

[This is another way of constructing R from Q.]

14.6 The contraction mapping theorem

If f is a mapping of a set X into itself, then an element x of X is a *fixed point* of f if $f(x) = x$. As we shall see, fixed points frequently have interesting properties.

A mapping $f : (X, d) \to (X, d)$ of a metric space into itself is a *contraction mapping* of (X, d) if there exists $0 \le K < 1$ such that $d(f(x), f(y)) \le Kd(x, y)$ for all $x, y \in X$; that is, f is a Lipschitz mapping with constant strictly less than 1. The fact that the constant K is *strictly* less than 1 is of fundamental importance.

Theorem 14.6.1 (The contraction mapping theorem) *If f is a contraction mapping of a non-empty complete metric space (X, d) then f has a unique fixed point x_∞.*

Proof   Let K be the Lipschitz constant of f. Let x_0 be any point of X. Define the sequence $(x_n)_{n=0}^\infty$ recursively by setting $x_{n+1} = f(x_n)$. Thus $x_n = f^n(x_0)$, and

$$d(x_n, x_{n+1}) \le Kd(x_{n-1}, x_n) \le K^2 d(x_{n-2}, x_{n-1}) \le \cdots \le K^n d(x_0, x_1).$$

We show that $(x_n)_{n=0}^\infty$ is a Cauchy sequence. Suppose that $\epsilon > 0$. There exists $n_0 \in N$ such that $K^n < (1 - K)\epsilon/(d(x_0, x_1) + 1)$ for $n \ge n_0$. If $n > m \ge n_0$ then

$$\begin{aligned}
d(x_m, x_n) &\le d(x_m, x_{m+1}) + d(x_{m+1}, x_{m+2}) + \cdots + d(x_{n-1}, x_n) \\
&\le K^m d(x_0, x_1) + K^{m+1} d(x_0, x_1) + \cdots + K^{n-1} d(x_0, x_1) \\
&\le K^m d(x_0, x_1)/(1 - K) < \epsilon.
\end{aligned}$$

Since (X, d) is complete, there exists $x_\infty \in X$ such that $x_n \to x_\infty$ as $n \to \infty$. Since f is continuous, $x_{n+1} = f(x_n) \to f(x_\infty)$ as $n \to \infty$, and so $x_\infty = f(x_\infty)$: x_∞ is a fixed point of f. If y is any fixed point of f then $d(y, x_\infty) = d(f(y), f(x_\infty)) \le Kd(y, x_\infty)$; hence $d(y, x_\infty) = 0$, and $y = x_\infty$; x_∞ is the unique fixed point of f. $\qquad\square$

Three points are worth making about this proof. First, we start with *any* point x_0 of X, and obtain a sequence which converges to the unique fixed point x_∞; further, $d(x_0, x_\infty) \leq d(x_0, x_1)/(1 - K)$. Secondly, $d(x_{n+1}, x_\infty) = d(f(x_n), f(x_\infty)) \leq K d(x_n, x_\infty)$, so that $d(x_n, x_\infty) \leq K^n d(x_0, x_\infty)$; the convergence is exponentially fast. Thirdly, the condition that $d(f(x), f(y)) < d(x, y)$ for $x \neq y$ is not sufficient for f to have a fixed point. The function $f(x) = x + e^{-x} : [0, \infty) \to [0, \infty)$ does not have a fixed point, but satisfies the condition; if $0 \leq y < x < \infty$ then, by the mean-value theorem, $f(x) - f(y) = (1 - e^{-c})(x - y)$ for some $x < c < y$, so that $|f(x) - f(y)| < |x - y|$.

Corollary 14.6.2 *Suppose that g is a mapping from X to X which commutes with f: $f \circ g = g \circ f$. Then x_∞ is a fixed point of g.*

Proof $f(g(x_\infty)) = g(f(x_\infty)) = g(x_\infty)$; $g(x_\infty)$ is a fixed point of f, and so $g(x_\infty) = x_\infty$. □

We can strengthen the contraction mapping in the following way.

Corollary 14.6.3 *Suppose that $h : (X, d) \to (X_d)$ is a mapping of a complete metric space into itself, and suppose that h^k is a contraction mapping for some $k \in \mathbf{N}$. Then h has a unique fixed point.*

Proof Let $f = h^k$. Then f has a unique fixed point x_∞. As $f \circ h = h \circ f = f^{k+1}$, x_∞ is a fixed point of h. If y is a fixed point of h then $f(y) = h^k(y) = y$, so that $y = x_\infty$; x_∞ is the unique fixed point of f. □

Suppose that (X, d) and (Y, ρ) are metric spaces and that $f : X \times Y \to Y$ is continuous. Can we solve the equation $y = f(x, y)$ for each $x \in X$? In other words, is there a function $\phi : X \to Y$ such that $\phi(x) = f(x, \phi(x))$ for each $x \in X$? If so, is it unique? Is it continuous?

Our first application of the contraction mapping theorem gives sufficient conditions for these questions to have a positive answer. It can be thought of as a contraction mapping theorem with a continuous parameter.

Theorem 14.6.4 (The Lipschitz implicit function theorem) *Suppose that (X, d) is a metric space, that (Y, ρ) is a complete metric space and that $f : X \times Y \to Y$ is continuous. If there exists $0 < K < 1$ such that $\rho(f(x, y), f(x, y')) \leq K \rho(y, y')$ for all $x \in X$ and $y, y' \in Y$ then there exists a unique mapping $\phi : X \to Y$ such that $\phi(x) = f(x, \phi(x))$ for each $x \in X$. Further, ϕ is continuous.*

Proof The proof of existence and uniqueness follows easily from the contraction mapping theorem. If $x \in X$ then the mapping $f_x : Y \to Y$ defined by $f_x(y) = f(x, y)$ is a contraction mapping, which has a unique fixed point $\phi(x)$. Then $f(x, \phi(x)) = f_x(\phi(x)) = \phi(x)$.

It remains to show that ϕ is continuous. Suppose that $x \in X$ and that $\epsilon > 0$. There exists $\delta > 0$ such that if $d(x, z) < \delta$ then

$$\rho(\phi(x), f(z, \phi(x))) = \rho(f(x, \phi(x)), f(z, \phi(x))) < (1 - K)\epsilon.$$

If $d(x, z) < \delta$, then

$$\rho(\phi(x), \phi(z)) \leq \rho(\phi(x), f(z, \phi(x)) + \rho(f(z, \phi(x)), f(z, \phi(z)))$$
$$\leq (1 - K)\epsilon + K\rho(\phi(x), \phi(z)),$$

so that $\rho(\phi(x), \phi(z)) \leq \epsilon$.   □

Our next application, which uses Corollary 14.6.3, gives a proof of the existence and uniqueness of solutions of certain ordinary differential equations.

Theorem 14.6.5 *Suppose that $M \geq 0$ and that $L \geq 0$. Suppose that H is a continuous real-valued function on the triangle*

$$T = \{(x, y) \in \mathbf{R}^2 : 0 \leq x \leq b, |y| \leq Mx\},$$

that $|H(x, y)| \leq M$ and that $|H(x, y) - H(x, y')| \leq L|y - y'|$, for $(x, y) \in T$ and $(x, y') \in T$. Then there exists a unique continuously differentiable function f on $[0, b]$ such that $f(0) = 0$, $(x, f(x)) \in T$ for $x \in [0, b]$ and

$$\frac{df}{dx}(x) = H(x, f(x)) \text{ for all } x \in [0, b].$$

Proof   If f is any solution, then

$$|f(x)| = |f(x) - f(0)| = |\int_0^x H(t, f(t))dt| \leq \int_0^x |H(t, f(t))|dt \leq Mx,$$

for $0 \leq x \leq b$, so that the graph of f is contained in T. The second condition is a Lipschitz condition, which is needed to enable us to use the contraction mapping theorem.

First let us observe that the fundamental theorem of calculus shows that solving this differential equation is equivalent to solving an integral equation. If f is a solution, then, as above,

$$f(x) = f(x) - f(0) = \int_0^x f'(t) \, dt = \int_0^x H(t, f(t)) \, dt \text{ for all } x \in [0, b].$$

Conversely, if f is a continuous function which satisfies this integral equation, then $f(0) = 0$ and the function $J(f)(x) = \int_0^x H(t, f(t)) \, dt$ is differentiable, with continuous derivative $H(x, f(x))$. Thus $f'(x) = H(x, f(x))$ for $x \in [0, b]$.

Let $X = \{g \in C[0, b] : |g(x)| \leq Mx \text{ for } x \in [0, b]\}$. X is a closed subset of the Banach space $(C[0, b], \|.\|_\infty)$, and so is a complete metric space under the metric defined by the norm. We define a mapping $J : X \to C[0, b]$ by setting

$$J(g)(x) = \int_0^x H(t, g(t)) \, dt, \text{ for } x \in [0, b].$$

Then $J(g)$ is a continuous function on $[0, b]$ and

$$|J(g)(x)| \leq \int_0^x M \, dt = Mx,$$

so that $J(g) \in X$. We now show by induction that, for each $n \in \mathbf{Z}^+$,

$$|J^n(g)(x) - J^n(h)(x)| \leq \frac{L^n x^n}{n!} \|g - h\|_\infty, \text{ for } g, h \in X \text{ and } 0 \leq x \leq b.$$

The result is certainly true for $n = 0$. Suppose that it is true for n. Then

$$|J^{n+1}(g)(x) - J^{n+1}(h)(x)| \leq \int_0^x |H(t, J^n(g)(t)) - H(t, J^n(h)(t))| \, dt$$

$$\leq \int_0^x L|J^n(g)(t) - J^n(h)(t)| \, dt$$

$$\leq \int_0^x \frac{L^{n+1}}{n!} \|g - h\|_\infty \, t^n \, dt$$

$$= \frac{L^{n+1} x^{n+1}}{(n+1)!} \|g - h\|_\infty.$$

Thus

$$\|J^n(g) - J^n(h)\|_\infty \leq \frac{L^n b^n}{n!} \|g - h\|_\infty.$$

Now $L^n b^n / n! \to 0$ as $n \to \infty$, and so there exists $k \in \mathbf{N}$ such that $L^k b^k / k! < 1$. Thus J^k is a contraction mapping of X. We apply Corollary 14.6.3. J has a unique fixed point f, and f is the unique solution of the integral equation. $\qquad\square$

The next application of the contraction mapping theorem is an inverse function theorem.

Theorem 14.6.6 (The Lipschitz inverse function theorem) *Suppose that U is an open subset of a Banach space $(E, \|.\|)$ and that $g : U \to E$ is a Lipschitz mapping with constant $K < 1$. Let $f(x) = x + g(x)$. If the closed neighbourhood $M_\epsilon(x)$ of x is contained in U then*

$$M_{(1-K)\epsilon}(f(x)) \subseteq f(M_\epsilon(x)) \subseteq M_{(1+K)\epsilon}(f(x)).$$

The mapping f is a homeomorphism of U onto $f(U)$, f^{-1} is a Lipschitz mapping with constant $1/(1 - K)$, and $f(U)$ is an open subset of E.

Proof   Since $x - y = (f(x) - f(y)) - (g(x) - g(y))$

$$\|x - y\| \le \|f(x) - f(y)\| + \|g(x) - g(y)\| \le \|f(x) - f(y)\| + K \|x - y\|,$$

so that $\|f(x) - f(y)\| \ge (1 - K) \|x - y\|$. Thus f is one-one, and f^{-1} is a Lipschitz mapping with constant $1/(1 - K)$.

Suppose that $x \in U$ and that the closed neighbourhood $M_\epsilon(x)$ is contained in U. Then

$$\|f(x) - f(y)\| \le \|x - y\| + \|g(x) - g(y)\| \le (1 + K) \|x - y\|,$$

so that $f(M_\epsilon(x)) \subseteq M_{(1+K)\epsilon}(f(x))$.

Suppose that $y \in M_{(1-K)\epsilon}(f(x))$. Let $h(z) = y - g(z)$, for $z \in M_\epsilon(x)$. We shall show that h is a contraction mapping of $M_\epsilon(x)$. First, if $z \in M_\epsilon(x)$ then

$$\|h(z) - x\| = \|y - x - g(z)\| = \|y - f(x) + g(x) - g(z)\|$$
$$\le \|y - f(x)\| + \|g(x) - g(z)\| \le (1 - K)\epsilon + K\epsilon = \epsilon,$$

so that $h(M_\epsilon(x)) \subseteq M_\epsilon(x)$. Secondly,

$$\|h(z) - h(w)\| = \|g(z) - g(w)\| \le K \|z - w\|,$$

so that h is a contraction mapping of $M_\epsilon(x)$. Since $M_\epsilon(x)$ is closed, it is complete, and so h has a unique fixed point v. Then $v = y - g(v)$, so that $y = f(v) \in f(M_\epsilon(x))$. Thus $M_{(1-K)\epsilon}(f(x)) \subseteq f(M_\epsilon(x))$. Consequently, $f(U)$ is an open subset of E. $\square$

We shall use this theorem later to prove a differentiable inverse function theorem (Theorem 17.4.1).

We can apply this result to linear operators on $(E, \|.\|)$. In this case, however, it is more natural to proceed directly. Suppose that $S, T \in L(E)$.

Then the composed map $S \circ T \in L(E)$. Further,

$$\|S \circ T\| = \sup\{\|S(T(x))\| : \|x\| \le 1\}$$
$$\le \|S\| \sup\{\|T(x)\| : \|x\| \le 1\} = \|S\| \cdot \|T\| \,.$$

We set $T^0 = I$. Then $\|T^n\| \le \|T\|^n$ for all $n \in \mathbf{Z}^+$. We use this inequality to prove the following.

Theorem 14.6.7 *Suppose that $(E, \|.\|)$ is a Banach space, that $T \in L(E)$ and that $\|T^k\| < 1$ for some $k \in \mathbf{N}$. Then $\sum_{n=0}^{\infty} T^n$ converges absolutely in $L(E)$. If $S = \sum_{n=0}^{\infty} T^n$ then $(I - T)S = S(I - T) = I$, so that $I - T$ is invertible, with inverse S.*

Proof Since $\|T^{nk}\| \le \|T^k\|^n$ for $n \in \mathbf{N}$, $\sum_{n=0}^{\infty} \|T^{nk}\| < \infty$. Thus

$$\sum_{n=0}^{\infty} \|T^n\| = \sum_{j=0}^{k-1} \sum_{n=0}^{\infty} \left\|T^j T^{nk}\right\| \le \sum_{j=0}^{k-1} \sum_{n=0}^{\infty} \|T^j\| \cdot \left\|T^{nk}\right\|$$

$$= \left(\sum_{j=0}^{k-1} \|T^j\|\right) \left(\sum_{n=0}^{\infty} \left\|T^{nk}\right\|\right) < \infty,$$

and so $\sum_{n=0}^{\infty} T^n$ converges absolutely in A, to S, say. In particular, $T^n \to 0$ as $n \to \infty$. Let $S_n = \sum_{j=0}^{n} T^j$. Then

$$(I - T)S_n = S_n(I - T) = I - T^{n+1} \to I \text{ as } n \to \infty.$$

But $(I - T)S_n \to (I - T)S$ and $S_n(I - T) \to S(I - T)$ as $n \to \infty$, and so $(I - T)S = S(I - T) = I$. $\square$

Corollary 14.6.8 *If $\|T\| < I$ then $I - T$ is invertible, and*

$$\|(I - T)^{-1}\| < 1/(1 - \|T\|).$$

Proof We can take $k = 1$. Then

$$\|S\| \le \sum_{n=0}^{\infty} \|T^n\| \le \sum_{n=0}^{\infty} \|T\|^n = \frac{1}{1 - \|T\|}.$$

$\square$

The series $\sum_{n=0}^{\infty} T^n$ is called the *Neumann series*.

Corollary 14.6.9 *Let $GL(E)$ be the set of invertible elements of $L(E)$. Then $GL(E)$ is an open subset of $L(E)$, and the mapping $S \to S^{-1}$ is a homeomorphism of $GL(E)$ onto itself.*

The set $GL(E)$ is a group under composition. It is called the *general linear group.*

Proof   Suppose that $S \in GL(E)$. Let $\alpha = \left\| S^{-1} \right\|^{-1}$. Suppose that $\|U\| < \alpha/2$. Then $\left\| US^{-1} \right\| \leq \|U\|/\alpha < \frac{1}{2}$, so that $I + US^{-1}$ is invertible and $\left\| (I + US^{-1})^{-1} \right\| < 2$. Then $S + U = (I + US^{-1})S$ is invertible, with inverse $S^{-1}(I + US^{-1})^{-1}$, so that $GL(E)$ is an open subset of $L(E)$. Further, $\left\| (S+U)^{-1} \right\| \leq \left\| S^{-1} \right\| \cdot \left\| (I + US^{-1})^{-1} \right\| < 2/\alpha$. Now $(S+U)^{-1} - S^{-1} = -(S+U)^{-1}US^{-1}$, so that

$$\left\| (S+U)^{-1} - S^{-1} \right\| \leq 2 \|U\|/\alpha^2,$$

and $(S+U)^{-1} \to S^{-1}$ as $U \to 0$. Thus the mapping $S \to S^{-1}$ is continuous on $GL(E)$. Since $(S^{-1})^{-1} = S$, it follows that the mapping $S \to S^{-1}$ is a homeomorphism of $GL(E)$ onto itself.   □

Corollary 14.6.10 *Suppose that E and F are Euclidean spaces and that $1 \leq k \leq d = \dim F$. The set $L_k(E, F)$ of linear mappings in $L(E, F)$ of rank greater than or equal to k is an open subset of $L(E, F)$. In particular, the set $L_d(E, F)$ of surjective mappings in $L(E, F)$ is an open subset of $L(E, F)$.*

Proof   Suppose that $T \in L_k(E, F)$. Let N be the null-space of T, let $E_1 = N^\perp$, and let $j : E_1 \to E$ be the inclusion mapping. Let $F_1 = T(E) = T(E_1)$ and let $P : F \to F_1$ be the orthogonal projection of F onto F_1. Then $T_1 = P \circ T \circ j$ is a linear isomorphism of E_1 onto F_1. Let $\delta = 1/\left\| T_1^{-1} \right\|$. If $S_1 \in L(E_1, F_1)$ and $\|S_1 - T_1\| < \delta$ then $\left\| T_1^{-1} \circ S_1 - I \right\| < 1$, so that $T_1^{-1}S_1 \in GL(E_1)$. Hence S_1 is a linear isomorphism of E_1 onto F_1. If $S \in L(E, F)$ and $\|S - T\| < \delta$ then $\|(P \circ S \circ j) - T_1\| < \delta$, so that $P \circ S \circ j$ is a linear isomorphism of E_1 onto F_1, and therefore has rank k. Thus $\mathrm{rank}(S) \geq \mathrm{rank}(P \circ S \circ j) \geq k$.   □

Let us apply Theorem 14.6.7 to some linear integral equations. First suppose that K is a bounded uniformly continuous real-valued function on the square $[a, b] \times [a, b]$, that $g \in C([a, b])$ and that λ is a parameter in $\mathbf{R}$. We seek a solution $f \in C([a, b])$ to the *Fredholm integral equation*

$$f(x) = g(x) + \lambda \int_a^b K(x, y)f(y)\, dy \text{ for } x \in [a, b].$$

K is the *kernel* of the equation. K defines an element of $L(C([a,b]))$: if $f \in C([a,b])$ let

$$T_K(f)(x) = \int_a^b K(x,y)f(y)\,dy \text{ for } x \in [a,b].$$

First we show that $T_K(f) \in C([a,b])$. Suppose that $\epsilon > 0$. Let $\eta = \epsilon/(b-a)(\|f\|_\infty + 1)$. There exists $\delta > 0$ such that

$$|K(x,y) - K(x',y')| < \eta \text{ if } |x - x'| < \delta \text{ and } |y - y'| < \delta.$$

If $|x - x'| < \delta$ then

$$|T_K(f)(x) - T_K(f)(x')| \le \int_a^b |K(x,y) - K(x',y)|\,|f(y)|\,dy$$
$$\le \eta\,\|f\|_\infty\,(b-a) < \epsilon.$$

Thus $T_K(f)$ is continuous on $[a,b]$.

Further

$$|T_K(f)(x)| \le \int_a^b |K(x,y)f(y)|\,dy \le (b-a)\,\|K\|_\infty\,\|f\|_\infty,$$

where $\|K\|_\infty = \sup\{|K(x,y)| : (x,y) \in [a,b] \times [a,b]\}$. Consequently $T_K \in L(C([a,b]))$, and $\|T_K\| \le (b-a)\,\|K\|_\infty$. Thus if $|\lambda|(b-a)\,\|K\|_\infty < 1$ then $\|\lambda T_K\| < 1$, and so the continuous linear operator $I - \lambda T_K$ is invertible. Thus for each $g \in C([a,b])$ there exists a unique $f \in C([a,b])$ such that $(I - \lambda T_K)f = g$; the Fredholm integral equation has a unique solution if $|\lambda|\,\|T_K\| < 1$.

Next, we consider the *Volterra integral equation*

$$f(x) = g(x) + \lambda \int_a^x K(x,y)f(y)\,dy \text{ for } x \in [a,b],$$

where K is a bounded uniformly continuous function on the triangle $T = \{(x,y) : a \le y \le x \le b\}$. If $f \in C([a,b])$ let

$$V_K(f)(x) = \int_a^x K(x,y)f(y)\,dy \text{ for } x \in [a,b].$$

Again, $V_K(f) \in C([a,b])$. We claim that

$$|V_K^n(f)(x)| \le \frac{(x-a)^n}{n!}\,\|K\|_\infty^n\,\|f\|_\infty \text{ for } x \in [a,b] \text{ and } n \in \mathbf{Z}^+,$$

where $\|K\|_\infty = \sup\{|K(x,y)| : (x,y) \in T\}$. We prove this by induction. The result is certainly true when $n = 0$. Suppose that it holds for n. Then

$$|V_K^{n+1}(f)(x)| = \left| \int_a^x K(x,y) V_K^n(f)(y)\, dy \right| \le \int_a^x |K(x,y) V_K^n(f)(y)|\, dy$$

$$\le \|K\|_\infty \int_a^x |V_K^n(f)(y)|\, dy \le \frac{\|K\|_\infty^{n+1}}{n!} \|f\|_\infty \int_a^x (y-a)^n\, dy$$

$$= \frac{(x-a)^{n+1}}{(n+1)!} \|K\|_\infty^{n+1} \|f\|_\infty .$$

In particular, $\|V_K^n\| \le (b-a)^n \|K\|_\infty^n /n!$, and so $|\lambda|^n \|V_K^n\| \to 0$ as $n \to \infty$. Thus $|\lambda|^n \|V_K^n\| < 1$ for large enough n, and so $1 - \lambda V_K$ is invertible for all $\lambda \in \mathbf{R}$; the Fredholm integral equation has a unique solution, for all $\lambda \in \mathbf{R}$.

A concluding remark: as we shall see in the next chapter, it is enough to assume that the kernels in the Fredholm and Volterra equations are continuous, since this implies that they are bounded and uniformly continuous.

Exercises

14.6.1 Give an example of a surjective contraction mapping f on an incomplete metric space (X, d) with no fixed point.

14.6.2 Suppose that f, g are contractions of a complete metric space (X, d). Show that there exists unique points x_0 and y_0 in X such that $x_0 = g(y_0)$ and $y_0 = f(x_0)$.

14.6.3 Suppose that $h \in C([a,b])$. If $f \in C([a,b])$, let $l_h(f) = \int_a^b f(x)h(x)\, dx$. Show that if $C([a,b])$ is given the uniform norm then l_h is a continuous linear functional, and $\|l_h\|' = \int_a^b |h(x)|\, dx$. [Consider approximating sums to the integral.]

14.6.4 Suppose that K is the kernel of a Fredholm operator on $C([a,b])$. Show that $\|T_K\| = \sup\{\int_a^b |K(x,y)|\, dy : x \in [a,b]\}$.

14.6.5 Verify that if $f \in C([a,b])$ then $V_K(f)$ is continuous.

14.7 *Baire's category theorem*

(This section can be omitted on a first reading.)

We now prove Baire's category theorem, which is a straightforward extension of Osgood's theorem to complete metric spaces.

Theorem 14.7.1 (Baire's category theorem)   *If (U_n) is a sequence of dense open subsets of a complete metric space (X, d) then $\cap_{n=1}^\infty U_n$ is dense in X.*

Proof   Suppose that V is a non-empty open subset of X. We must show that $V \cap (\cap_{n=1}^{\infty} U_n)$ is not empty. Since U_1 is dense in X, there exists $c_1 \in V \cap U_1$. Since $V \cap U_1$ is open, there exists $0 < \epsilon_1 \leq 1/2$ such that

$$N_{\epsilon_1}(c_1) \subseteq M_{\epsilon_1}(c_1) \subseteq V \cap U_1.$$

We now iterate the argument; for each $n \in \mathbf{N}$ there exist

$$c_n \in N_{\epsilon_{n-1}}(c_{n-1}) \cap U_n \text{ and } 0 < \epsilon_n < 1/2^n$$

such that

$$N_{\epsilon_n}(c_n) \subseteq M_{\epsilon_n}(c_n) \subseteq N_{\epsilon_{n-1}}(c_{n-1}) \cap U_n.$$

The sequence $(N_{\epsilon_n}(c_n))_{n=1}^{\infty}$ is decreasing, so that if $m, p \geq n$ then $c_m \in M_{\epsilon_n}(c_n)$ and $c_p \in M_{\epsilon_n}(c_n)$, so that

$$d(c_m, c_p) \leq d(c_m, c_n) + d(c_n, c_p) < 2/2^n;$$

thus $(c_n)_{n=1}^{\infty}$ is a Cauchy sequence in (X, d). Since (X, d) is complete, it converges to an element c of X. Suppose that $n \in \mathbf{N}$. Since $c_m \in M_{\epsilon_n}(c_n)$ for $m \geq n$ and since $M_{\epsilon_n}(c_n)$ is closed, $c \in M_{\epsilon_n}(c_n) \subseteq U_n$. Thus $c \in \cap_{n=1}^{\infty} U_n$. Further, $c \in M_{\epsilon_1}(c_1) \subseteq V$, and so $c \in V$. $\square$

Note that the proof uses the axiom of dependent choice. This cannot be avoided: if Baire's category theorem is true for all complete metric spaces then the axiom of dependent choice must hold (Exercise 14.7.4). On the other hand, the theorem can be proved for separable complete metric spaces without using the axiom of dependent choice (Exercise 14.7.5).

The following corollary is particularly useful.

Corollary 14.7.2   *Suppose that $(C_n)_{n=1}^{\infty}$ is a sequence of closed subsets of a complete metric space (X, d) whose union is X. Then there exists n such that C_n has a non-empty interior.*

Proof   Let $U_n = X \setminus C_n$. Then $(U_n)_{n=1}^{\infty}$ is a sequence of open sets and $\cap_{n=1}^{\infty} U_n$ is empty, and so is certainly not dense in X. Thus there exists U_n which is not dense in X; that is C_n has a non-empty interior. $\square$

It is sometimes useful to have a local version of this corollary. This depends upon the important observation that the hypotheses and conclusions of the theorem are topological ones, so that Baire's category theorem applies to topologically complete metric spaces; in particular, by Theorem 14.1.13, it applies to open subsets of complete metric spaces, and to G_δ subsets of complete metric spaces.

Corollary 14.7.3 *Suppose that $(C_n)_{n=1}^{\infty}$ is a sequence of closed subsets of a complete metric space (X, d) whose union contains a non-empty open set W. Then there exists n such that $C_n \cap W$ has a non-empty interior.*

Proof The sets $C_n \cap W$ are closed subsets of W whose union is W, and so there exists n and a non-empty open subset V of W such that $V \subseteq C_n \cap W$. Since W is open in X, it follows that V is open in X. $\qquad\square$

Baire proved his theorem (for $\mathbf{R}^n$) independently of Osgood. It was included in his doctoral thesis, published in 1899. Why is the word 'category' used? This is a matter of terminology. A subset A of a topological space is said to be *nowhere dense* if its closure has an empty interior. It is said to be of the *first category* in X if it is the union of a sequence of nowhere dense sets, and is said to be of the *second category* in X if it is not of the first category in X. Thus Corollary 14.7.2 states that a complete metric space is of the second category in itself.

Let us now turn to some applications of the theorem.

Proposition 14.7.4 *Suppose that F is a set of continuous mappings from a complete metric space (X, d) into a metric space (Y, ρ), with the property that $F(x) = \{f(x) : f \in F\}$ is bounded, for each $x \in X$. Then there exists a non-empty open subset U of X and a positive number K such that $\mathrm{diam}\,(F(x)) \leq K$ for each $x \in U$.*

Proof If $f, g \in F$ then the function $x \to \rho(f(x), g(x))$ is continuous on X and so the set $\{x \in X : \rho(f(x), g(x)) \leq n\}$ is closed. Consequently the set

$$C_n = \{x \in X : \mathrm{diam}\,(F(x)) \leq n\}$$
$$= \cap_{f,g \in F}\{x \in X : \rho(f(x), g(x)) \leq n\}$$

is closed. By hypothesis, $X = \cup_{n=1}^{\infty} C_n$, and so there exists n such that C_n has a non-empty interior. $\qquad\square$

The corresponding result for continuous linear mappings is more useful.

Theorem 14.7.5 (The principle of uniform boundedness) *Suppose that A is a set of continuous linear mappings from a Banach space $(E, \|.\|_E)$ into a normed space $(F, \|.\|_F)$ with the property that $A(x) = \{T(x) : T \in A\}$ is bounded, for each $x \in E$. Then $\sup\{\|T\| : T \in A\}$ is finite.*

Proof Let $C_n = \{x \in E : \sup\{\|T(x)\| : T \in A\} \leq n\}$. Then $(C_n)_{n=1}^{\infty}$ is a sequence of closed sets whose union is E, and so there exists n such that

C_n has a non-empty interior. Thus there exists $x_0 \in E$ and $\epsilon > 0$ such that $M_\epsilon(x_0) \subseteq C_n$. Let $K = 1/\epsilon$. If $T \in A$ and $\|x\|_E \leq 1$ then

$$\|T(x)\|_F = K \|T(\epsilon x)\|_F = K \|T(x_0 + \epsilon x) - T(x_0)\|_F$$
$$\leq K(\|T(x_0 + \epsilon x)\|_F + \|T(x_0)\|_F) \leq 2Kn,$$

so that $\|T\| \leq 2Kn$.   □

The contrapositive is equally useful.

Theorem 14.7.6 (The principle of condensation of singularities) *Suppose that D is an unbounded set of continuous linear mappings from a Banach space $(E, \|.\|_E)$ into a normed space $(F, \|.\|_F)$. If $x \in E$, let $D(x) = \{T(x) : T \in D\}$. Then $H = \{x \in E : D(x)$ is unbounded$\}$ is of the second category in E.*

Proof   For each $n \in \mathbf{N}$, the set $G_n = \{x \in E : \sup_{T \in D} \|T(x)\| \leq n\}$ is a closed nowhere dense subset of E, so that $\cup_{n \in \mathbf{N}} G_n$ is of the first category in E. By Baire's category theorem, H cannot be of the first category in E.   □

The principle of uniform boundedness has the following consequence.

Theorem 14.7.7 (The Banach–Steinhaus theorem) *Suppose that $(T_n)_{n=1}^\infty$ is a sequence of continuous linear mappings from a Banach space $(E, \|.\|_E)$ into a normed space $(F, \|.\|_F)$, and that $T_n(x)$ converges, to $T(x)$, say, as $n \to \infty$, for each $x \in E$. Then T is a continuous linear mapping from E into F.*

Proof   The mapping T is certainly linear. For each $x \in X$, the set $\{T_n(x) : n \in \mathbf{N}\}$ is bounded. By the principle of uniform boundedness, there exists K such that $\|T_n\| \leq K$ for all $n \in \mathbf{N}$. If $x \in E$ then $\|T(x)\|_F = \lim_{n \to \infty} \|T_n(x)\|_F \leq K \|x\|_E$, so that T is continuous.   □

(Terminology varies; many authors call the principle of uniform boundedness the Banach–Steinhaus theorem.)

We now combine the Baire category theorem with Corollary 14.4.2 to prove some of the most powerful results of Banach space theory.

Theorem 14.7.8 (The open mapping theorem) *Suppose that T is a surjective continuous linear mapping of a Banach space $(E, \|.\|_E)$ onto a Banach space $(F, \|.\|_F)$. If U is open in E then $T(U)$ is open in F.*

Proof   Let B_E be the unit ball in E, B_F the unit ball in F. Let $A_n = \overline{T(nB_E)}$, for $n \in \mathbf{N}$. Then $A_n = nA_1$, A_1 is convex (Corollary 11.2.2) and $A_1 = -A_1$. Now $F = T(\cup_{n=1}^\infty nB_E) = \cup_{n=1}^\infty T(nB_E) \subseteq \cup_{n=1}^\infty A_n$ so that

$F = \cup_{n=1}^{\infty} A_n$. By Baire's category theorem there exists n so that A_n has a non-empty interior. Since the mapping $y \to y/n$ is a homeomorphism of F, A_1 has a non-empty interior. Thus there exist $y_0 \in A_1$ and $\epsilon > 0$ such that $M_\epsilon(y_0) \subseteq A_1$. If $\|y\|_F \le \epsilon$ then $y_0 + y \in A_1$ and $y_0 - y \in A_1$. Since $A_1 = -A_1$, $-y_0 + y \in A_1$, and since A_1 is convex $y = \frac{1}{2}((y_0 + y) + (-y_0 + y)) \in A_1$, so that $A_1 = \overline{T(B_E)} \supset \epsilon B_F$. The result now follows from Corollary 14.4.2.   □

Corollary 14.7.9 (The isomorphism theorem)   *If T is a bijective continuous linear mapping of a Banach space $(E, \|.\|_E)$ onto a Banach space $(F, \|.\|_F)$, then T^{-1} is continuous, so that T is a homeomorphism.*

Recall that a continuous mapping from a topological space to a T_1 topological space has a closed graph.

Corollary 14.7.10 (The closed graph theorem)   *If T is a linear mapping of a Banach space $(E, \|.\|_E)$ into a Banach space $(F, \|.\|_F)$ which has a closed graph, then T is continuous.*

Proof   The graph G_T of T is a closed linear subspace of the Banach space $(E, \|.\|_E) \times (F, \|.\|_F)$, and so is a Banach space, under the norm $\|(x, T(x))\| = \|x\|_E + \|T(x)\|_F$. If $(x, T(x)) \in G_T$ let $R((x, T(x))) = T(x)$ and let $L((x, T(x))) = x$. R is a norm-decreasing linear mapping of G_T into F, and is therefore continuous. L is a bijective norm-decreasing linear mapping of the Banach space G_T onto the Banach space E; it is continuous, and so L^{-1} is continuous, by the isomorphism theorem. Thus $T = R \circ L^{-1}$ is continuous.   □

This theorem says the following. Suppose that T is a linear mapping of a Banach space $(E, \|.\|_E)$ into a Banach space $(F, \|.\|_F)$ with the property that whenever $(x_n)_{n=1}^{\infty}$ is a sequence in E for which $x_n \to x$ and $T(x_n) \to y$ as $n \to \infty$, then $T(x_n) \to T(x)$ as $n \to \infty$. Then if $(x_n)_{n=1}^{\infty}$ is a sequence in E for which $x_n \to x$ as $n \to \infty$, then $T(x_n) \to T(x)$ as $n \to \infty$. The gain may appear to be slight, but this is a powerful theorem.

The general principle of convergence ensures that the uniform limit of continuous functions is continuous, but the same is not true for functions which are the pointwise limit of continuous functions. On the other hand, as we shall see, not every function on a complete metric space is the pointwise limit of continuous functions. Baire used his category theorem to establish properties of such limits.

We shall restrict attention to real-valued functions defined on a complete metric space (X, d); the results extend easily to functions taking values in a separable metric space. Suppose that f is a function on X. Recall that if A is a subset of X the the oscillation $\Omega(f, A)$ of f on A is defined as $\Omega(f, A) = \sup\{|f(x) - f(y)| : x, y \in A\}$. If $x \in X$ and $\delta > 0$, we set $\Omega_\delta(f)(x) = \Omega(f, N_\delta(x))$. Then $\Omega_\delta(f)(x)$ is an increasing function of δ taking values in $[0, \infty]$. We set $\Omega(f)(x) = \inf\{\Omega_\delta(f)(x) : \delta > 0\}$. Then it is easy to see that f is continuous at x if and only if $\Omega(f)(x) = 0$.

Proposition 14.7.11 *If f is a real-valued function on a metric space (X, d) and $\epsilon > 0$ then the set $U_\epsilon = \{x \in X : \Omega(f)(x) < \epsilon\}$ is open in (X, d).*

Proof   Suppose that $x \in U_\epsilon$. There exists $\delta > 0$ such that $\Omega_\delta(f)(x) < \epsilon$. If $y \in N_\delta(x)$, there exists $\eta > 0$ such that $N_\eta(y) \subseteq N_\delta(x)$. Then

$$\Omega(f)(y) \leq \Omega_\eta(f)(y) \leq \Omega_\delta(f)(x) < \epsilon,$$

so that $y \in U_\epsilon$. Thus $N_\delta(x) \subseteq U_\epsilon$, and U_ϵ is open.   □

Corollary 14.7.12 *Suppose that f is a real-valued function on a complete metric space (X, d) for which the set $U_\epsilon = \{x \in X : \Omega(f)(x) < \epsilon\}$ is dense in X, for each $\epsilon > 0$. Then the set C of points of continuity of f is dense in (X, d).*

Proof   For $C = \cap_{n=1}^\infty U_{1/n}$, and so the result follows from Baire's category theorem.   □

Theorem 14.7.13 *Suppose that f is the pointwise limit of a sequence $(f_n)_{n=1}^\infty$ of continuous functions on a complete metric space (X, d). Then the set C of points of continuity of f is dense in (X, d).*

Proof   We show that the conditions of Corollary 14.7.12 are satisfied. Suppose that $\epsilon > 0$. If $j \in \mathbf{Z}$ let $a_j = j\epsilon/4$ and let $b_j = a_j + \epsilon/2$, so that $\mathbf{R} = \cup_{j \in \mathbf{Z}}(a_j, b_j)$. Suppose that V is a non-empty open subset of X. Recall (Theorem 14.1.13) that V is topologically complete; there is a complete metric on V which defines the subspace topology of V. Let

$$A_{n,j} = \{x \in V : f_m(x) \in [a_j, b_j] \text{ for } m \geq n\}$$
$$= \cap_{m \geq n}\{x \in V : f_m(x) \in [a_j, b_j]\}.$$

Then $A_{n,j}$ is a closed subset of V, and $V = \cup\{A_{n,j} : n \in \mathbf{N}, j \in \mathbf{Z}\}$. Note that if $x \in A_{n,j}$ then $f(x) \in [a_j, b_j]$. By Baire's category theorem, there exist n, j such that $A_{n,j}$ has a non-empty interior in V. Since V is open in X, $A_{n,j}$ has a non-empty interior in X. Thus there exist $x \in V$ and $\eta > 0$ such that $N_\eta(x) \subseteq A_{n,j}$. Then $\Omega(f)(x) \leq \Omega_\eta(f)(x) \leq \epsilon/2 < \epsilon$, so that $x \in V \cap U_\epsilon$, and U_ϵ is dense in (X, d).   □

In fact, we can say more.

Proposition 14.7.14 *Suppose that f is a real-valued function on a complete metric space (X, d) and that the set C of points of continuity of f is dense in (X, d). Then the set D of points of discontinuity of f is of the first category in X.*

Proof   For $D = \cup_{n=1}^{\infty} B_n$, where $B_n = \{x \in X : \Omega(f)(x) \geq 1/n\}$, and each B_n is closed and nowhere dense. $\square$

Corollary 14.7.15   *If (X, d) has no isolated points, then C is uncountable.*

Proof   If not, then C is the union of countably many singleton sets, each of which is nowhere dense. Thus $X = \cup_{n=1}^{\infty} B_n \cup C$ is the countable union of closed nowhere dense sets, giving a contradiction. $\square$

We end this section with a remarkable result of the Catalan mathematician Ferran Sunyer y Balaguer.

Theorem 14.7.16   *Suppose that f is an infinitely differentiable function on $(0, 1)$ with the property that for each $x \in (0, 1)$ there exists $n \in \mathbf{Z}^+$ such that $f^{(n)}(x) = 0$. Then f is a polynomial function.*

Proof   Let $A_n = \{x \in (0, 1) : f^{(n)}(x) = 0\}$, and let E_n be the interior of A_n. Let $E = \cup_{n=0}^{\infty} E_n$, and let $F = (0, 1) \setminus E$. Since each A_n is closed and since if $[a, b] \subseteq (0, 1)$ then $[a, b] = \cup_{n=0}^{\infty}(A_n \cap [a, b])$, it follows from Baire's category theorem that E is dense in $(0, 1)$. In particular, there exists $n \in \mathbf{N}$ such that E_n is not empty. Note that if $m > n$ then $f^{(m)}(x) = 0$ for $x \in E_n$, so that $E_n \subseteq E_m$. Note also that, by continuity, $f^{(m)}(x) = 0$ for $x \in \overline{E}_n$. We shall show that there exists n such that $E_n = (0, 1)$. Suppose not. If $E_m \neq \emptyset$ then E_m is the union of countably many disjoint open intervals, the *constituent intervals* of E_m.

Now E is open, and is the union of countably many disjoint non-empty open intervals, the *constituent intervals* of E. Suppose that I is one of them, and that $x \in I$. Then there exists a least $m \in \mathbf{Z}^+$ for which $x \in E_m$, and x is in a constituent interval I_m of E_m. Then $I_m \subseteq I$. We show that $I_m = I$. If not, one of the endpoints of I_m is in I. Without loss of generality, we can suppose that it is a right-hand endpoint b. Since the sequence $(E_n)_{n=1}^{\infty}$ is increasing, and $b \notin E_m$, there exists a least integer $p > m$ such that $b \in E_p$.

Since $I_m \subseteq E_m \subseteq E_{p-1}$, it follows that b is a right-hand endpoint of a constituent interval of E_{p-1}. Consequently, $f^{p-1}(b) = 0$. Since E_p is open, there exists $c > b$ such that $(b, c) \subseteq E_p$. If $x \in (b, c)$ then

$$f^{(p-1)}(x) = f^{(p-1)}(b) + \int_b^x f^{(p)}(t)\, dt = 0,$$

so that $(b, c) \subseteq E_{p-1}$. Consequently $b \in E_{p-1}$, contradicting the minimality of p. Thus $I_m = I$. Consequently, the constituent intervals of E_{m+1} are either constituent intervals of E_m, or are intervals disjoint from E_m.

The set $F = (0, 1) \setminus E$ is a closed nowhere-dense subset of $(0, 1)$. It is also a perfect subset of $(0, 1)$. For if b were an isolated point of F there would be two disjoint open intervals in E with b as end-point. Thus there would exist $(a, b) \subseteq E_m$ and $(b, c) \subseteq E_n$, for some $0 \le a < b < c \le 1$ and $m, n \in \mathbf{N}$. But then $b \in (a, c) \subseteq E_{\max(m,n)} \subseteq E$, giving a contradiction.

We now apply Baire's category theorem again, this time to the sequence $(A_n \cap F)_{n=1}^\infty$ of closed subsets of F. It follows from Baire's category theorem that there exist $n \in \mathbf{N}$, $x \in F$ and $\eta > 0$ such that $N_\eta(x) \cap F \subseteq A_n$. If $y \in N_\eta(x) \cap F$, y is not an isolated point of F, and so there exists a sequence $(y_j)_{j=1}^\infty$ in $(N_\eta(x) \cap F) \setminus \{y\}$ which converges to y. Consequently

$$f^{(n+1)}(y) = \lim_{j \to \infty} \frac{f^{(n)}(y_j) - f^{(n)}(y)}{y_j - y} = 0.$$

Thus $N_\eta(x) \cap F \subseteq A_{n+1}$. Iterating the argument, $N_\eta(x) \cap F \subseteq \cup_{p \ge n} A_p$.

Suppose now that $z \in N_\eta(x) \cap E$. Then z is in one of the constituent intervals I of E and one of its end-points, b say, is in $N_\eta \cap F$. We can suppose, without loss of generality, that $b < z$. Further, there is a least integer p such that I is a constituent interval of E_p. Suppose, if possible, that $p > n$. Then, arguing as above, if $w \in I$ then

$$f^{(p-1)}(w) = f^{(p-1)}(b) + \int_b^w f^{(p)}(t)\, dt = 0,$$

so that I is a constituent interval of E_{p-1}, contradicting the minimality of p. Thus $p \le n$, so that $z \in E_n$. Hence $N_\eta(x) \subseteq E_n$, contradicting the fact that $x \in F$. $\square$

Corollary 14.7.17 *Suppose that f is an infinitely differentiable function on $\mathbf{R}$ with the property that for each $x \in \mathbf{R}$ there exists $n \in \mathbf{Z}^+$ such that $f^{(n)}(x) = 0$. Then f is a polynomial function.*

Proof For f is a polynomial function on each bounded open interval of **R**, and two polynomial functions which are equal on an interval must be defined by the same polynomial.   □

Exercises

14.7.1 Use Baire's category theorem to show that a perfect subset of a complete metric space is uncountable.

14.7.2 Show that the real line is not the union of a set of proper non-trivial disjoint open intervals.

14.7.3 Let G_n be the set of functions f in $C([0,1])$ for which there exists $0 \leq x \leq 1$ for which $|f(x) - f(y)| \leq n|x - y|$ for all $y \in [0,1]$. Show that G_n is a closed subset of $C([0,1])$. Show that G_n is nowhere dense. Deduce that the set of continuous functions on $[0,1]$ which are nowhere differentiable is of the second category in $C([0,1])$.

14.7.4 This exercise shows that if Baire's category theorem is true, then the axiom of dependent choice must hold. Suppose that X is a non-empty set, and that ϕ is a mapping from X into the set of non-empty subsets of X. Let $X_n = X$ for $n \in \mathbf{Z}^+$, and let $P = \prod_{n=0}^{\infty} X_n$. Give each X_n the discrete metric, and give P a uniform product metric d.

(a) Show that (P, d) is a complete metric space.

(b) If $n \in \mathbf{Z}^+$, let $V_n = \{f \in P : \text{there exists } k > n \text{ with } f_k \in \phi(f_n)\}$. Show that V_n is open and dense in (P, d).

(c) If Baire's category theorem is true, there exists $f \in \cap_{n=0}^{\infty} V_n$. If $n \in \mathbf{Z}^+$, let $j(n) = \inf\{k : k > n, f_k \in \phi_n\}$. Use recursion to show there exists an increasing sequence $(c_n)_{n=0}^{\infty}$ such that $c_0 = 0$ and $f(c_{n+1}) \in \phi_{c_n}$ for $n \in \mathbf{N}^+$.

(d) Show that the axiom of dependent choice holds.

14.7.5 Suppose that (X, d) is a separable complete metric space. Prove Baire's category theorem for (X, d) without using the axiom of dependent choice. Let $S = \{s_1, s_2, \ldots\}$ be a countable dense subset of (X, d), and let $r_1, r_2, \ldots$ be an enumeration of the positive rational numbers. Show that at each stage there is a least $j(n)$ such that if $c_n = s_{j(n)}$ then $c_n \in N_{\epsilon_{n-1}}(c_{n-1}) \cap U_n \cap S$, and a least $k(n)$ such that if $\epsilon_n = r_{k(n)}$ then

$$N_{\epsilon_n}(c_n) \subseteq M_{\epsilon_n}(c_n) \subseteq N_{\epsilon_{n-1}}(c_{n-1}) \cap U_n.$$

In particular, Osgood's theorem does not need the axiom of dependent choice.

14.7.6 Suppose that f is a continuous function on $\mathbf{T}$. The n-th Fourier coefficient $\hat{f}_n$ of f is defined as

$$\hat{f}_n = \frac{1}{2\pi} \int_{-\pi}^{\pi} e^{-int} f(e^{it}) \, dt.$$

Let

$$S_n(f)(t) = \sum_{j=-n}^{n} \hat{f}_j e^{ijt}.$$

In Volume I, Section 9.5, we constructed an example of a continuous function on $\mathbf{T}$ whose Fourier series is unbounded at 0. In this exercise, we show that the set of functions for which this is true is of the second category in $C(\mathbf{T})$.

(a) Show that

$$S_n(f)(0) = \frac{1}{2\pi} \int_{-\pi}^{\pi} D_n(t) f(e^{it}) \, dt,$$

where

$$D_n(0) = 2n+1 \quad \text{and} \quad D_n(t) = \sum_{j=-n}^{n} e^{ijt} = \frac{\sin(n+\frac{1}{2})t}{\sin t/2} \quad \text{otherwise.}$$

(b) If $0 \le t \le \pi$, let $f_n(e^{it}) = \sin(n+\frac{1}{2})t$, and if $-\pi \le t \le 0$, let $f_n(e^{it}) = -\sin(n+\frac{1}{2})t$. Let $t_j = j\pi/(2n+1)$. Show that

$$S_n(f_n)(0) = \frac{1}{\pi} \sum_{j=1}^{2n+1} \int_{t_{j-1}}^{t_j} \frac{\sin^2(n+\frac{1}{2})t}{\sin t/2} \, dt$$

$$\ge \frac{1}{\pi} \sum_{j=1}^{2n+1} \frac{2}{t_j} \int_{t_{j-1}}^{t_j} \sin^2(n+\frac{1}{2})t \, dt = \frac{1}{\pi} \sum_{j=1}^{2n+1} \frac{1}{j}.$$

(c) Let $\phi_n(f) = S_n(f)(0)$. Deduce that $(\phi_n)_{n=0}^{\infty}$ is a sequence of continuous linear functionals on $C(\mathbf{T})$ which is unbounded in norm.

(d) Use the principle of condensation of singularities to show that the set of functions f in $C(\mathbf{T})$ for which the sequence $(S_n(f)(0))_{n=1}^{\infty}$ is unbounded is of the second category in $C(\mathbf{T})$.

14.7.7 Suppose that T is a linear mapping of a normed space $(E, \|.\|_E)$ into a normed space $(F, \|.\|_F)$. Show that T has a closed graph if and only whenever $x_n \to 0$ in E and $T(x_n) \to y$ in F then $y = 0$.

14.7.8 Suppose that T is a linear mapping from a Hilbert space H into itself for which $\langle T(x), y \rangle = \langle x, T(y) \rangle$ for all $x, y \in H$. Show that T is continuous.

14.7.9 Let ω be the vector space of all real sequences, and let ϕ be the linear subspace of all sequences with finitely many non-zero terms. A *Banach sequence space* $(E, \|.\|_E)$ is a Banach space $(E, \|.\|_E)$, where E is a linear subspace of ω which contains ϕ with the property that if $(x^{(n)})_{n=1}^\infty$ is a sequence in E for which $\|x^{(n)}\| \to 0$ as $n \to \infty$ then $x_j^{(n)} \to 0$ as $n \to \infty$ for each $j \in \mathbf{N}$. Show that if $(E, \|.\|_E)$ and $(F, \|.\|_F)$ are Banach sequence spaces and $E \subseteq F$ then the inclusion mapping $E \to F$ is continuous.

14.7.10 Suppose that $\|.\|$ is a complete norm on the space $C_b(X)$ of bounded continuous real-valued functions on a topological space (X, τ) with the property that if $\|f_n\| \to 0$ as $n \to \infty$ then $f_n(x) \to 0$ as $n \to \infty$ for each $x \in X$. Show that the norm $\|.\|$ is equivalent to the uniform norm $\|.\|_\infty$.

14.7.11 Give an example of a norm $\|.\|$ on $C([0, 1])$ with the property that if $\|f_n\| \to 0$ as $n \to \infty$ then $f_n(x) \to 0$ as $n \to \infty$ for each $x \in [0, 1]$ which is not equivalent to the uniform norm $\|.\|_\infty$.

15

Compactness

15.1 Compact topological spaces

Two of the most powerful results that we met when considering functions of a real variable were the Bolzano–Weierstrass theorem and the Heine–Borel theorem. Both of these involve topological properties, and we now consider these properties for topological spaces. We shall see that they give rise to three distinct concepts; in Section 15.4, we shall see that these three are the same for metric spaces.

We begin with *compactness*; this is the most important of the three properties. It is related to the Heine–Borel theorem, and the definition is essentially the same as for subsets of the real line. If A is a subset of a set X and $\mathcal{B}$ is a set of subsets of X then we say that $\mathcal{B}$ *covers* A, or that $\mathcal{B}$ is a *cover* of A, if $A \subseteq \cup_{B \in \mathcal{B}} B$. A subset $\mathcal{C}$ of $\mathcal{B}$ is a *subcover* if it covers A. A cover $\mathcal{B}$ is *finite* if the set $\mathcal{B}$ has finitely many members. If (X, τ) is a topological space, then a cover $\mathcal{B}$ is *open* if each $B \in \mathcal{B}$ is an open set. A topological space (X, τ) is *compact* if every open cover of X has a finite subcover. A subset A of a topological space (X, τ) is *compact* if it is compact, with the subspace topology. If U is a subset of A which is open in the subspace topology, there exists an open subset V of X such that $U = V \cap A$, and so A is a compact subset of X if and only if every cover of A by open subsets of X has a finite subcover.

The Heine–Borel theorem states that a subset of $\mathbf{R}$ is compact if and only if it is closed and bounded.

We can formulate the definition of compactness in terms of closed sets: this version is quite as useful as the 'open sets' version. Recall that a set $\mathcal{F}$ of subsets of a set X has the *finite intersection property* if whenever $\{F_1, \ldots, F_n\}$ is a finite subset of $\mathcal{F}$ then $\cap_{j=1}^n F_j$ is non-empty.

Theorem 15.1.1 *A topological space (X, τ) is compact if and only if whenever $\mathcal{F}$ is a set of closed subsets of X with the finite intersection property then the total intersection $\cap_{F \in \mathcal{F}} F$ is non-empty.*

Proof This is just a matter of taking complements. Suppose that $\cap_{F \in \mathcal{F}} F = \emptyset$. Then $\{C(F) : F \in \mathcal{F}\}$ is an open cover of X, and so there is a finite subcover $\{C(F_1), \cdots, C(F_n)\}$. Thus

$$X = C(F_1) \cup \ldots \cup C(F_n) = C(F_1 \cap \ldots \cap F_n),$$

so that $F_1 \cap \ldots \cap F_n = \emptyset$, contradicting the finite intersection property.

The converse is as easy, and is left to the reader as an exercise.   □

We have the following 'local' corollary.

Corollary 15.1.2 *Suppose that $\mathcal{C}$ is a set of closed subsets of a compact topological space (X, τ) and that $\cap_{C \in \mathcal{C}} C$ is contained in an open set U. Then there exists a finite subset $\mathcal{F}$ of $\mathcal{C}$ such that $\cap_{C \in \mathcal{F}} C \subseteq U$.*

Proof Let $\mathcal{C}_1 = \mathcal{C} \cup \{X \setminus U\}$. Then $\mathcal{C}_1$ is a set of closed subsets of X, and $\cap_{C \in \mathcal{C}_1} C = \emptyset$, and so $\mathcal{C}_1$ fails to have the finite intersection property. There exists a finite subset $\mathcal{F}$ of $\mathcal{C}$ such that $(\cap_{C \in \mathcal{F}} C) \cap (X \setminus U) = \emptyset$: that is, $\cap_{C \in \mathcal{F}} C \subseteq U$.   □

Proposition 15.1.3 *Suppose that (X, τ) is a topological space and that A is a subset of X.*

   (i) If (X, τ) is compact and A is closed, then A is compact.

   (ii) If (X, τ) is Hausdorff and A is compact, then A is closed.

   (iii) If (X, τ) is compact and Hausdorff then it is normal.

Proof (i) Suppose that $\mathcal{F}$ is a set of closed subsets of A with the finite intersection property. Since A is closed, the sets in $\mathcal{F}$ are closed in X. Since (X, τ) is compact, $\cap \{C : C \in \mathcal{F}\}$ is not empty.

(ii) Suppose that $x \notin A$. We shall show that there are disjoint open sets U and V with $A \subseteq U$ and $x \in V$. For each $a \in A$ there exist disjoint open subsets U_a and V_a of X with $a \in U_a$ and $x \in V_a$. The sets $\{U_a : a \in A\}$ form an open cover of A, and so there is a finite subset F of A such that $\{U_a : a \in F\}$ is a finite subcover of A. Then $U = \cup \{U_a : a \in F\}$ and $V = \cap \{V_a : a \in F\}$ are disjoint open sets, and $A \subseteq U$, $x \in V$. Thus $x \notin \bar{A}$, so that $A = \bar{A}$, and A is closed.

(iii) Suppose that A and B are disjoint closed subsets of X. We repeat the argument used in (ii). For each $b \in B$, there exist disjoint open subsets U_b and V_b of X with $A \subseteq U_b$ and $b \in V_b$. The sets $\{V_b : b \in B\}$ form an open cover of B, and so there is a finite subset G of B such that $\{V_b : b \in G\}$ is a finite subcover of B. Then $U = \cap\{U_b : b \in G\}$ and $V = \cup\{V_b : b \in G\}$ are disjoint open sets, and $A \subseteq U$, $B \subseteq V$. $\qquad\square$

Compact spaces which are not Hausdorff are less well behaved. For example, if X is an infinite set with the cofinite topology τ_f then (X, τ_f) is compact, and so are all of its subsets. Some authors include the Hausdorff property in their definition of compactness, and we shall concentrate our attention on such spaces.

Proposition 15.1.4 *Suppose that f is a continuous mapping from a topological space (X, τ) into a topological space (Y, σ). If A is a compact subset of X then $f(A)$ is a compact subset of Y.*

Proof Suppose that $\mathcal{U}$ is an open cover of $f(A)$. If $U \in \mathcal{U}$ then $f^{-1}(U)$ is open, since f is continuous. Thus $\{f^{-1}(U) : U \in \mathcal{U}\}$ is an open cover of A. Since A is compact, there is a finite subcover $\{f^{-1}(U_1), \ldots, f^{-1}(U_n)\}$. Then $\{U_1, \ldots, U_n\}$ is a finite subcover of $f(A)$. $\qquad\square$

Corollary 15.1.5 *Suppose that f is a continuous real-valued function on a compact space (X, τ). Then f is bounded on A, and attains its bounds: there exist $y \in X$ with $f(y) = \sup_{x \in X} f(x)$ and $z \in X$ with $f(z) = \inf_{x \in X} f(x)$.*

Proof For $f(X)$ is a compact subset of $\mathbf{R}$, and so is bounded and closed, by Theorem 5.4.4 of Volume I. $\qquad\square$

Proposition 15.1.6 *Suppose that f is a continuous mapping from a compact topological space (X, τ) onto a Hausdorff topological space (Y, σ), and that g is a mapping from (Y, σ) into a topological space (Z, ρ). Then g is continuous if and only if $g \circ f$ is continuous.*

Proof If g is continuous then certainly $g \circ f$ is continuous. Conversely, suppose that $g \circ f$ is continuous. Suppose that C is a closed subset of Z. Then $(g \circ f)^{-1}(C)$ is closed in X, and is therefore compact, by Proposition 15.1.3 (i). Thus $g^{-1}(C) = f((g \circ f)^{-1}(C))$ is compact, by Proposition 15.1.4, and is therefore closed, by Proposition 15.1.3 (ii). Thus g is continuous. $\qquad\square$

Corollary 15.1.7 *If f is a continuous bijection from a compact topological space (X, τ) onto a Hausdorff topological space (Y, σ), then f is a homeomorphism.*

Proof Take $g = f^{-1}$. $\qquad\square$

The topology of a compact Hausdorff space has a certain minimal property.

Corollary 15.1.8 *Suppose that (X, τ) is a compact Hausdorff space, and that σ is a Hausdorff topology on X which is coarser than τ. Then $\sigma = \tau$.*

Proof   Apply the corollary to the continuous identity mapping from (X, τ) to (X, σ).   □

Theorem 15.1.9 *The product of finitely many compact spaces is compact.*

Proof   A standard induction argument shows that it is enough to prove that the product of two compact spaces (X, τ) and (Y, σ) is compact. Suppose that $\mathcal{U}$ is an open cover of $X \times Y$. If $P = (x, y) \in X \times Y$, there exists $U_P \in \mathcal{U}$ with $P \in U_P$. Since U_P is open, there exist open neighbourhoods V_P of x and W_P of y such that $V_P \times W_P \subseteq U_P$. It is then clearly sufficient to show that finitely many of the sets $V_P \times W_P$ cover $X \times Y$.

Suppose that $x \in X$. The cross-section $C_x = \{(x, y) : y \in Y\}$ is homeomorphic to Y, and is therefore compact. It is covered by the collection $\{V_{(x,y)} \times W_{(x,y)} : y \in Y\}$ of open sets, and is therefore covered by a finite subset $\{V_{(x,y_j)} \times W_{(x,y_j)} : 1 \leq j \leq n\}$. Let $Q_x = \cap_{j=1}^{n} V_{(x,y_j)}$. Then Q_x is an open neighbourhood of x, and $Q_x \times Y \subseteq \cup_{j=1}^{n} V_{(x,y_j)} \times W_{(x,y_j)}$. The sets $\{Q_x : x \in X\}$ cover X. Since (X, τ) is compact, there is a finite subcover $\{Q_{x_1}, \ldots, Q_{x_n}\}$. Then the sets $\{Q_{x_1} \times Y, \ldots, Q_{x_n} \times Y\}$ cover $X \times Y$. Since each of then is covered by finitely many sets $V_P \times W_P$, $X \times Y$ is covered by finitely many sets $V_P \times W_P$.   □

Corollary 15.1.10 *A subset A of $\mathbf{R}^d$ or $\mathbf{C}^d$ is compact if and only if it is closed and bounded.*

The proof of Theorem 15.1.9 is rather awkward, and only deals with the product of finitely many spaces. In fact, a careful use of the axiom of choice can be used to prove the following.

Theorem 15.1.11 (Tychonoff's theorem)   *If $(X_\alpha, \tau_\alpha)_{\alpha \in A}$ is a family of compact topological spaces, then $\prod_{\alpha \in A} X_\alpha$ is compact in the product topology.*

In particular, $P(X)$, with the Bernoulli topology, is compact.

To prove this, 'sequences' are replaced by 'filters'. This involves introducing a fair amount of machinery. A proof is given in Appendix D.

Exercises

15.1.1 Show that the union of finitely many compact subsets of a topological space is compact.

15.1.2 Show that the intersection of a collection of compact subsets of a Hausdorff topological space is compact.

15.1.3 Give an example of two compact subsets of a T_1 topological space whose intersection is not compact.

15.1.4 Suppose that (X_1, τ_1) and (X_2, τ_2) are Hausdorff topological spaces and that (X_2, τ_2) is compact. Show that if A is a closed subset of $X_1 \times X_2$ then $\pi_1(A)$ is closed in (X_1, τ_1).

15.1.5 Suppose that f is a mapping from a topological space (X_1, τ_1) into a compact topological space (X_2, τ_2) whose graph G_f is a closed subset of $X_1 \times X_2$. Show that f is continuous. Can the condition that (X_2, d_2) is compact be dropped?

15.1.6 Suppose that G is a closed subgroup of $(\mathbf{R}^d, +)$ which does not contain a line (if $x \in \mathbf{R}^d \setminus \{0\}$, then $l_x = \{\alpha x : \alpha \in \mathbf{R}\}$ is not contained in G). Suppose that $x \in S^{d-1} = \{x \in \mathbf{R}^d : \|x\| = 1\}$. By considering $l_x \cap G$, show that there exist $r > 0$ and $\epsilon > 0$ such that if $y \in N_\epsilon(x) \cap S^{d-1}$ and $0 < \alpha < r$, then $\alpha y \notin G$. Use the compactness of S^{d-1} to show that G is a discrete subset of $\mathbf{R}^n$.

15.1.7 Let $B_n([0,1])$ be the collection of subsets of $[0,1]$ with at most n elements. Show (without appealing to Tychonoff's theorem) that $B_n([0,1])$ is a compact subset of $P([0,1])$, with the Bernoulli topology. (Hint: induction on n.)

15.2 Sequentially compact topological spaces

We now make a definition inspired by the Bolzano–Weierstrass theorem. We say that a topological space (X, τ) is *sequentially compact* if whenever $(x_n)_{n=1}^\infty$ is a sequence in X then there exists a subsequence $(x_{n_k})_{k=1}^\infty$ and an element $x \in X$ such that $x_{n_k} \to x$ as $k \to \infty$. We say that a subset A of X is sequentially compact, if it is sequentially compact, with the subspace topology: if $(a_n)_{n=1}^\infty$ is a sequence in A then there exists a subsequence $(a_{n_k})_{k=1}^\infty$ and an element $a \in A$ such that $a_{n_k} \to a$ as $k \to \infty$.

Thus the Bolzano–Weierstrass theorem implies that a subset of $\mathbf{R}$ is sequentially compact if and only if it is closed and bounded; that is, if and only if it is compact.

Proposition 15.2.1 *Suppose that (X, τ) is a sequentially compact topological space.*

(i) A closed subset A of X is sequentially compact.

(ii) If f is a continuous mapping of (X, τ) into a topological space (Y, σ) then $f(X)$ is sequentially compact.

Proof   (i) Suppose that $(a_n)_{n=1}^\infty$ is a sequence in A. There is a subsequence $(a_{n_k})_{k=1}^\infty$ and an element $x \in X$ such that $a_{n_k} \to x$ as $k \to \infty$. Since A is closed, $x \in A$.

(ii) Suppose that $(y_n)_{n=1}^\infty$ is a sequence in $f(X)$. For each $n \in \mathbf{N}$ there exists $x_n \in X$ such that $f(x_n) = y_n$. Then there exist a subsequence $(x_{n_k})_{k=1}^\infty$ and an element $x \in X$ such that $x_{n_k} \to x$ as $k \to \infty$. Then $y_{n_k} = f(x_{n_k}) \to f(x)$ as $k \to \infty$. $\qquad \square$

Theorem 15.2.2   *Suppose that (X, τ) is a finite product $\prod_{j=1}^n (X_j, \tau_j)$ or a countably infinite product $\prod_{j=1}^\infty (X_j, \tau_j)$ of sequentially compact topological spaces. Then (X, τ) is sequentially compact.*

Proof   We consider the countably infinite case: the finite case is easier. We use a diagonal argument, as in the proof of the Bolzano–Weierstrass theorem. Suppose that $(x^{(n)})_{n=1}^\infty$ is a sequence in X. There exists a subsequence $(y^{(1k)})_{k=1}^\infty$ and an element y_1 of X_1 such that $y_1^{(1k)} \to y_1$ as $k \to \infty$. Inductively, for each $j \in \mathbf{N}$ we can find a subsequence $(y^{(jk)})_{k=1}^\infty$ of $(y^{(j-1,k)})_{k=1}^\infty$ and an element y_j of such that $y_j^{(jk)} \to y_j$ as $k \to \infty$. Then $(y^{(kk)})_{k=1}^\infty$ is a subsequence of $(x^{(n)})_{n=1}^\infty$, and $y_j^{(kk)} \to y_j$ as $k \to \infty$, for each $j \in \mathbf{N}$. Thus if we set $y = (y_j)_{j=1}^\infty$ then $y^{(kk)} \to y$ in (X, τ) as $k \to \infty$. $\qquad \square$

Corollary 15.2.3   *The Hilbert cube is sequentially compact.*

Let us give two examples, related to these results.

Example 15.2.4   An uncountable product of sequentially compact topological spaces which is compact, but not sequentially compact.

Let S be the Bernoulli sequence space $\Omega(\mathbf{N})$; S is the set of all sequences taking the values 0 and 1, and is an uncountable set. Let $X = \Omega(S)$, with the product topology. Then (X, τ) is compact, by Tychonoff's theorem. We show that (X, τ) is not sequentially compact. For $n \in \mathbf{N}$ and $s \in S$ let $x_s^{(n)} = s_n$. Then $(x^{(n)})_{n=1}^\infty$ is a sequence in (X, τ). We shall show that it has no convergent subsequences. Suppose that $(x^{(n_k)})_{n=1}^\infty$ is a subsequence. Define an element s of S by setting $s_{n_k} = 1$ if k is even, and setting $s_n = 0$ otherwise. Then $x_s^{(n_k)}$ takes each of the values 0 and 1 infinitely often, and so does not converge.

Example 15.2.5   A sequentially compact subset C of a Hausdorff topological space (X, τ) which is not closed, and is therefore not compact.

Take (X, τ) the space of the preceding example; it is a Hausdorff space. Let

$$C = \{x \in X : \{s \in S : x_s = 1\} \text{ is countable}\}.$$

C is a dense proper subset of X, and so is not closed. Suppose that $(x^{(n)})_{n=1}^{\infty}$ is a sequence in C. For $n \in \mathbf{N}$, let $S_n = \{s \in S : x_s^{(n)} = 1\}$ and let $S_\infty = \cup_{n=1}^{\infty} S_n$. Then S_∞ is countable, and if $s \in S \setminus S_\infty$ then $x_s^{(n)} = 0$ for all $n \in \mathbf{N}$. A diagonal argument just like that of Theorem 15.2.2 shows that there is a subsequence $(x^{(n_k)})_{k=1}^{\infty}$ such that $x_s^{(n_k)}$ converges, to l_s, say, as $k \to \infty$, for each $s \in S_\infty$. Thus if we set $l_s = 0$ for $s \in S \setminus S_\infty$ then $l \in C$ and $x_{n_k} \to l$ as $k \to \infty$.

We now introduce a topological property that is rather weaker than sequential compactness. We need a definition. If (x_n) is a sequence in a topological space (X, τ), and $x \in X$, then x is a *limit point of the sequence* if whenever N is a neighbourhood of x and $n \in \mathbf{N}$, there exists $m \geq n$ such that $x_m \in \mathbf{N}$. A topological space (X, τ) is *countably compact* if every sequence in (X, τ) has a limit point.

Proposition 15.2.6 *A sequentially compact topological space (X, τ) is countably compact.*

Proof Suppose that $(x^{(n)})_{n=1}^{\infty}$ is a sequence in X. There exists $x \in X$ and a subsequence $(x^{(n_k)})_{k=1}^{\infty}$ which converges to x. Then x is a limit point of the sequence $(x^{(n)})_{n=1}^{\infty}$. □

Proposition 15.2.7 *Suppose that (X, τ) is a countably compact topological space.*

(i) A closed subset A of X is countably compact.

(ii) If f is a continuous mapping of (X, τ) into a topological space (Y, σ) then $f(X)$ is countably compact.

Proof The proof is very similar to the proof of Proposition 15.2.1, and the details are left to the reader. □

Proposition 15.2.8 *A first countable topological space (X, τ) is sequentially compact if and only if it is countably compact.*

Proof It is enough to show that if (X, τ) is countably compact, then it is sequentially compact. Suppose that $(x_n)_{n=1}^{\infty}$ is a sequence in X. It has a limit point l, and l has a decreasing base of neighbourhoods $(B_k)_{k=1}^{\infty}$. There then exists a subsequence $(x_{n_k})_{k=1}^{\infty}$ such that $x_{n_k} \in B_k$ for $k \in \mathbf{N}$. Then $x_{n_k} \to l$ as $k \to \infty$, so that (X, τ) is sequentially compact. □

What is the relationship between compactness and countable compactness?

Proposition 15.2.9 *A topological space (X, τ) is countably compact if and only if every cover of X by a sequence $(O_n)_{n=1}^{\infty}$ of open sets has a finite subcover.*

Proof Suppose that $(O_n)_{n=1}^{\infty}$ is an open cover of X, and that there is no finite subcover. Then for each n there exists $x_n \in X \setminus (\cup_{j=1}^{n} O_j)$. We show that the sequence $(x_n)_{n=1}^{\infty}$ has no limit point in X, so that (X, τ) is not countably compact. If $x \in X$, then $x \in O_n$ for some $n \in \mathbf{N}$, so that O_n is an open neighbourhood of x. Since $x_j \notin O_n$ for $j \geq n$, it follows that x is not a limit point of the sequence $(x_n)_{n=1}^{\infty}$. Thus the condition is necessary.

Conversely, suppose that every cover of X by a sequence $(O_n)_{n=1}^{\infty}$ of open sets has a finite subcover. Taking complements, this implies that if $(F_n)_{n=1}^{\infty}$ is a sequence of closed sets with the finite intersection property then $\cap_{n=1}^{\infty} F_n \neq \emptyset$. Suppose that $(x_n)_{n=1}^{\infty}$ is a sequence in X. Let $T_n = \{x_j : j \geq n\}$, and let $F_n = \bar{T}_n$. Then $(F_n)_{n=1}^{\infty}$ is a decreasing sequence of non-empty closed sets, and so has the finite intersection property. Thus there exists $x \in \cap_{n=1}^{\infty} F_n$. If $n \in \mathbf{N}$ and $N \in \mathcal{N}_x$, then $N \cap T_n \neq \emptyset$, so that there exists $m \geq n$ such that $x_m \in N$: x is a limit point of the sequence $(x_n)_{n=1}^{\infty}$. $\square$

Corollary 15.2.10 *If $(U_n)_{n=1}^{\infty}$ is an increasing sequence of open subsets of a countably compact topological space (X, τ) whose union is X, then there exists $n_0 \in \mathbf{N}$ such that $U_{n_0} = X$.*

Corollary 15.2.11 *A compact topological space is countably compact.*

In fact, countable compactness is sufficient for many problems concerning sequences of functions, as the next result shows. Recall that a sequence of continuous real-valued functions on a closed interval which converges pointwise to a continuous function need not converge uniformly. Things improve if the convergence is monotone.

Theorem 15.2.12 (Dini's theorem) *Suppose that $(f_n)_{n=1}^{\infty}$ is an increasing sequence of continuous real-valued functions on a countably compact topological space (X, τ) which converges pointwise to a continuous function f. Then $f_n \to f$ uniformly as $n \to \infty$.*

Proof Suppose that $\epsilon > 0$. Let $U_n = \{x \in X : f_n(x) > f(x) - \epsilon\}$. Since the function $f - f_n$ is continuous, U_n is open. $(U_n)_{n=1}^{\infty}$ is an increasing sequence, and $\cup_{n \in \mathbf{N}} U_n = X$, since $f_n(x) \to f(x)$ for each $x \in X$; $\{U_n : n \in \mathbf{N}\}$ is an open cover of X. Since $(U_n)_{n=1}^{\infty}$ is an increasing sequence, there exists $n_0 \in \mathbf{N}$ such that $U_{n_0} = X$ (Corollary 15.2.10). If $n \geq n_0$ and $x \in X$ then

$$0 \leq f(x) - f_n(x) \leq f(x) - f_{n_0}(x) < \epsilon,$$

so that $\|f - f_n\|_\infty \leq \epsilon$; $f_n \to f$ uniformly as $n \to \infty$. $\qquad\square$

15.3 Totally bounded metric spaces

We now consider what happens when we restrict attention to metric spaces. We need to introduce one further idea. If (X, d) is an unbounded metric space then the function $d'(x, y) = \min(d(x, y), 1)$ is a metric on X which is uniformly equivalent to d, and X is bounded under this metric. Thus boundedness is not a uniform property. We introduce a stronger boundedness property that is preserved under uniform homeomorphisms. Suppose that $\epsilon > 0$. A subset F of a metric space (X, d) is an ϵ-*net* if $\cup_{x \in F} N_\epsilon(x) = X$; every point of X is within ϵ of a point of F. (X, d) is *totally bounded*, or *precompact*, (the two names are used equally frequently, but we shall prefer the former) if, for every $\epsilon > 0$, there exists a finite ϵ-net; X can be covered by finitely many open neighbourhoods of radius ϵ. In other terms, (X, d) is totally bounded if and only if, for every $\epsilon > 0$, X is the union of finitely many subsets of diameter at most ϵ. A subset A of X is totally bounded if it is totally bounded with the subset metric. This concept is not a topological one. For example, the subset $(-\pi/2, \pi/2)$ is clearly totally bounded under the usual metric, but is not totally bounded under the metric $\rho(x, y) = |\tan x - \tan y|$, since $\tan$ defines an isometry of $((-\pi/2, \pi/2), \rho)$ onto $\mathbf{R}$, with its usual metric, and the latter is certainly not totally bounded.

Proposition 15.3.1 *A totally bounded subset A of a metric space (X, d) is bounded.*

Proof Take $\epsilon = 1$. There exists a finite subset F of A such that $A \subseteq \cup_{x \in F} N_1(x)$. If $y_1, y_2 \in A$ then there exist $x_1, x_2 \in F$ such that $y_1 \in N_1(x_1)$ and $y_2 \in N_2(x_2)$. By the triangle inequality,

$$d(y_1, y_2) \leq d(y_1, x_1) + d(x_1, x_2) + d(x_2, y_2) \leq \mathrm{diam}\,(F) + 2,$$

so that A is bounded. $\qquad\sqcup$

Boundedness is not a uniform property, but total boundedness is.

Proposition 15.3.2 *Suppose that A is a totally bounded subset of a metric space (X, d) and that f is a uniformly continuous mapping of (X, d) into a metric space (Y, ρ). Then $f(A)$ is a totally bounded subset of (Y, ρ).*

Proof   Suppose that $\epsilon > 0$. Then there exists $\delta > 0$ such that if $d(x, y) < \delta$ then $\rho(f(x), f(y)) < \epsilon$. Let F be a finite δ-net in A. Then $f(F)$ is a finite ϵ-net in $f(A)$.   □

Corollary 15.3.3   *If d and d' are uniformly equivalent metrics on X, then (X, d) is totally bounded if and only if (X, d') is.*

Proposition 15.3.4   *A totally bounded metric space (X, d) is second countable, and is therefore separable.*

Proof   For each $n \in \mathbf{N}$ there exists a finite $1/n$-net F_n in X. Let $\mathcal{U}_n = \{N_{1/n}(x) : x \in F_n\}$ and let $\mathcal{U} = \cup_{n \in \mathbf{N}}\mathcal{U}_n$. Then $\mathcal{U}$ is a countable collection of open subsets of X. Let us show that it is a base for the topology. Suppose that O is an open subset of X and that $x \in X$. There exists $\delta > 0$ such that $N_\delta(x) \subseteq O$. Choose n so that $1/n < \delta$. Then there exists $U_x = N_{1/2n}(y) \in \mathcal{U}_{2n}$ such that $x \in U_x$. If $z \in U_x$ then $d(z, x) \leq d(z, y) + d(y, x) < 1/n$ so that $z \in O$. Thus $x \in U_x \subseteq O$, so that $O = \cup\{U \in \mathcal{U} : U \subseteq O\}$, and $\mathcal{U}$ is a base for the topology.   □

Proposition 15.3.5   *Suppose that S is a dense totally bounded metric subspace of a metric space (X, d). Then (X, d) is totally bounded.*

Proof   Suppose that $\epsilon > 0$. There exists a finite subset F of S such that $S = \cup\{N_{\epsilon/2} \cap S : f \in F\}$. If $x \in X$ there exists $s \in S$ with $d(x, s) < \epsilon/2$ and there exists $f \in F$ with $d(s, f) < \epsilon/2$. Thus $d(x, f) < \epsilon$, and F is a finite ϵ-net in X.   □

Corollary 15.3.6   *A metric space (X, d) is totally bounded if and only if its completion is totally bounded.*

Total boundedness can be characterized in terms of Cauchy sequences.

Theorem 15.3.7   *A metric space (X, d) is totally bounded if and only if every sequence in X has a Cauchy subsequence.*

Proof   Suppose first that (X, d) is totally bounded, and that $(x_n)_{n=1}^\infty$ is a sequence in X. We use a diagonal argument to obtain a Cauchy subsequence. There exists a finite cover $\{A_1, \ldots, A_k\}$ of X by sets of diameter at most 1. By the pigeonhole principle, there exists j such that $x_n \in A_j$ for infinitely many n. That is, there exists a subsequence $(y_{1,n})_{n=1}^\infty$ of $(x_n)_{n=1}^\infty$ such that $d(y_{1,m}, y_{1,n}) \leq 1$ for $m, n \in \mathbf{N}$. Repeating the argument, there exists a subsequence $(y_{2,n})_{n=1}^\infty$ of $(y_{1,n})_{n=1}^\infty$ such that $d(y_{2,m}, y_{2,n}) \leq 1/2$ for $m, n \in \mathbf{N}$, and, iterating the argument, for each j there exists a subsequence $(y_{j+1,n})_{n=1}^{\infty}$ of $(y_{j,n})_{n=1}^{\infty}$ such that $d(y_{j+1,m}, y_{j+1,n}) \leq 1/(j+1)$ for $m, n \in \mathbf{N}$. The sequence $(y_{j,j})_{j=1}^{\infty}$ is then a Cauchy subsequence of $(x_n)_{n=1}^{\infty}$.

Suppose next that (X, d) is not totally bounded; there exists $\epsilon > 0$ such that there is no finite ϵ-net in X. Choose $x_1 \in X$. Then $N_\epsilon(x_1) \neq X$, and so there exists $x_2 \in X$ with $d(x_1, x_2) \geq \epsilon$. Iterating this argument, there exists a sequence $(x_n)_{n=1}^{\infty}$ such that for each $n \in \mathbf{N}$, $x_{n+1} \notin \cup_{j=1}^{n} N_\epsilon(x_j)$. Thus if $m \neq n$ then $d(x_m, x_n) \geq \epsilon$, and so the sequence $(x_n)_{n=1}^{\infty}$ has no Cauchy subsequence. $\qquad \square$

Exercise

15.3.1 Show that a subset A of a metric space (X, d) is totally bounded if and only if whenever $\epsilon > 0$ there exists a finite subset G of X such that $A \subseteq \cup_{x \in G} N_\epsilon(x)$.

15.4 Compact metric spaces

Things work extremely well for metric spaces.

Theorem 15.4.1 *Suppose that (X, d) is a metric space. The following are equivalent:*

(i) (X, d) is compact;

(ii) (X, d) is sequentially compact;

(iii) (X, d) is countably compact;

(iv) (X, d) is complete and totally bounded.

Proof We have seen that (i) implies (iii) (Corollary 15.2.11), and that (ii) and (iii) are equivalent (Proposition 15.2.8).

Let us show that (ii) and (iv) are equivalent. Suppose first that (X, d) is sequentially compact, and suppose that $(x_n)_{n=1}^{\infty}$ is a Cauchy sequence in X. Then $(x_n)_{n=1}^{\infty}$ has a convergent subsequence, and so by Proposition 14.1.1 $(x_n)_{n=1}^{\infty}$ is convergent. Thus (X, d) is complete. Since every sequence has a convergent subsequence, which is a Cauchy subsequence, (X, d) is totally bounded, by Theorem 15.3.7. Conversely, suppose that (X, d) is complete and totally bounded, and that $(x_n)_{n=1}^{\infty}$ is a sequence in X. Since (X, d) is totally bounded, there is a Cauchy subsequence $(x_{n_k})_{k=1}^{\infty}$, and this subsequence converges, since (X, d) is complete.

Finally let us show that (ii) and (iv) imply that (X, d) is compact. We need a lemma, of interest in its own right.

Lemma 15.4.2 *If $\mathcal{O}$ is an open cover of a countably compact metric space (X, d), there exists $\delta > 0$ such that for each $x \in X$ there exists $O \in \mathcal{O}$ for which $N_\delta(x) \subseteq O$.*

Proof Suppose not. Then for each $n \in \mathbf{N}$ there exists $x_n \in X$ for which $N_{1/n}(x_n)$ is not contained in any $O \in \mathcal{O}$. Let x be a limit point of the sequence $(x_n)_{n=1}^\infty$. Then $x \in O$, for some $O \in \mathcal{O}$. Since O is open, there exists $\epsilon > 0$ such that $N_\epsilon(x) \subseteq O$. Since x is a limit point of the sequence, there exists $n > 2/\epsilon$ such that $x_n \in N_\epsilon(x)$. If $y \in N_{1/n}(x_n)$, then $d(y, x_n) < \epsilon/2$, so that $d(y, x) \leq d(y, x_n) + d(x_n, x) < \epsilon$. Hence $N_{1/n}(x_n) \subseteq N_\epsilon(x) \subseteq O$, giving a contradiction.   □

A number δ which satisfies the conclusion of the lemma is called a *Lebesgue number* of the cover.

Suppose now that $\mathcal{O}$ is an open cover of (X, d). Let $\delta > 0$ be a Lebesgue number of the cover. Since (X, d) is totally bounded, there exists a finite δ-net F in X. For each $x \in F$ there exists $O_x \in \mathcal{O}$ such that $N_\delta(x) \subseteq O_x$. Then $X = \cup_{x \in F} N_\delta(x) = \cup_{x \in F} O_x$, so that $\{O_x : x \in F\}$ is a finite subcover of X.   □

Note that neither of the conditions of (iv) is a topological condition, but that together they are equivalent to topological conditions.

Let us bring some earlier results together.

Corollary 15.4.3 *A compact metric space is second countable, and is therefore separable.*

Proof Proposition 15.3.4.   □

Corollary 15.4.4 *The completion of a totally bounded metric space is compact.*

Proof Proposition 15.3.5.   □

This explains the terminology 'precompact'.

Corollary 15.4.5 *A finite or countably infinite product of compact metric spaces, with a product metric, is a compact metric space.*

Proof Theorem 15.2.2.   □

Corollary 15.4.6 *The Hilbert cube and the Bernoulli sequence space $\Omega(\mathbf{N})$ are compact.*

Proof Corollary 15.4.5.   □

We can characterize compactness in terms of the Hilbert cube, and in terms of the Bernoulli sequence space.

Corollary 15.4.7 *A metric space (X, d) is compact if and only if it is homeomorphic to a closed subspace of the Hilbert cube.*

Proof If (X, d) is compact, it is second countable (Corollary 15.4.3), and therefore there is a homeomorphism f of X onto a metric subspace $f(X)$ of the Hilbert cube, by Urysohn's metrization theorem (Theorem 13.5.6). But $f(X)$ is compact, and so it is a closed subset of the Hilbert cube. Conversely, if (X, d) is homeomorphic to a closed subspace of the Hilbert cube, it must be compact. □

Theorem 15.4.8 *A metric space is compact if and only if there is a continuous surjective mapping of the Bernoulli sequence space $\Omega(\mathbf{N})$ onto X.*

Proof Since $\Omega(\mathbf{N})$ is compact, the condition is sufficient.

Suppose that (X, d) is compact. We give $\Omega(\mathbf{N})$ the product metric $\rho(y, z) = \sum_{j=1}^{\infty} |y_j - x_j|/3^j$. If $y \in \Omega(\mathbf{N})$ and $j \in \mathbf{N}$, let $C_j(y) = N_{1/3^j}(y)$. Then

$$C_j(y) = \{z \in \Omega(\mathbf{N}) : z_i = y_i \text{ for } 1 \leq i \leq j\}.$$

Such a set is called a *j-cylinder set*. Let $\mathcal{C}_j$ be the set of j-cylinder sets: $|\mathcal{C}_j| = 2^j$.

We now show that there is a strictly increasing sequence $(s_k)_{k=1}^{\infty}$ in $\mathbf{N}$, and a sequence $(f_k : \mathcal{C}_{s_k} \to X)_{k=1}^{\infty}$ of mappings, such that

(i) $f_k(\mathcal{C}_{s_k})$ is a $1/2^k$-net in (X, d), and

(ii) if $y \in \Omega(\mathbf{N})$ then $d(f_k(\mathcal{C}_{s_k}(y)), f_{k+1}(\mathcal{C}_{s_{k+1}}(y))) \leq 1/2^k$.

We first define s_1 and f_1. There exists a finite $1/2$-net F_1 in (X, d). Choose s_1 so that $2^{s_1} \geq |F_1|$. Since $|\mathcal{C}_{s_1}| = 2^{s_1}$, there is a surjective mapping of $\mathcal{C}_{s_1}$ onto F_1.

Suppose now that $s_1, \ldots, s_k$ and $f_1, \ldots f_k$ have been defined. For each $C \in \mathcal{C}_{s_k}$ there is a $1/2^{k+1}$-net $F_{k+1}(C)$ in $N_{1/2^k}(f_k(C))$. Choose $n_{k+1} > n_k$ so that

$$2^{n_{k+1}} \geq \max\{|F_{k+1}(C)| : C \in \mathcal{C}_{s_k}\}.$$

Let $s_{k+1} = s_k + n_{k+1}$, and let $\mathcal{C}_{k+1}(C)$ be the set of s_{k+1}-cylinder sets contained in C: there are $2^{n_{k+1}}$ of them. There is therefore a surjective mapping $f_{k+1,C}$ from $\mathcal{C}_{k+1}(C)$ onto $F_{k+1}(C)$. Letting C vary, and combining the mappings $f_{k+1,C}$, we obtain a mapping of $\mathcal{C}_{s_{k+1}}$ into (X, d) which satisfies (i) and (ii).

If $y \in \Omega(\mathbf{N})$ and $k \in \mathbf{N}$, let $g_k(y) = f_k(C_{s_k}(y))$. Then each g_k is a continuous mapping of $\Omega(\mathbf{N})$ into (X, d), and $d(g_k(y), g_{k+1}(y)) \leq 1/2^k$. Thus if $k < l$ then $d(g_k(y), g_l(y)) \leq 2/2^k$. It therefore follows from the general principle of uniform convergence that the mappings g_k converge uniformly to a continuous function g mapping $\Omega(\mathbf{N})$ to (X, d), and that $d(g(y), g_k(y)) \leq 2/2^k$ for $y \in \Omega(\mathbf{N})$ and $k \in \mathbf{N}$.

It remains to show that g is surjective. Since $\Omega(\mathbf{N})$ is compact, $g(\Omega(\mathbf{N}))$ is compact, and is therefore closed in X. It is therefore sufficient to show that $g(\Omega(\mathbf{N}))$ is dense in X. Suppose that $x \in X$ and that $\epsilon > 0$. There exists k such that $1/2^k < \epsilon/3$, and there exists $y \in \Omega(\mathbf{N})$ such that $d(x, g_k(y)) < 1/2^k$. Thus $d(x, g(y)) \leq d(x, g_k(y)) + d(g_k(y), g(y)) \leq 3/2^k < \epsilon$. $\square$

Corollary 15.4.9 *A metric space is compact if and only if there is a continuous surjective mapping of the Cantor set C onto X.*

Proof   For the Cantor set is homeomorphic to $\Omega(\mathbf{N})$. $\square$

The next result is particularly important.

Theorem 15.4.10 *If f is a continuous map from a compact metric space (X, d) into a metric space (Y, ρ) then f is uniformly continuous.*

Proof   Suppose not. Then there exists $\epsilon > 0$ for which we can find no suitable $\delta > 0$. Thus for each $n \in \mathbf{N}$ there exist x_n, x'_n in X with $d(x_n, x'_n) < 1/n$ and $\rho(f(x_n), f(x'_n)) \geq \epsilon$. By sequential compactness there exists a subsequence $(x_{n_k})_{k=1}^{\infty}$ which converges to an element $x \in X$ as $k \to \infty$. Since $d(x_{n_k}, x'_{n_k}) \to 0$ as $k \to \infty$, $x'_{n_k} \to x$, as well. Since f is continuous at x, $f(x_{n_k}) \to f(x)$ and $f(x'_{n_k}) \to f(x)$ as $k \to \infty$, so that $\rho(f(x_{n_k}), f(x'_{n_k})) \to 0$ as $k \to \infty$. As $\rho(f(x_{n_k}), f(x'_{n_k})) \geq \epsilon$ for all $k \in \mathbf{N}$, we have a contradiction. $\square$

Isometries of compact metric spaces behave well.

Theorem 15.4.11 *Suppose that f is an isometry of a compact metric space into itself. Then f is surjective.*

Proof   Suppose not. Then $f(X)$ is a proper closed subset of X. Suppose that $x \in X \setminus f(X)$, and let $\delta = d(x, f(X))$. If $n \in \mathbf{N}$ then $f^n(x) \in f(X)$, and so $d(f^n(x), x) \geq \delta$. Since d is an isometry, if $k \in \mathbf{N}$ then $d(f^{n+k}(x), f^k(x)) = d(f^k(f^n(x)), f^k(x)) \geq \delta$, so that the sequence $(f^n(x))_{n=1}^{\infty}$ has no convergent subsequence. $\square$

Exercises

15.4.1 Suppose that A and B are disjoint subsets of a metric space (X, d), and that A is compact and B is closed. Show that

$$d(A, B) = \inf\{d(a, b) : a \in A, b \in B\} > 0.$$

15.4.2 Suppose that (X, d) is a compact metric space. Let

$$C = \{x \in X : x \text{ has a countable neighbourhood}\}$$

Show that C is an open subset of X and that $P = X \setminus C$ is perfect. By considering sets $P_n = \{x \in X : d(x, P) < 1/n\}$, or otherwise, show that C is countable: X is the union of a countable set and a perfect set.

15.4.3 Suppose that (X, d) is a compact metric space, and that $(f_n)_{n=0}^\infty$ is a sequence of continuous functions from (X, d) into a metric space (Y, ρ), with the property that $d(f_n(x), f_0(x))$ is a decreasing null sequence, for each $x \in X$. Show that f_n converges uniformly to f_0 as $n \to \infty$. (This generalizes Dini's theorem.)

15.4.4 Give an example of a sequence $(U_n)_{n=1}^\infty$ of open subsets of $\mathbf{R}$ whose union contains the rationals and whose complement is infinite.

15.4.5 Show that an open cover of a separable metric space has a countable subcover. (The previous exercise shows that some care is needed.)

15.4.6 Suppose that (X, d) is a compact metric space, and that f is a mapping of X into itself which satisfies $d(f(x), f(y)) \geq d(x, y)$ for all $x, y \in X$. By considering the sequence $((f^n(x), f^n(y)))_{n=1}^\infty$ in $X \times X$, show that if $x, y \in X$ and $\epsilon > 0$ then there exists $n \in \mathbf{N}$ such that $d(x, f^n(x)) < \epsilon$ and $d(y, f^n(y)) < \epsilon$. Show that f is an isometry, and give another proof that f is surjective.

15.5 Compact subsets of $C(K)$

Suppose that (K, d) is a compact topological space. Let $C(K)$ denote the (real or complex) vector space of continuous (real or complex) functions on K. If $f \in C(K)$ then $f(K)$ is compact, and is therefore closed and bounded. Thus $C(K) = C_b(K)$, and $C(K)$ is a Banach space under the uniform norm $\|f\|_\infty = \sup\{|f(x)| : x \in K\}$. What are the compact subsets of $C(K)$? Since $C(K)$ is complete, a subset of $C(K)$ is compact if and only if it is closed and totally bounded, and so it is enough to characterize the totally bounded subsets of $C(K)$.

In order to do this, we need another definition. Suppose that (X, τ) is a topological space, that (Y, d) is a metric space, and that $x \in X$. A set A of mappings from (X, τ) to (Y, d) is *equicontinuous* at x if, given $\epsilon > 0$, there exists $N \in \mathcal{N}_x$ such that $d(a(x), a(y)) < \epsilon$ for all $a \in A$ and $y \in N$. The set A is *equicontinuous* on X if it is equicontinuous at each point of X.

Theorem 15.5.1 (The Arzelà–Ascoli theorem) *Suppose that (K, τ) is a compact topological space, that $(C(K), \|.\|_\infty)$ is the (real or complex) Banach space of continuous (real or complex) functions on K and that $A \subseteq C(K)$. Then A is totally bounded if and only if A is bounded in norm and equicontinuous on K.*

Proof   Suppose that A is totally bounded. Then A is bounded, by Proposition 15.3.1. Let us show that A is equicontinuous. Suppose that $x \in K$ and that $\epsilon > 0$. There exists a finite $\epsilon/3$-net F in A. For each $f \in F$ there exists $N_f \in \mathcal{N}_x$ such that if $y \in N_f$ then $|f(x) - f(y)| < \epsilon/3$. Let $N = \cap_{f \in F} N_f$: N is a neighbourhood of x. Now if $a \in A$ there exists $f \in F$ such that $\|a - f\|_\infty < \epsilon/3$. Thus if $y \in N$ then

$$|a(x) - a(y)| \le |a(x) - f(x)| + |f(x) - f(y)| + |f(y) - a(y)|$$
$$\le \epsilon/3 + \epsilon/3 + \epsilon/3 = \epsilon.$$

Since this holds for all $a \in A$, A is equicontinuous at x.

Conversely, suppose that A is bounded and equicontinuous. Suppose that $\epsilon > 0$. If $x \in K$, there exists $N(x)$ in $\mathcal{N}_x$ such that if $y \in N(x)$ then $|a(x) - a(y)| < \epsilon/4$ for all $a \in A$. Since (K, τ) is compact, there exists a finite subset $Y = \{y_1, \ldots y_n\}$ of K such that $K = \cup_{m=1}^n N(y_m)$. We use Y to define a linear mapping of $C(K)$ into $\mathbf{R}^n$ (or $\mathbf{C}^n$): if $f \in C(K)$, we set $T(f) = (f(y_1), \ldots f(y_n))$. We give $\mathbf{R}^n$ (or $\mathbf{C}^n$) the supremum norm; $\|(x_1, \ldots, x_n)\|_\infty = \max\{|x_m| : 1 \le m \le n\}$. Then

$$\|T(f)\|_\infty = \sup_{1 \le m \le n} |f(x_m)| \le \|f\|_\infty,$$

so that $T(A)$ is bounded in $\mathbf{R}^n$ (or $\mathbf{C}^n$). It is therefore totally bounded, by Theorem 15.1.10, and so there exists a finite subset F of A such that $T(F)$ is an $\epsilon/4$-net in $T(A)$. We shall show that F is an ϵ-net in A, so that A is totally bounded. If $a \in A$ there exists $f \in F$ such that $\|(T(a) - T(f)\|_\infty < \epsilon/4$; that is, $|a(y_m) - f(y_m)| < \epsilon/4$ for $1 \le m \le n$. If $x \in K$ there exists y_m such that $x \in N(y_m)$; then $|a(x) - a(y_m)| < \epsilon/4$ and $|f(x) - f(y_m)| < \epsilon/4$. Putting these inequalities together,

$$|a(x) - f(x)| \leq |a(x) - a(y_m)| + |a(y_m) - f(y_m)| + |f(y_m) - f(x)|$$
$$\leq \epsilon/4 + \epsilon/4 + \epsilon/4 = 3\epsilon/4.$$

Since this holds for all $x \in K$, $\|a - f\|_\infty \leq 3\epsilon/4 < \epsilon$, and $a \in N_\epsilon(f)$. Thus F is an ϵ-net in A. $\square$

Corollary 15.5.2 *A is compact if and only if it is closed, bounded and equicontinuous on K.*

It is possible to characterize the compact subsets of $C(K)$ locally.

Suppose that (X, τ) is a topological space and that Y is a subset of X. The *restriction mapping* π_Y from $C_b(X)$ to $C_b(Y)$ is defined by setting $\pi_Y(f)(y) = f(y)$, for $y \in Y$. π_Y is a norm-decreasing linear mapping from $(C_b(X), \|.\|_\infty)$ to $(C_b(Y), \|.\|_\infty)$, and so is continuous.

Theorem 15.5.3 *Suppose that $M_1, \ldots, M_n$ are closed subsets of a compact topological space (K, τ) and that $K = \cup_{j=1}^n M_j$. Then a subset A of $(C(K), \|.\|_\infty)$ is compact if and only if $\pi_{M_j}(A)$ is compact in $(C(M_j), \|.\|_\infty)$ for $1 \leq j \leq n$.*

Proof If A is compact, then $\pi_{M_j}(A)$ is compact, for $1 \leq j \leq n$, since the mappings π_{M_j} are continuous.

Conversely, suppose the condition is satisfied. Give the product space $\prod_{j=1}^n C(M_j)$ the norm

$$\|(f_1, \ldots, f_n)\|_\infty = \max_{1 \leq j \leq n} \|f_j\|_\infty .$$

If $f \in C(K)$, let $\pi(f) = (\pi_{M_1}(f), \ldots, \pi_{M_1}(f))$. Then

$$\|\pi(f)\|_\infty = \max_{1 \leq j \leq n} \|\pi_{M_j}(f)\|_\infty = \max_{1 \leq j \leq n} (\sup_{x \in M_j} |f(x)|) = \|f\|_\infty .$$

Thus π is an isometric linear mapping of $(C(K), \|.\|_\infty)$ into $\prod_{j=1}^n (C(M_j), \|.\|_\infty)$. In particular, $\pi(C(K))$ is closed in $\prod_{j=1}^n C(M_j)$. It follows from Corollary 15.4.5 that $\prod_{j=1}^n \pi_{M_j}(A)$ is compact in $\prod_{j=1}^n (C(M_j), \|.\|_\infty)$. Since $\pi(A) = (\prod_{j=1}^n \pi_{M_j}(A)) \cap \pi(C(K))$, it follows that $\pi(A)$ is compact. Since π is an isometry, A is compact. $\square$

Exercises

15.5.1 Suppose that $(f_n)_{n=1}^\infty$ is an equicontinuous sequence of mappings from a topological space (X, τ) into a metric space (Y, σ), and that

$f_n(x)$ converges pointwise to $f(x)$ for each x in X. Show that f is continuous.

15.5.2 Suppose that $(f_n)_{n=1}^\infty$ is an equicontinuous sequence of mappings from a topological space (X, τ) into a complete metric space (Y, σ), and that $f_n(d)$ converges to $f(d)$ for each d in a dense subset D of X. Show that $f_n(x)$ converges in Y for each $x \in X$.

15.5.3 Suppose that (K, τ) is a compact topological space and that $(f_n)_{n=1}^\infty$ is an equicontinuous sequence in $C(K)$, which converges pointwise to a function f. Show that f_n converges uniformly to f.

15.5.4 Suppose that $(f_n)_{n=1}^\infty$ is an increasing sequence of continuous real-valued functions on a compact topological space (X, τ) which converges pointwise to a continuous function f. Show that the sequence $(f_n)_{n=1}^\infty$ is totally bounded in $C(K)$. Use this to give another proof of Dini's theorem (for functions on a compact topological space).

15.5.5 Suppose that (X, d) and (Y, ρ) are metric spaces. A set A of mappings from (X, d) to (Y, ρ) is *uniformly equicontinuous* if, given $\epsilon > 0$, there exists $\delta > 0$ such that if $d(x_1, x_2) < \delta$ then $\rho(a(x_1), a(x_2)) < \epsilon$, for all $a \in A$. Thus each $a \in A$ is uniformly continuous, and we can control all the elements of A simultaneously. Suppose that (X, d) is compact. Show that an equicontinuous set of mappings A from X to Y is uniformly equicontinuous.

15.5.6 Suppose that $A \subseteq C[0, 1]$ is equicontinuous and that $\{a(x) : a \in A\}$ is bounded, for some $x \in [0, 1]$. Show that A is bounded in norm, and is therefore totally bounded.

15.5.7 Prove the following vector-valued version of the Arzelà–Ascoli theorem. Suppose that (K, τ) is a compact topological space, that $(E, \|.\|)$ is a normed space and that A is a subset of $C(K; E), \|.\|_\infty)$, the (real or complex) normed space of continuous functions on K taking values in E. Show that A is totally bounded if and only if A is bounded, $\{a(x) : a \in A\}$ is totally bounded in E for each $x \in K$, and A is equicontinuous on K.

15.6 *The Hausdorff metric*

(This section can be omitted on a first reading.)

We now give an example of an interesting class of metric spaces. Suppose that (X, d) is a metric space. If A is a non-empty subset of (X, d), the *open ϵ-neighbourhood* $N_\epsilon(A)$ is defined to be

$$N_\epsilon(A) = \{x \in X : d(x, A) < \epsilon\} = \cup_{x \in A} N_\epsilon(x);$$

since $N_\epsilon(A)$ is the union of open sets, it is open. Since $d(x, A) = d(x, \overline{A})$, it follows that $N_\epsilon(A) = N_\epsilon(\overline{A})$. For this reason, we concentrate on the non-empty closed subsets of X. Let $\mathrm{Con}(X)$ be the set of non-empty bounded closed subsets of (X, d); $\mathrm{Con}(X)$ is the *configuration space* of X.

Theorem 15.6.1 *Suppose that* $\mathrm{Con}(X)$ *is the* configuration *space of a metric space* (X, d). *If* $A, B \in \mathrm{Con}(X)$ *and* $x \in X$, *let*

$$d_H(A, B) = \sup_{x \in X} |d(x, A) - d(x, B)|.$$

(i) Let $e(A, B) = \sup_{a \in A} d(a, B)$. *Then*

$$d_H(A, B) = \max(e(A, B), e(B, A)) < \infty.$$

(ii) d_H *is a metric on* $\mathrm{Con}(X)$.
(iii) $d_H(A, B) = \inf\{\epsilon > 0 : A \subseteq N_\epsilon(B) \text{ and } B \subseteq N_\epsilon(A)\}$.

Proof   (i) First, $e(A, B) = \sup_{a \in A} d(a, B) - \sup_{a \in A} d(a, A) \le d_H(A, B)$, and similarly, $e(B, A) \le d_H(A, B)$, so that

$$\max(e(A, B), e(B, A)) \le d_H(A, B).$$

Conversely, suppose that $x \in X$ and that $\epsilon > 0$. There exists $b \in B$ such that $d(x, b) < d(x, B) + \epsilon/2$ and there exists $a \in A$ such that $d(b, a) < d(b, A) + \epsilon/2$. Then

$$d(x, A) \le d(x, a) \le d(x, b) + d(b, a) \le d(x, B) + d(b, A) + \epsilon$$
$$\le d(x, B) + e(B, A) + \epsilon.$$

Since $\epsilon > 0$ is arbitrary,

$$d(x, A) - d(x, B) \le e(B, A) \le \mathrm{diam}\,(A \cup B) < \infty,$$

and similarly $d(x, B) - d(x, A) \le e(A, B) < \infty$. Thus

$$d_H(A, B) \le \max(e(A, B), e(A, B)).$$

(ii) Clearly $d_H(A, B) = d_H(B, A)$ and $d_H(A, A) = 0$. Suppose that $A \ne B$; without loss of generality we can suppose that $A \setminus B \ne \emptyset$. If $a \in A \setminus B$ then

$$d_H(A, B) \ge d(a, B) - d(a, A) = d(a, B) > 0.$$

If $x \in X$, and $A, B, C \in \mathrm{Con}(X)$ then

$$|d(x, A) - d(x, C)| \le |d(x, A) - d(x, B)| + |d(x, B) - d(x, C)|$$
$$\le d_H(A, B) + d_H(B, C);$$

thus $d_H(A, C) \leq d_H(A, B) + d_H(B, C)$, and so d_H is a metric on $\mathrm{Con}(X)$.

(iii) This follows immediately from (i). $\square$

The metric d_H is the *Hausdorff metric* on $\mathrm{Con}(X)$; it measures how far apart A and B are.

Let $\mathrm{Con}_n(X)$ denote the set of subsets of X with n elements; $\mathrm{Con}_n(X)$ is the *n-point configuration space* of X. Let $\mathrm{Con}_F(X) = \cup_{n=1}^{\infty} \mathrm{Con}_n(X)$; $\mathrm{Con}_F(X)$ is the *finite configuration space* of X.

Proposition 15.6.2   *A closed bounded subset A of (X, d) is totally bounded if and only if it is in the closure of $\mathrm{Con}_F(X)$.*

Proof   If A is totally bounded and $\epsilon > 0$ there exists a finite subset F such that $A \subseteq \cup_{x \in F} N_\epsilon(x)$. Thus $e(A, F) \leq \epsilon$. But we can clearly suppose, by discarding terms if necessary, that $N_\epsilon(x) \cap A \neq \emptyset$, for each $x \in F$, and then $e(F, A) < \epsilon$; consequently $d_H(A, F) < \epsilon$, and A is in the closure of $\mathrm{Con}_F(X)$.

Conversely, suppose that A is in the closure of $\mathrm{Con}_F(X)$, and that $\epsilon > 0$. Then there exists a finite set F such that $d_H(A, F) < \epsilon$. Thus $A \subseteq N_\epsilon(F)$, and A is totally bounded. $\square$

The mapping $i_H : x \to \{x\} : (X, d) \to (\mathrm{Con}(X), d_H)$ is an isometry; further, $\mathrm{Con}_1(X) = i_H(X)$ is a closed subset of (x_H, d_H), since if $A \in \mathrm{Con}(X)$ has two distinct points a and a' then

$$e(A, \{b\}) \geq \max(d(a, b), d(a', b)) \geq d(a, a')/2,$$

by the triangle inequality. Note also that if Y is a closed subset of (X, d) then the natural inclusion: $(\mathrm{Con}(Y), d_H) \to (\mathrm{Con}(X), d_H)$ is an isometry.

Properties of (X, d) are reflected in properties of $(\mathrm{Con}(X), d_H)$.

Theorem 15.6.3   *Suppose that (X, d) is a bounded metric space. The following are equivalent.*

(i) (X, d) is totally bounded.

(ii) $(\mathrm{Con}(X), d_H)$ is totally bounded.

(iii) $(\mathrm{Con}(X), d_H)$ is separable.

Proof   Suppose that (X, d) is totally bounded, and suppose that $\epsilon > 0$. Then there exists a finite subset F of X such that $X = N_\epsilon(F)$. Suppose that $A \in \mathrm{Con}(X)$. Let $F_A = \{x \in F : N_\epsilon(x) \cap A \neq \emptyset\}$. Then F_A is not empty, and $d_H(A, F_A) < \epsilon$. Thus if $C(F)$ is the collection of the $2^{|F|} - 1$ non-empty subsets of F then $\mathrm{Con}(X) = N_\epsilon(C(F))$. Thus (i) implies (ii). Since (X, d) is isometric to a subset of $(\mathrm{Con}(X), d_H)$, (ii) implies (i).

Since a totally bounded metric space is separable, (ii) implies (iii). Suppose that (X, d) is not totally bounded. Then, as in Theorem 15.3.7, there exists $\epsilon > 0$ such that there is no finite ϵ-net in (X, d), and there therefore exists an infinite sequence $(x_n)_{n=1}^\infty$ such that $d(x_m, x_n) \geq \epsilon$ for $m \neq n$. Let $S = \{x_n : n \in \mathbf{N}\}$, with the subspace metric. Then any subset of S is closed in (X, d), and if A and B are distinct non-empty subsets of S then $d_H(A, B) \geq \epsilon$. Since there are uncountably many such sets, $(\mathrm{Con}(X), d_H)$ is not separable. Thus (iii) implies (i). □

Theorem 15.6.4 *If (X, d) is a metric space, $(\mathrm{Con}(X), d_H)$ is complete if and only if (X, d) is complete.*

Proof   The condition is necessary, since (X, d) is isometric to a closed metric subspace of $(\mathrm{Con}(X), d_H)$. Suppose that (X, d) is complete, and that $(A_n)_{n=1}^\infty$ is a Cauchy sequence in $(\mathrm{Con}(X), d_H)$. By Proposition 14.1.1, it is enough to show that $(A_n)_{n=1}^\infty$ has a convergent subsequence. There exists a subsequence $(B_k)_{k=1}^\infty = (A_{n_k})_{k=1}^\infty$ such that $d_H(B_k, B_{k+1}) < 1/2^k$, for $k \in \mathbf{N}$. We shall show that the sequence $(B_k)_{k=1}^\infty$ converges.

Let $B = \cap_{k=1}^\infty M_{2/2^k}(B_k)$, where $M_\epsilon(A) = \{x \in X : d(x, A) \leq \epsilon\}$. Since the mapping $x \to d(x, A)$ is continuous, $M_\epsilon(A)$ is closed, and so B is closed. Further, $B \subseteq N_{3/2^k}(B_k)$, for $k \in \mathbf{N}$. We show that B is non-empty, and that $B_k \subseteq N_{3/2^k}(B)$, for each $k \in \mathbf{N}$. Suppose that $k \in \mathbf{N}$ and that $x_k \in B_k$. First, for $1 \leq l < k$ there exists $x_l \in B_l$ with $d(x_l, x_k) \leq 2/2^l$. Secondly, an inductive argument shows that for each $l > k$ there exists $x_l \in B_l$ such that $d(x_{l-1}, x_l) \leq 1/2^{l-1}$. If $k \leq l < m$ then

$$d(x_l, x_m) \leq \sum_{j=l+1}^m d(x_{j-1}, x_j) \leq \sum_{j=l+1}^m \frac{1}{2^{j-1}} < \frac{2}{2^l},$$

so that $(x_l)_{l=1}^\infty$ is a Cauchy sequence in (X, d). Let $x = \lim_{l\to\infty} x_l$. Then $d(x_l, x) \leq 2/2^l$ for $l \geq k$, and

$$d(x_l, x) \leq d(x_l, x_k) + d(x_k, x) \leq 1/2^l + 2/2^k \leq 2/2^l$$

for $1 \leq l < k$. Thus $x \in B$, and so B is not empty. Further, $x_k \in N_{3/2^k}(x) \subseteq N_{3/2^k}(B)$; this holds for any $x_k \in B_k$, and so $B_k \subseteq N_{3/2^k}(B)$. Hence $d_H(D, B_k) \leq 3/2^k$, and so $B_k \to B$ as $k \to \infty$. □

Combining this with Theorem 15.6.3, we have the following.

Corollary 15.6.5 *If (X, d) is a metric space, $(\mathrm{Con}(X), d_H)$ is compact if and only if (X, d) is compact.*

Exercises

15.6.1 Suppose that (X, d) is a metric space.

(a) Let $D_n(X) = \cup_{j=1}^n \mathrm{Con}_j(X)$. Show that $D_n(X)$ is closed in $(\mathrm{Con}(X), d_H)$ and give an example to show that $\mathrm{Con}_n(X)$ need not be closed.

(b) Let ρ be a product metric on the product X^n. Show that the mapping $s : (X^n, \rho) \to (\mathrm{Con}(X), d_H)$ defined by

$$s(x_1, \ldots, x_n) = \{x_1, \ldots, x_n\}$$

is continuous, and that $s(X^n) = D_n(X)$. What is $s^{-1}(C_n(X))$?.

(c) Define a relation on X^n by setting $x \sim y$ if there is a permutation $\sigma \in \Sigma_n$ such that $y_j = x_{\sigma(j)}$ for $1 \le j \le n$. Show that this is an equivalence relation. Let $W_n(X)$ be the quotient space of equivalence classes; $W_n(X)$ is the *n-point weighted configuration space* of X. Define a function d_W on $W_n(X) \times W_n(X)$ by setting $d_W(a, b) = \inf\{\rho(x, y) : x \in a, y \in b\}$. Show that this is a metric on $W_n(X)$, and that the quotient mapping $q : (X^n, \rho) \to (W_n(X), d_W)$ is continuous.

(d) Show that there is a unique mapping $\widetilde{s} : W_n(x) \to D_n(X)$ such that $s = \widetilde{s} \circ q$, and show that $\widetilde{s} : (W_n(x), d_W) \to (D_n(X), d_H)$ is continuous.

15.6.2 Suppose that (X, d) is a complete metric space and that $\mathcal{F}$ is a finite set of contraction mappings of (X, d). If $A \in \mathrm{Con}(X)$, let $U(A) = \cup\{T(A) : T \in \mathcal{F}\}$. Show that U is a contraction mapping of $(\mathrm{Con}(X), d_H)$.

Show that there is a unique $A \in \mathrm{Con}(X)$ for which $A = \cup\{T(A) : T \in \mathcal{F}\}$, and that if $B \in \mathrm{Con}(X)$ then $U^n(B) \to A$ as $n \to \infty$. A is called the *attractor* of $\mathcal{F}$.

Let $X = [0, 1]$, Suppose that $0 \le \alpha, \beta < 1$. If $x \in [0, 1]$, let $L_\alpha(x) = \alpha x$ and let $R_\beta(x) = 1 - \beta x$. What is the attractor of $\{L_\alpha, R_\beta\}$ when $\alpha = \beta = 1/3$? What is the attractor when $\alpha = \beta = 0$? What is the attractor when $0 < \alpha + \beta < 1$? What is the attractor when $1 \le \alpha + \beta < 2$?

15.7 Locally compact topological spaces

The Bolzano–Weierstrass and Heine–Borel theorems are useful in proving results in analysis on $\mathbf{R}$, even though $\mathbf{R}$ is not compact. Every point in $\mathbf{R}$ has a compact neighbourhood, and $\mathbf{R}$ is the union of an increasing sequence

$([-n,n])_{n=1}^{\infty}$ of compact sets. We can consider topological spaces for which similar phenomena hold. A topological space space (X,d) is said to be *locally compact* if each point x of X has a base of neighbourhoods consisting of compact sets: if U is open and $x \in U$ then there exist an open set V and a compact set K such that $x \in V \subseteq K \subseteq U$. Thus $\mathbf{R}$ is locally compact. So are $\mathbf{R}^d$ and $\mathbf{C}^d$, with product metrics. More generally, if (X_1,d_1) and (X_2,d_2) are locally compact, then so is $X_1 \times X_2$, with the product topology. A compact Hausdorff topological space is locally compact. Any set X, with the discrete topology, is locally compact.

As usual, we shall concentrate attention on Hausdorff spaces.

Proposition 15.7.1 *A Hausdorff topological space (X,τ) is locally compact if and only if each point has a compact neighbourhood.*

Proof The condition is certainly necessary. Suppose that it is satisfied, that $x \in X$ and that K is a compact neighbourhood of x. Let V be the interior of K, so that $x \in V \subseteq K$. Suppose that U is an open neighbourhood of x in X. Then $U \cap V$ is an open neighbourhood of x in K, for the subspace topology. Since K is compact and Hausdorff, it is normal, and so there exist a subset W of K, open in the subspace topology, and a subset L of K, closed in the subspace topology, such that $x \in W \subseteq L \subseteq U \cap V$. Since K is compact and Hausdorff, L is compact. Since W is open in K there exists an open subset Y of X such that $W = Y \cap K$. But then $W \subseteq Y \cap V \subseteq Y \cap K = W$, so that $W = Y \cap V$ is open in X and $x \in W \subseteq L \subseteq U$; (X,τ) is locally compact. $\qquad\square$

Corollary 15.7.2 *Suppose that Y is a subspace of a locally compact Hausdorff topological space (X,τ).*
  (i) If Y is closed, then Y is locally compact.
  (ii) If Y is open, then Y is locally compact.

Proof (i) If $y \in Y$ and M is a compact neighbourhood of y in X, then $M \cap Y$ is a compact neighbourhood of y in Y.
  (ii) If $y \in Y$ then Y is an open neighbourhood of y in X, and so there exists a compact neighbourhood K of y in X with $K \subseteq Y$. Then $K = K \cap Y$ is a compact neighbourhood of y in Y. $\qquad\square$

Proposition 15.7.3 *If (X,τ) is a Hausdorff topological space and Y is a dense subspace of X which is locally compact in the subspace topology then Y is an open subset of X.*

Proof Suppose that $y \in Y$. There exists an open subset U of Y and a compact subset K of Y such that $y \in U \subseteq K$. Let $W = X \setminus K$. There exists an open subset V of X such that $U = V \cap Y$. We show that $V \cap W$ is empty. Since K is compact and X is Hausdorff, K is closed in X, and so $V \cap W$ is open in X. Suppose that $V \cap W$ is not empty. Since Y is dense in X, there exists $z \in Y \cap (V \cap W) = U \cap W$. But $U \cap W \subseteq K \cap W = \emptyset$. Thus $V \cap W = \emptyset$, so that $V \subseteq K$. Thus $U = V \cap K = V$, and U is open in X. Thus Y is open in X. $\qquad\square$

Theorem 15.7.4 *Suppose that (X, τ) is a topological space. There exists a compact topological space space (X_∞, τ_∞), a point x_∞ of X_∞ (the* point at infinity*) and a homeomorphism j of (X, τ) onto the subspace $X_\infty \setminus \{x_\infty\}$ of X_∞.*

Proof We adjoin a point x_∞ to X, and set $X_\infty = X \cup \{x_\infty\}$. We define a topology τ_∞ on X_∞ by saying that $U \in \tau_\infty$ if

- $U \cap X \in \tau$, and
- if $x_\infty \in U$ then $X \setminus U$ is compact.

Let us verify that this defines a topology. First, $\emptyset$ and X_∞ are in τ_∞. Suppose that $U_1, U_2 \in \tau_\infty$. Then $U_1 \cap U_2 \cap X \in \tau$; if $x_\infty \in U_1 \cap U_2$ then $x_\infty \in U_1$ and $x_\infty \in U_2$, so that $X \setminus U_1$ and $X \setminus U_2$ are compact, and $X \setminus (U_1 \cap U_2) = (X \setminus U_1) \cup (X \setminus U_2)$ is compact. Thus finite intersections of sets in τ_∞ are in τ_∞. Finally suppose that $\mathcal{U} \subseteq \tau_\infty$. Let $W = \cup_{U \in \mathcal{U}} U$. Then $W \cap X = \cup_{U \in \mathcal{U}} (U \cap X) \in \tau$. If $x_\infty \in W$ then there exists $U_0 \in \mathcal{U}$ such that $x_\infty \in U_0$. Then $X \setminus W = (X \setminus U_0) \cap (\cap_{U \in \mathcal{U}} (X \setminus U))$. Since $X \setminus U_0$ is compact and $\cap_{U \in \mathcal{U}} (X \setminus U)$ is closed, $X \setminus W$ is compact. Thus arbitrary unions of sets in τ_∞ are in τ_∞, and τ_∞ is a topology.

Next we show that (X_∞, τ_∞) is compact. Suppose that $\mathcal{U}$ is an open cover of X_∞. There exists $U_0 \in \mathcal{U}$ such that $x_\infty \in U_0$, so that $X \setminus U_0$ is compact. Finitely many elements $U_1, \ldots, U_n$ of $\mathcal{U}$ cover $X \setminus U_0$, and so $U_0, U_1, \ldots U_n$ cover X.

The natural inclusion mapping $j : (X, \tau) \to (X_\infty, \tau_\infty)$ clearly defines a homeomorphism of (X, τ) onto the subspace $X_\infty \setminus \{x_\infty\}$. $\qquad\square$

A *compactification* of a topological space (X, τ) is a compact topological space $(\widetilde{X}, \widetilde{\tau})$, together with a homeomorphism j of (X, τ) onto a dense subspace of $(\widetilde{X}, \widetilde{\tau})$. The space (X_∞, τ_∞) is called the *one-point compactification* of (X, τ).

The one point compactification is only really useful when it is Hausdorff. When does this happen?

Theorem 15.7.5 *The one point compactification (X_∞, τ_∞) of a topological space (X, τ) is Hausdorff if and only if (X, τ) is locally compact and Hausdorff.*

Proof If (X_∞, τ_∞) is Hausdorff, then (X, τ) is Hausdorff. Further, (X_∞, τ_∞) is regular, and so is locally compact. If $x \in X$ then a subset N of X is a neighbourhood of x in X if and only if it is a neighbourhood of x in X_∞, and so (X, τ) is locally compact.

Conversely, suppose that (X, τ) is locally compact and Hausdorff. First, suppose that x and y are distinct points of X. Then, since (X, τ) is Hausdorff, then there are disjoint open sets U and V in τ such that $x \in U$ and $y \in V$. Since $U, V \in \tau_\infty$, they separate x and y in (X_∞, τ_∞). Secondly, suppose that $x \in X$. Since (X, τ) is locally compact, there exist an open set W and a compact set K in (X, τ) such that $x \in W \subseteq K$. Since (X, τ) is Hausdorff, K is closed, and so $X_\infty \setminus K \in \tau_\infty$. Thus W and $X_\infty \setminus K$ separate x and x_∞. $\qquad \square$

What can we say about the compact subsets of a locally compact space?

Proposition 15.7.6 *Suppose that K is a compact subset of a locally compact space (X, τ). Then there exists an open set U and a compact set L such that $K \subseteq U \subseteq L$.*

Proof For each $x \in K$ there exists an open set U_x and a compact set L_x such that $x \in U_x \subseteq L_x$. Then $\{U_x : x \in K\}$ is an open cover of K, which has a finite subcover $\{U_x : x \in F\}$. Take $U = \cup_{x \in F} U_x$ and $L = \cup_{x \in F} L_x$. $\qquad \square$

A topological space (X, τ) is *σ-compact* if there is a sequence $(K_n)_{n=1}^\infty$ of compact subsets of X such that $X = \cup_{n=1}^\infty K_n$. If so, let $J_n = \cup_{j=1}^n K_j$; then $(J_n)_{n=1}^\infty$ is an increasing sequence of compact subsets of X whose union is X.

Corollary 15.7.7 *If (X, τ) is a σ-compact locally compact topological space then $X = \cup_{n=1}^\infty L_n$, where $(L_n)_{n=1}^\infty$ is an increasing sequence of compact sets, with L_n contained in the interior L_{n+1}° of L_{n+1}, for $n \in \mathbf{N}$. If K is a compact subset of X then $K \subseteq L_n$, for some $n \in \mathbf{N}$.*

Proof We construct the sequence $(L_n)_{n=1}^\infty$ inductively. Suppose that $X = \cup_{n=1}^\infty K_n$, where the sets K_n are compact. Set $L_1 = K_1$. Having constructed L_n there is an open set U_{n+1} and a compact subset L_{n+1} such that $L_n \cup K_n \subseteq U_{n+1} \subseteq L_{n+1}$. If K is compact, it is covered by the increasing sequence $(L_n^\circ)_{n=1}^\infty$ of open sets, so that $K \subseteq L_n^\circ \subseteq L_n$, for some n. $\qquad \square$

A sequence $(L_n)_{n=1}^{\infty}$ of compact subsets of a σ-compact locally compact space (X, τ) which satisfies the conclusions of this corollary is called a *fundamental sequence of compact sets*.

What can we say about metrizable locally compact spaces?

Theorem 15.7.8 *Suppose that (X, τ) is a metrizable locally compact topological space. The following are equivalent.*

(i) (X, τ) is separable.

(ii) (X, τ) is σ-compact.

(iii) (X, τ) is second countable.

(iv) The one-point compactification (X_∞, τ_∞) of (X, τ) is metrizable.

(v) (X, τ) is homeomorphic to an open subset of a compact metric space.

Proof   Let d be a metric on X which defines the topology τ.

Suppose that (i) holds. Let $(x_r)_{r=1}^{\infty}$ be a dense sequence of elements of X. For each r let

$$C_r = \{n \in \mathbf{N} : M_{1/n}(x_r) = \{x \in X : d(x, x_r) \le 1/n\} \text{ is compact}\}.$$

We shall show that $X = \cup_{r=1}^{\infty}\left(\cup_{n \in C_r} M_{1/n}(x_r)\right)$, so that (ii) holds. Suppose that $x \in X$. There exists n such that $M_{1/n}(x)$ is compact. There exists x_r such that $d(x, x_r) \le 1/2n$. Then $M_{1/2n}(x_r) \subseteq M_{1/n}(x)$, and so $2n \in C_r$. Since $x \in M_{1/2n}(x_r)$, $X = \cup_{r=1}^{\infty}\left(\cup_{n \in C_r} M_{1/n}(x_r)\right)$.

Suppose that (ii) holds. By Corollary 15.7.7, there exists an increasing sequence of compact sets (K_n) with $K_n \subseteq K_{n+1}^{\circ} \subseteq K_{n+1}$, for which $X = \cup_{n=1}^{\infty} K_n$. Each K_n is second countable, and so therefore is each K_n°. Let $\mathcal{U}_n$ be a countable basis for the topology of K_n°. Suppose that U is a non-empty open subset of X. Then $U \cap K_n^{\circ}$ is the union of countably many subsets in $\mathcal{U}_n$, and so $U = \cup_{n=1}^{\infty}(U \cap K_n^{\circ})$ is the union of countably many sets in $\mathcal{U} = \cup_{n=1}^{\infty}\mathcal{U}_n$; $\mathcal{U}$ is a countable basis for the topology of (X, τ). Thus (iii) holds.

Since a second countable space is separable, (i), (ii) and (iii) are equivalent. Suppose that they hold. Let the sequence $(K_n)_{n=1}^{\infty}$ and $\mathcal{U}$ be as in the previous paragraph. Then $\mathcal{U} \cup \{X_\infty \setminus K_n : n \in \mathbf{N}\}$ is a countable basis for the topology τ_∞. Thus (X_∞, τ_∞) is metrizable, by Urysohn's metrization theorem: (iv) holds.

Since (X, τ) is homeomorphic to an open subset of (X_∞, τ_∞), (iv) implies (v), and (v) implies (i), since a compact metric space is separable (Corollary 15.4.3), and a subspace of a separable metric space is separable (Corollary 13.5.4). $\square$

Exercises

15.7.1 Construct a *two-point compactification* of **R**, and a compactification of $\mathbf{R}^2$ which is homeomorphic to the closed disc $\{(x, y) \in \mathbf{R}^2 : x^2 + y^2 \leq 1\}$.

15.7.2 Give an example of a countable product of locally compact spaces which is not locally compact.

15.7.3 Suppose that (X, d) is a separable metric space. Show that (X, d) is locally compact if and only if there is continuous mapping of $C \setminus \{0\}$ (where C is Cantor's ternary set) onto X.

15.7.4 Let $X = \{x \in \mathcal{H}_2 : \sup_{j \in \mathbf{N}} |x_j/j| < 1\}$, with the subspace metric. Show that X is the union of an increasing sequence of compact sets, but is not locally compact.

15.7.5 Show that the intersection of two locally compact topological subspaces of a topological space is locally compact.

15.7.6 Give an example of two locally compact subsets of **R** whose union is not locally compact.

15.7.7 Give an example of a separable locally compact metric space which is not complete.

15.7.8 Show that if (X, d) is a separable locally compact metric space then there is an equivalent metric ρ on X such that (X, ρ) is complete.

15.7.9 Suppose that (X, d) is a separable locally compact metric space. Use the metric d to construct a metric d_∞ on the one-point compactification X_∞ which defines the topology τ_∞, avoiding the use of Urysohn's metrization theorem.

15.8 Local uniform convergence

Suppose that (X, τ) is a σ-compact locally compact topological space, and that $(E, \|.\|)$ is a Banach space. Let $C(X, E)$ denote the space of continuous functions on X taking values in E. (An important example is the case where X is an open subset of the complex plane **C**, and $E = \mathbf{C}$, and it is good to keep this example in mind.) Functions in $C(X, E)$ need not be bounded, and uniform convergence is too strong for many purposes. If $(f_k)_{k=1}^\infty$ is a sequence in $C(X, E)$ and $f \in C(X, E)$ then we say that f_k converges to f *uniformly on compact sets*, or *locally uniformly*, as $k \to \infty$ if $f_k(x) \to f(x)$ uniformly on K, for each compact subset K of X. Let us show that this convergence can be defined by a suitable metric topology on $C(X, E)$. Let $(L_n)_{n=1}^\infty$ be a fundamental sequence of compact subsets of X.

If K is a compact subset of X, let $\pi_K : C(X, E) \to C(K, E)$ be the restriction mapping (so that $\pi_K(f) = f_{|K}$), and let $\pi_n = \pi_{L_n}$. If $m \le n$ let $\pi_{m,n}$ be the restriction mapping from $C(L_n, E)$ to $C(L_m, E)$. If $f \in C(X, E)$, let $\pi(f) = (\pi_n(f))_{n=1}^\infty$; π is an injective linear mapping of $C(X, E)$ into the product space $\prod_{n=1}^\infty C(L_n, E)$. We give each space $C(L_n, E)$ its uniform norm, and we give the product $\prod_{n=1}^\infty C(L_n, E)$ the product topology τ, defined by a suitable complete product metric ρ.

Proposition 15.8.1 *With the notation above, $\pi(C(X, E))$ is a closed linear subspace of $\prod_{n=1}^\infty C(L_n, E)$.*

Proof   The mapping π is linear, and so $\pi(C(X, E))$ is a linear subspace of $\prod_{n=1}^\infty C(L_n, E)$. If $m \le p$ then

$$G_{m,p} = \{(g, h) \in C(L_m, E) \times C(L_p, E) : \pi_{m,p}(h) = g\}$$

is the graph of the continuous mapping $\pi_{m,p}$, and is therefore a closed linear subspace of $C(L_m, E) \times C(L_p, E)$. Consequently

$$H_{m,p} = \{(g_n)_{n=1}^\infty \in \prod_{n=1}^\infty C(L_n, E) : \pi_{m,p}(g_p) = g_m\}$$

is a closed linear subspace of $\prod_{n=1}^\infty C(L_n, E)$. Since

$$\pi(C(X, E)) = \cap_{m=1}^\infty \left(\cap_{p \ge m} H_{m,p}\right),$$

$\pi(C(X, E))$ is a closed subset of $\prod_{n=1}^\infty C(L_n, E)$. $\square$

Let us set $d(f, g) = \rho(\pi(f), \pi(g))$. Then d is a complete metric on $C(X, E)$; for example, as in Section 13.3, we can take

$$d(f, g) = \sum_{n=1}^\infty \frac{\sup_{x \in L_n} \|f(x) - g(x)\|}{2^n (1 + \sup_{x \in L_n} \|f(x) - g(x)\|)}$$

$$\text{or } d(f, g) = \sum_{n=1}^\infty \min(2^{-n}, \sup_{x \in L_n} \|f(x) - g(x)\|).$$

We call the corresponding topology on $C(K, E)$ the *topology of local uniform convergence*.

Proposition 15.8.2 *Suppose that (X, τ) is a σ-compact locally compact topological space, and that $(E, \|.\|)$ is a Banach space. Let $(L_n)_{n=1}^\infty$ be a fundamental sequence of compact subsets of X. Suppose that $(f_k)_{k=1}^\infty$ is a sequence in $C(X, E)$, and that $f \in C(X, E)$. The following are equivalent.*

(i) $f_k(x) \to f(x)$ *uniformly on compact sets, as* $k \to \infty$.

(ii) $f_k(x) \to f(x)$ *uniformly on each* L_n, *as* $k \to \infty$.

(iii) For each $x \in X$ *there exists a compact neighbourhood* N_x *of* x *such that* $f_k(x) \to f(x)$ *uniformly on* N_x.

(iv) $f_k(x) \to f(x)$ *in the topology of local uniform convergence.*

Proof Certainly (i) implies (ii) and (ii) implies (iii), since if $x \in X$ then there exists $n \in \mathbf{N}$ such that $x \in L_n^\circ$, so that L_n is a compact neighbourhood of x. If (iii) holds and if K is a compact subset of X, then for each $x \in K$ there exists a compact neighbourhood N_x of x such that $f_k(x) \to f(x)$ uniformly on N_x. The collection of sets $\{N_x^\circ : x \in K\}$ is then an open cover of K, and so there is a finite subcover $\{N_x^\circ : x \in F\}$. Since $f_k \to f$ uniformly on $\cup_{x \in F} N_x$, it follows that $f_k \to f$ uniformly on K. Thus (iii) implies (i). Finally, the properties of the product topology imply that (ii) and (iv) are equivalent. $\qquad\square$

Theorem 15.8.3 (The general principle of local uniform convergence) *Suppose that* E *is a Banach space and that* (X, τ) *is a* σ-*compact locally compact topological space. With the terminology as above, if* $(f_k)_{k=1}^\infty$ *is a sequence in* $C(X, E)$ *then* f_k *converges locally uniformly to a function* f *in* $C(X, E)$ *if and only if* $(\pi_n(f_k))_{k=1}^\infty$ *is a Cauchy sequence in* $(C(L_n, E), \|.\|_\infty)$, *for each* $n \in \mathbf{N}$.

Proof This follows immediately from the facts that convergent sequences are Cauchy sequences, and that the spaces $(C(L_n, E), \|.\|_\infty)$ are complete. $\qquad\square$

We can characterize compact subsets of $C(X, E)$. A subset A of $C(X, E)$ is *locally uniformly bounded* if $\pi_K(A)$ is bounded in $(C(K, E), \|.\|_\infty)$, for each compact subset K of X.

Theorem 15.8.4 (The local Arzelà–Ascoli theorem) *Suppose that* (X, τ) *is a* σ-*compact locally compact topological space and that* d *is a complete metric on* $C(X)$ *defining the topology of local uniform convergence. A closed subset* A *of* $C(X)$ *is compact if and only if it is locally uniformly bounded and is equicontinuous at each point of* X.

Proof Of course we use the Arzelà–Ascoli theorem. Let $(L_n)_{n=1}^\infty$ be a fundamental sequence of compact subsets of X, and for each n let $\pi_n : C(X) \to C(L_n)$ be the restriction mapping.

If A is compact, then $\pi_n(A)$ is compact, and is therefore uniformly bounded on L_n, for each $n \in \mathbf{N}$; thus A is locally uniformly bounded. If

$x \in X$ and if $x \in L_n^\circ$ then $\pi_n(A)$ is equicontinuous in $C(L_n)$. Since x is an interior point of L_n, it follows that A is equicontinuous at x.

Conversely, suppose that the conditions are satisfied. We show that A is sequentially compact. Suppose that $(f_k)_{k=1}^\infty$ is a sequence in A. The conditions imply that $\pi_n(A)$ is totally bounded in $C(L_n)$, for each $n \in \mathbf{N}$. A diagonal argument, then show that there exists a subsequence $(f_{k_j})_{j=1}^\infty$ such that $(\pi_n(f_{k_j}))_{j=1}^\infty$ is a Cauchy sequence in $(C(L_n), \|.\|_\infty)$, for each $n \in \mathbf{N}$. Since $(C(L_n), \|.\|_\infty)$ is complete, it therefore follows that, for each $n \in \mathbf{N}$, $(\pi_n(f_{k_j}))_{j=1}^\infty$ converges in norm to an element $f^{(n)}$, say, of $C(L_n)$. Since $f^{(n)}(x) = f^{(m)}(x)$ for $m < n$ and $x \in L_m$, there exists a function f on X such that $f(x) = f^{(n)}(x)$ for $x \in L_n$, for each $n \in \mathbf{N}$. If $x \in X$ then $x \in L_n^\circ$ for some n, from which it follows that $f \in C(X)$. Thus $(f_{k_j})_{j=1}^\infty$ converges locally uniformly to f. Since A is closed, $f \in A$. $\qquad\square$

Exercise

15.8.1 Suppose that (X, τ) is a σ-compact locally compact topological space, and that $(E, \|.\|)$ is a Banach space. Let $(L_n)_{n=1}^\infty$ be a fundamental sequence of compact subsets of X. Suppose that $A \subseteq C(X, E)$. Show that the following are equivalent.

  (i) A is locally uniformly bounded.
  (ii) $\pi_n(A)$ is bounded in $C(L_n, E), \|.\|_\infty)$, for each $n \in \mathbf{N}$.
  (iii) For each $x \in X$ there exists a compact neighbourhood N_x of x such that $\pi_{N_x}(A)$ is bounded in $C(N_x, E), \|.\|_\infty)$.

15.9 Finite-dimensional normed spaces

Suppose that $(E, \|.\|)$ is a normed space. When is E locally compact? Since the mappings $x \to x + a : E \to E$ and $x \to \lambda x : E \to E$ $(\lambda \neq 0)$ are homeomorphisms, $(E, \|.\|)$ is locally compact if and only if the closed unit ball $B_E = \{x \in E : \|x\| \leq 1\}$ is compact.

The space $\mathbf{R}^d$, with its usual Euclidean norm, is locally compact. We also have the following easy result.

Proposition 15.9.1 *Any linear operator T from $\mathbf{R}^d$, with its usual Euclidean norm, into a normed space $(E, \|.\|)$ is continuous.*

Proof If $x = (x_1, \ldots, x_d) = x_1 e_1 + \cdots x_d e_d \in \mathbf{R}^d$ then, using the triangle inequality and the Cauchy–Schwarz inequality,

$$\|T(x)\| = \|x_1 T(e_1) + \cdots x_d T(e_d)\| \leq |x_1| . \|T(e_1)\| + \cdots + |x_d| . \|T(e_d)\|$$

$$\leq \left(\sum_{j=1}^d \|T(e_j)\|^2 \right)^{1/2} \left(\sum_{j=1}^d |x_j|^2 \right)^{1/2} = K \|x\|_2 ,$$

where $K = (\sum_{j=1}^d \|T(e_j)\|^2)^{1/2}$. $\qquad\square$

Theorem 15.9.2 *Any two norms on a finite-dimensional normed space E are equivalent.*

Proof Since a finite-dimensional complex normed space of dimension d can be considered as a real normed space of dimension $2d$, we can suppose without loss of generality that E is a real vector space.

Let $(f_1, \ldots, f_d)$ be a basis for E. Define a bijective linear mapping $T : \mathbf{R}^d \to E$ by setting $T(x) = \sum_{j=1}^d x_j f_j$, and define $\|T(x)\|_2 = \|x\|$, where $\|.\|$ is the Euclidean norm on $\mathbf{R}^d$. Then $\|.\|_2$ is a norm on E, and T is an isometry of $\mathbf{R}^d$ onto $(E, \|.\|_2)$. Since $\mathbf{R}^d$ is locally compact, and complete, so also is $(E, \|.\|_2)$. It is sufficient to show that any norm $\|.\|_1$ on E is equivalent to $\|.\|_2$. By Proposition 15.9.1, the identity mapping $(E, \|.\|_2) \to (E, \|.\|_1)$ is continuous. Thus the function $f : (E, \|.\|_2) \to \mathbf{R}$ defined by $f(x) = \|x\|_1$ is continuous on $(E, \|.\|_2)$. The set $S = \{x \in E : \|x\|_2 = 1\}$ is compact, and so f attains its minimum on S at a point s_0 of S. Since $s_0 \neq 0$, $m = f(s_0) > 0$. Thus if $\|x\|_2 = 1$ then $\|x\|_1 \geq m$. We now use a standard homogeneity argument. If $x \in E$ and $x \neq 0$, then $x/\|x\|_2 \in S$, and so

$$\left\| \frac{x}{\|x\|_2} \right\|_1 = \frac{\|x\|_1}{\|x\|_2} \geq m.$$

Thus $\|x\|_2 \leq \|x\|_1 / m$ for all $x \in E$, so that the identity mapping $(E, \|.\|_1) \to (E, \|.\|_2)$ is continuous. $\qquad\square$

Corollary 15.9.3 *(i) Any finite-dimensional normed space is locally compact, and complete.*

(ii) Any linear operator from a finite-dimensional normed space into a normed space is continuous.

(iii) Any finite-dimensional subspace of a normed space $(E, \|.\|)$ is closed in E.

Theorem 15.9.4 *Any locally compact normed space* $(E, \|.\|)$ *is finite-dimensional.*

Proof Since B_E is compact, it is totally bounded. There therefore exists a finite subset F of B_E such that

$$B_E \subseteq F + B_E/2 = \{f + x : f \in F, \|x\| \le 1/2\}.$$

Let G be the linear span of F. G is a finite-dimensional linear subspace of E; we shall show that $G = E$. Now $B_E \subseteq G + B_E/2$, and so $B_E \subseteq G + (G + B_E/2)/2 = G + B_E/4$. Iterating this argument, we see that $B_E \subseteq G + B_E/2^n$, for $n \in \mathbf{N}$. Thus if $x \in B_E$ and $n \in \mathbf{N}$, there exists $g_n \in G$ such that $\|x - g_n\| \le 1/2^n$. Hence $g_n \to x$ as $n \to \infty$, and so, since G is closed in E, $x \in \overline{G} = G$. Consequently, $B_E \subseteq G$, and so $E = \mathrm{span}\,(B_E) \subseteq G$. $\qquad\square$

Exercises

15.9.1 Suppose that F is a proper closed linear subspace of a normed space $(E, \|.\|)$. Suppose that $y \in E \setminus F$ and that $0 < \epsilon < 1$. There exists $f \in F$ such that $\|y - f\| < d(y, F)/(1 - \epsilon)$. Let $x = (y - f)/\|y - f\|$. Show that $\|x\| = 1$ and that $d(x, F) > 1 - \epsilon$.

15.9.2 Suppose that F is a finite-dimensional linear subspace of a normed space $(E, \|.\|)$ and that $y \in E \setminus F$. Show that there exists $f \in F$ such that $\|y - f\| = d(y, F)$. Show that there exists $x \in E$ with $\|x\| = d(x, F) = 1$.

15.9.3 Suppose that $(E, \|.\|)$ is an infinite-dimensional normed space. Show that there exists a sequence $(x_n)_{n=1}^{\infty}$ of unit vectors in E with $\|x_j - x_k\| \ge 1$ for $j, k \in \mathbf{N}$ with $j \ne k$. Give another proof that $(E, \|.\|)$ is not locally compact.

15.9.4 If $x \in l_{\infty}$, let $\phi(x) = \sum_{n=1}^{\infty} x_n/2^n$, and let $H = \{x \in l_{\infty} : \phi(x) = 1\}$. Let $H_0 = H \cap c_0$.
   (a) Show that $\phi : l_{\infty} \to \mathbf{R}$ is a continuous linear mapping.
   (b) Show that H is closed in l_{∞}.
   (c) Show that there exists a unique $h \in H$ such that $d_{\infty}(0, h) = d_{\infty}(0, H)$.
   (d) Show that H_0 is closed in c_0.
   (e) Show that there is no element h_0 of H_0 such that $d_{\infty}(0, h_0) = d_{\infty}(0, H_0)$.

15.9.5 Suppose that F is a closed linear subspace of a normed space $(E, \|.\|)$ and that $y \in E \setminus F$. Let $G = \mathrm{span}\,(F, y)$. Suppose that $z \in \overline{G}$ and that $f_n + \lambda_n b \to z$ as $n \to \infty$. Suppose that $(\lambda_n)_{n=1}^{\infty}$ does not converge. Show that there is $\epsilon > 0$ and a subsequence $(\lambda_{n_k})_{k=1}^{\infty}$ such that

$|\lambda_{n_{k+1}} - \lambda_{n_k}| \geq \epsilon$. Show that $(a_{n_{k+1}} - a_{n_k})/(\lambda_{n_k} - \lambda_{n_{k+1}}) \to y$ as $n \to \infty$, giving a contradiction. Show that if $\lambda_n \to \lambda$ as $n \to \infty$ then $z \in G$. Deduce that G is closed. Deduce that if D is a finite-dimensional subspace of E then $F + D$ is closed.

16

Connectedness

16.1 Connectedness

In Section 5.3 of Volume I we introduced the notion of connectedness of subsets of the real line, and showed that a non-empty subset of $\mathbf{R}$ is connected if and only if it is an interval. The notion extends easily to topological spaces. A topological space *splits* if $X = F_1 \cup F_2$, where F_1 and F_2 are disjoint non-empty closed subsets of (X, τ). The decomposition $X = F_1 \cup F_2$ is a *splitting* of X. If X does not split, it is *connected*. A subset A of (X, τ) is connected if it is connected as a topological subspace of (X, τ). If $X = F_1 \cup F_2$ is a splitting, then $F_1 = C(F_2)$ and $F_2 = C(F_1)$ are open sets, and so X is the disjoint union of two non-empty sets which are both open and closed; conversely if U is a non-empty proper open and closed subset of X, $X = U \cup (X \setminus U)$ is a splitting of (X, τ). Thus (X, τ) is connected if and only if X and $\emptyset$ are the only subsets of X which are both open and closed.

Proposition 16.1.1 *Suppose that A is a connected subset of a topological space (X, τ) and that $X = F_1 \cup F_2$ is a splitting of X. Then either $A \subseteq F_1$ or $A \subseteq F_2$.*

Proof The sets $A \cap F_1$ and $A \cap F_2$ are disjoint open and closed subsets of A whose union is A; one of them must be empty and the other one be equal to A. $\square$

Proposition 16.1.2 *Suppose that (X, τ) and (Y, σ) are topological spaces and that $f : (X, \tau) \to (Y, \sigma)$ is continuous. If (X, τ) is connected, then so is the topological subspace $f(X)$ of Y.*

Proof If $f(X) = F_1 \cup F_2$ is a splitting of $f(X)$ then $X = f^{-1}(F_1) \cup f^{-1}(F_2)$ is a splitting of X. $\square$

Connectedness can be characterized in terms of continuous mappings. Let D_2 be the two point set $\{0, 1\}$, with the discrete topology. Then $D_2 = \{0\} \cup \{1\}$ is a splitting of D_2.

Corollary 16.1.3 *A topological space (X, τ) is connected if and only if there is no continuous surjective mapping of (X, τ) onto D_2.*

Proof   The proposition shows that the condition is necessary. On the other hand, if $X = F_0 \cup F_1$ is a splitting of X, and if $x \in F_i$, let $f(x) = i$. Then f is a continuous surjection of X onto D_2.   □

We also have an intermediate value theorem.

Corollary 16.1.4 (The intermediate value theorem) *Suppose that f : $(X, \tau) \to \mathbf{R}$ is a continuous real-valued function on a connected topological space (X, τ), and that $f(x) < v < f(x')$ (where $x, x' \in X$). Then there exists $y \in X$ such that $f(y) = v$.*

Proof $f(X)$ is a connected subset of $\mathbf{R}$, and so it is an interval. Since $f(x), f(x') \in f(X)$ and $f(x) < v < f(x')$, it follows that $v \in f(X)$.   □

Let us consider the connected subsets of a topological space (X, τ)

Proposition 16.1.5 *Suppose that $\mathcal{A}$ is a set of connected subsets of a topological space (X, τ) and that $\cap_{A \in \mathcal{A}} A \neq \emptyset$. Then $B = \cup_{A \in \mathcal{A}} A$ is connected.*

Proof   Let c be an element of $\cap_{A \in \mathcal{A}} A$. Suppose that F is an open and closed subset of B. We must show that either $F = B$ or that $F = \emptyset$. Suppose first that $c \in F$. If $A \in \mathcal{A}$, then $F \cap A$ is a non-empty open and closed subset of A. Since A is connected, $A = F \cap A \subseteq F$. Since this holds for all $A \in \mathcal{A}$, $B \subseteq F$, so that $F = B$. Secondly, suppose that $c \notin F$. If $A \in \mathcal{A}$, then $F \cap A$ is a open and closed subset of A which is not equal to A. Since A is connected, $F \cap A = \emptyset$. Since this holds for all $A \in \mathcal{A}$, $F \cap B = \emptyset$, so that $F = \emptyset$.   □

As an example, consider the open unit ball $U_E = \{x \in E : \|x\| < 1\}$ of a normed space. If $x \in U_E$, let $f_x : [0, 1] \to U_E$ be defined by $f(t) = tx$. Then f_x is continuous, so that $f_x([0, 1])$ is connected. Since $0 \in f_x([0, 1])$ for all $x \in U_E$, the set $U_E = \cup\{f_x([0, 1]) : x \in U_E\}$ is connected. It follows, by translation and scaling, that $N_\epsilon(x)$ is connected, for each $x \in E$ and $\epsilon > 0$.

Proposition 16.1.6 *Suppose that A is a dense connected subset of a topological space (X, τ). Then (X, τ) is connected.*

Proof   Suppose that F is a non-empty open and closed subset of X. Since F is open and A is dense in X, $F \cap A \neq \emptyset$. But $F \cap A$ is open and closed in A, and so $F \cap A = A$, since A is connected. Thus $A \subseteq F$. Since F is closed in X, $X = \overline{A} \subseteq F$, and so $F = X$. □

Corollary 16.1.7   *Suppose that A is a connected subset of a topological space (X, τ) and that $A \subset B \subset \overline{A}$. Then B is connected.*

Proof A is a dense subset of the subspace B of (X, τ). □

Theorem 16.1.8   *If $(X_\alpha, \tau_\alpha)_{\alpha \in A}$ is a family of connected topological spaces, then the topological product $X = \prod_{\alpha \in A}(X_\alpha, \tau_\alpha)$ is connected.*

Proof   Suppose that G is a non-empty open and closed subset of X, and that $x \in G$. We shall show that if $y \in X$ then $y \in G$, so that $G = X$, and X is connected. Suppose that $\beta \in A$. The cross-section $C_{x,\beta} = \{z \in X : z_\alpha = x_\alpha$ for $\alpha \neq \beta\}$ is homeomorphic to (X_β, τ_β), and is therefore connected. $G \cap C_{x,\beta}$ is an open and closed subset of $C_{x,\beta}$ containing x, and so $G \cap C_{x,\beta} = C_{x,\beta}$. In particular, let $w^{(\beta)}(\beta) = y_\beta$ and $w^{(\beta)}(\alpha) = x_\alpha$ for $\alpha \neq \beta$; then $w^{(\beta)} \in G$. Iterating this argument, if F is a finite subset of A, let $w^F(\beta) = y_\beta$ for $\beta \in F$ and $w^F(\alpha) = x_\alpha$ for $\alpha \notin F$. Then $w^F \in G$. Now suppose that N is a neighbourhood of y. Then there exists a finite subset F of N such that $N \supseteq \{z \in X : z_\beta = y_\beta$ for $\beta \in F\}$. In particular, $w^F \in N$, so that $N \cap G \neq \emptyset$. Thus $y \in \overline{G}$. Since G is closed, $y \in G$. □

We now consider the collection of connected subsets of a topological space (X, τ). We define a relation on X by setting $a \sim b$ if there is a connected subset A of (X, τ) which contains a and b. This is clearly symmetric and reflexive (take $A = \{a\}$), and is also transitive, by Proposition 16.1.5, and so it is an equivalence relation on X. The equivalence classes are called the *connected components* of (X, τ).

Proposition 16.1.9   *Suppose that E is a connected component of a topological space (X, τ), and that $x \in E$. Let*

$$G_x = \cup \{F \subseteq X : x \in F \text{ and } F \text{ is connected}\}.$$

Then $E = G_x$.

Proof   If $y \in E$ then there exists a connected subset F of X such that $\{x, y\} \subseteq F$, and so $y \in G_x$; thus $E \subseteq G_x$. On the other hand, G_x is connected, by Proposition 16.1.5, and so $G_x \subseteq E$. □

Corollary 16.1.10   *A connected component E of a topological space (X, τ) is closed and connected, and is a maximal connected subset of (X, τ).*

Proof E is connected, since G_x is connected. Since $\overline{E}$ is connected, $\overline{E} \subseteq G_x = E$, and so E is closed. If F is a connected subset containing E then $F \subseteq G_x$, and so $F = E$; E is a maximal connected subset of (X, τ). $\square$

Corollary 16.1.11 *If $X = G \cup H$ is a splitting of X, then either $E \subseteq G$ or $E \subseteq H$.*

Example 16.1.12 Two distinct connected components of a Hausdorff topological space which cannot be separated by a splitting.

We construct an example in $\mathbf{R}^2$. For $n \in \mathbf{N}$ let C_n be the circle $\{(x, y) : x^2 + y^2 = n/(n+1)\}$, let $W = \cup_{n=1}^{\infty} C_n$ and let

$$X = \{(1, 0)\} \cup \{(-1, 0)\} \cup W.$$

Then the circles C_n are connected components in X, and so therefore are the singleton sets $\{(1, 0)\}$ and $\{(-1, 0)\}$. Suppose that G is an open and closed subset of W which contains $(1, 0)$. Since G is open, there exists n_0 such that $(n/(n+1), 0)) \in G$ for $n \geq n_0$. Since each C_n is connected, $C_n \subseteq G$ for $n \geq n_0$, and so $(-n/(n+1), 0) \in G$ for $n \geq n_0$. Since G is closed, $(-1, 0) \in G$. Thus if $F_1 \cup F_2$ is a splitting of X, the two connected components $\{(1, 0)\}$ and $\{(-1, 0)\}$ must either both be in F_1 or both be in F_2.

Things work better for compact Hausdorff spaces. We need a preliminary lemma.

Lemma 16.1.13 *Suppose that $\mathcal{G}$ is a set of open and closed subsets of a compact Hausdorff space (X, τ), and that $J = \cap_{G \in \mathcal{G}} G$. If $J_1 \cup J_2$ is a splitting of J, then there is a splitting $F_1 \cup F_2$ of X, with $J_1 \subseteq F_1$ and $J_2 \subseteq F_2$.*

Proof The sets J_1 and J_2 are disjoint closed subsets of X, and (X, τ) is normal, and so there exist disjoint open subsets U_1 and U_2 of X with $J_1 \subseteq U_1$ and $J_2 \subseteq U_2$. The open sets U_1, U_2 and $\{X \setminus G; G \in \mathcal{G}\}$ cover X, and so there is a finite subcover $U_1, U_2, X \setminus G_1, \ldots, X \setminus G_n$. Let $F_1 = U_1 \cap (\cap_{j=1}^{n} G_j)$ and $F_2 = U_2 \cup (\cup_{j=1}^{n} (X \setminus G_j))$. Then $F_1 \cup F_2 = X$, F_1 and F_2 are open and disjoint, $J_1 \subseteq F_1$ and $J_2 \subseteq F_2$. $\square$

Theorem 16.1.14 *If C and D are distinct connected components of a compact Hausdorff space (X, τ), there exists a splitting $G \cup H$ with $C \subseteq G$ and $D \subseteq H$.*

Proof Let $\mathcal{G} = \{G : G \text{ is open and closed, and } C \subseteq G\}$, and let $J = \cap_{G \in \mathcal{G}} G$. First we show that J is connected. Suppose not, and suppose that $J = J_1 \cup J_2$ is a splitting of J. By Lemma 16.1.13, there exists a splitting $X = F_1 \cup F_2$ of X with $J_1 \subseteq F_1$ and $J_2 \subseteq F_2$. But C is connected, and so $C \subseteq F_1$

or $C \subseteq F_2$. Suppose, without loss of generality, that $C \subseteq F_1$. Then $F_1 \in \mathcal{G}$, so that $J \subseteq F_1$. Thus $J_2 \subseteq F_1 \cap F_2 = \emptyset$, giving a contradiction.

Thus J is connected. But $C \subseteq J$ and C is a maximal connected subset of X, and so $C = J$. Suppose that $d \in D$. Then there exists $G \in \mathcal{G}$ such that $d \notin G$. Since $H = X \setminus G$ is open and closed and D is connected, $D \subseteq H$. Thus the splitting $X = G \cup H$ has the required properties. $\square$

When are connected components open? A topological space is *locally connected* if for each $a \in X$ and each $N \in \mathcal{N}_a$ there exists a connected $M \in \mathcal{N}_a$ with $M \subseteq N$; that is, each point of X has a base of neighbourhoods consisting of connected sets. For example, a normed space is locally connected, since if $x \in E$ then the sets $\{N_\epsilon(x) : \epsilon > 0\}$ form a base of neighbourhoods of x consisting of connected sets. An open subset of a locally connected space is clearly locally connected.

Proposition 16.1.15   *If (X, τ) is a locally connected topological space then the connected components of X are open and closed.*

Proof   Suppose that E is a connected component of X. E is closed, by Corollary 16.1.10. If $a \in E$, there exists a connected neighbourhood N of a. Since E is the maximal connected subset of X containing a, $N \subseteq E$. Thus $a \in E^\circ$. Since this holds for all $a \in E$, E is open. $\square$

Thus the connected components form a partition of X into connected open and closed sets.

Corollary 16.1.16   *If U is an open subset of a normed space $(E, \|.\|)$, then the connected components of U are open subsets of E.*

Corollary 16.1.17   *If (X, τ) is a compact locally connected topological space then there are only finitely many connected components.*

Proof   For the connected components of X form an open cover of X. $\square$

Corollary 16.1.18   *If (X, τ) is a separable locally connected topological space then there are only countably many connected components.*

Proof   Let D be a countable dense subset of X. If $d \in D$ let C_d be the connected component to which d belongs. If C is a connected component, then $C \cap D$ is not empty, and so the mapping $d \to C_d$ is a surjection from the countable set D onto the set of connected components. $\square$

In particular, and this will be very important later, an open subset of $\mathbf{C}$ is the countable union of disjoint open connected sets.

Proposition 16.1.19 *If* $(X, \tau) = \prod_{i=1}^{n}(X_n, \tau_n)$ *is a finite product of locally connected topological spaces then* (X, τ) *is locally connected.*

Proof   If $x \in X$ then the sets

$$\{N = N_1 \times \cdots \times N_n : N_j \text{ is a connected neighbourhood of } x_j\}$$

form a base of neighbourhoods of x consisting of connected sets.   □

On the other hand, an infinite product of locally connected topological spaces need not be locally connected. If $X_n = \{0, 1\}$, with the discrete topology, for all $n \in \mathbf{N}$, then X_n is locally connected, while the connected components of $(X, \tau) = \prod_{n=1}^{\infty}(X_n, \tau_n)$ are the singleton sets. But we have the following.

Proposition 16.1.20 *If* $(X, \tau) = \prod_{\alpha \in A}(X_\alpha, \tau_\alpha)$ *is a product of connected locally connected topological spaces then* (X, τ) *is locally connected.*

Proof   This is left as an exercise for the reader.   □

Exercises

16.1.1 A point x in a connected topological space (X, τ) is a *splitting point* of X if the topological subspace $X \setminus \{x\}$ is not connected. Determine the splitting points of the spaces $(0, 1)$, $(0, 1]$, $[0, 1]$ and $[0, 1] \times [0, 1]$. Show that no two of them are homeomorphic.

16.1.2 Show that there is no continuous injective map from $\mathbf{R}^2$ into $\mathbf{R}$.

16.1.3 Suppose that A is a closed subset of $[0, 1] \times [0, 1]$ for which each cross-section $A \cap (\{x\} \times [0, 1])$ is a non-empty interval. Show that there exists $0 \leq x \leq 1$ such that $(x, x) \in A$.

16.1.4 Suppose that $\mathcal{A}$ is a set of non-empty subsets of a set S. $\mathcal{A}$ is *linked* if whenever $A, B \in \mathcal{A}$ then there exists a finite sequence $(A_0, \ldots A_n)$ such that $A_0 = A$, $A_n = B$ and $A_{j-1} \cap A_j \neq \emptyset$ for $1 \leq j \leq n$. Show that if $\mathcal{A}$ is a linked set of connected subsets of a topological space (X, τ) then $\cup_{A \in \mathcal{A}} A$ is connected.

16.1.5 Suppose that (X, d) is a compact metric space for which $\overline{N_\epsilon(x)} = M_\epsilon(x)$ for each $x \in X$ and $\epsilon > 0$. Show that the sets $N_\epsilon(x)$ and $M_\epsilon(x)$ are connected, for each $x \in X$ and $\epsilon > 0$. In particular, (X, d) is connected and locally connected.

16.1.6 A metric space (X, d) is *well-linked* if whenever $a, a' \in X$ and $\epsilon > 0$ there exists a finite sequence $(a_j)_{j=0}^{n}$ in X with $a_0 = a, a_n = a'$ and $d(a_{j-1}, a_j) < \epsilon$ for $1 \leq j \leq n$.
    (a) Show that a connected metric space is well-linked.

(b) Give an example of a well-linked metric space which is not connected.

(c) Show that a compact well-linked metric space is connected.

16.1.7 Show that a countable Hausdorff topological space with more than one point is not connected.

16.1.8 Prove Proposition 16.1.20.

16.1.9 Suppose that the compact metric space (X, d) has the property that if $x, y \in X$ then there exists $z \in X$ with $d(x, z) = d(y, z) = d(x, y)/2$. Show that (X, d) is connected.

16.1.10 Suppose that $(C_n)_{n=1}^\infty$ is a decreasing sequence of closed connected subsets of a compact topological space. Show that $\cap_{n=1}^\infty C_n$ is connected.

16.1.11 Give an example in $\mathbf{R}^2$ of a decreasing sequence $(C_n)_{n=1}^\infty$ of closed connected sets whose intersection is not connected.

16.2 Paths and tracks

Suppose that (X, τ) is a Hausdorff topological space. Let us consider the set $C_X([a, b])$ of continuous mappings from the closed interval $[a, b]$ into (X, τ). An element f of $C_X([a, b])$ is called a *path* in X. $f(a)$ is the *initial point* of the path, and $f(b)$ is its *final point*, and f is a path from a to b. The image $f([a, b])$ is called the *track* from $f(a)$ to $f(b)$, and is denoted by $[f]$. It is a compact connected subset of (X, d). It can be helpful, if not mathematical, to think of the interval $[a, b]$ as a time interval; we start at time a at $f(a)$; $f(t)$ denotes the point of the track that we have reached by time t, and we reach $f(b)$ at time b. We may retrace our footsteps, or cross the path at a point that we have reached before, and, as we shall see, many other strange things can happen. A path is *closed* if $f(a) = f(b)$; we return to our starting point.

Before going any further, let us give a word of warning: there is considerable variation in terminology. Different authors use words such as 'path', 'track', 'curve' and 'arc' with a variety of different meanings.

As a simple example, if a and b are elements in a normed space $(E, \|.\|)$ then the *linear path* $\sigma(a, b) : [0, 1] \to E$ from a to b is defined as $\sigma(a, b)(t) = (1 - t)a + tb$, for $t \in [0, 1]$; its track is denoted by $[a, b]$, and is called the *line segment* from a to b.

We can *juxtapose* two paths to obtain a new path: if $f : [a, b] \to X$ and $g : [c, d] \to X$ are paths, and $f(b) = g(c)$ we define $f \vee g$ to be the path from $[a, b + (d - c)]$ into X defined by $f \vee g(x) = f(x)$ for $x \in [a, b]$ and

$f \vee g(x) = g(x + (c - b))$ for $x \in [b, b + (d - c)]$. Thus $f \vee g$ has initial point $f(a)$ and final point $g(d)$, and $[f \vee g] = [f] \cup [g]$.

The juxtaposition of finitely many linear paths is called a *piecewise-linear* or *polygonal* path. If $\gamma = \sigma(v_0, v_1) \vee \ldots \vee \sigma(v_{k-1}, v_k)$ is a polygonal path, then the points $v_0, v_1, \ldots, v_k$ are called the *vertices* of γ, and the line segments $[v_{j-1}, v_j]$ are called the *edges* of γ. A polygonal path γ in $\mathbf{R}^d$ is called a *rectilinear path* if its edges are parallel to the axes: that is, if $1 \leq j \leq k$ then all but one of the coordinates of v_{j-1} and v_j are the same. A polygonal path in $\mathbf{R}^d$ is called a *dyadic* path if each coordinate of each vertex is a dyadic rational number. Dyadic rectilinear paths will be very useful, since we can apply counting arguments to sets of dyadic rectilinear paths.

We can *reverse* a path. If $f : [a, b] \to X$ is a path, we set $f^{\leftarrow}(t) = f(a + b - t)$ for $t \in [a, b]$. Then $f^{\leftarrow}$ is a path, the *reverse* of f, with initial point $f(b)$ and final point $f(a)$, and with $[f^{\leftarrow}] = [f]$.

We can define an equivalence relation on paths. If $f : [a, b] \to X$ and $g : [c, d] \to X$ are paths, we say that f and g are *similar paths*, or *equivalent paths*, if there exists a homeomorphism $\phi : [c, d] \to [a, b]$ such that $\phi(c) = a$, $\phi(d) = b$ and $g = f \circ \phi$. It is easy to see that this is an equivalence relation. Recall that $\phi : [c, d] \to [a, b]$ with $\phi(c) = a$ and $\phi(b) = d$ is a homeomorphism if and only if it is a strictly increasing continuous function. Similar paths have the same track, and if $g(s) = f \circ \phi$, we can think of $t = \phi(s)$ as a *change of variables* or *reparametrization*. For example, if $f : [a, b] \to X$ is a path, and we set $g(t) = f((1 - t)a + tb)$ for $t \in [0, 1]$ then $g : [0, 1] \to X$ is a path equivalent to f. Again, if $f : [0, 1] \to X$ is a path, and we set $g(t) = f(t^2)$ for $t \in [0, 1]$ then $g : [0, 1] \to X$ is a path equivalent to f. For many purposes, equivalent paths play the same role, and we shall frequently identify equivalent paths. Thus it is often convenient to consider paths defined on $[0, 1]$ or $[0, 2\pi]$.

We can change the initial point of a closed path. Suppose that $f : [a, b] \to X$ is a closed path and that $s \in [a, b]$. Define $f_s : [a, b] \to X$ by setting

$$f_s(t) = f(s - a + t) \text{ for } a \leq t \leq a + b - s,$$

$$\text{and } f_s(t) = f(t + s - b) \text{ for } a + b - s \leq t \leq b.$$

Then f_s is a closed path with initial point $f(s)$ and with the same track as f. Two closed paths $f : [a, b] \to X$ and $g : [c, d] \to X$ are *similar closed paths* if there exists $s \in [a, b]$ such that f_s and g are similar paths.

A path $f : [a, b] \to (X, \tau)$ is *simple* if f is an injective mapping from $[a, b]$ into X. A simple path is sometimes called an *arc*. In this case, since $[a, b]$ is compact and (X, τ) is Hausdorff, f is a homeomorphism of $[a, b]$ onto the track $[f]$. Suppose that $f : [a, b] \to X$ and $g : [c, d] \to X$ are simple paths with the same track, and with the same initial points and final points. Let $f^{-1} : [f] \to [a, b]$ be the inverse mapping, and let $\gamma = f^{-1} \circ g$. Then γ is a homeomorphism of $[a, b]$ onto $[c, d]$ with $\gamma(a) = c$ and $\gamma(b) = d$, and so f and g are equivalent.

A *simple closed path* $f : [a, b] \to X$ is a closed path whose restriction to $[a, b)$ is injective. Thus a simple closed path is not a simple path, since $f(a) = f(b)$, but otherwise f takes different values at different points of $[a, b]$. For example, the mapping $\kappa : [0, 2\pi] \to \mathbf{R}^2$ defined by $\kappa(t) = (\cos t, \sin t)$ is a simple closed path, with track the *unit circle* $\mathbf{T} = \{(x, y) \in \mathbf{R}^2 : \|(x, y)\| = 1\}$. More generally, if $w = (x, y) \in \mathbf{R}^2$ and $r > 0$, then the *circular path* $\kappa_r(w)$ is defined as

$$\kappa_r(w)(t) = (x + r \cos t, y + r \sin t) \text{ for } t \in [0, 2\pi];$$

we denote its track by $\mathbf{T}_r(w)$. Suppose that $f : [0, 2\pi] \to X$ is a simple closed path. Let $h = f \circ \kappa^{-1}$. Then h is a homeomorphism of $\mathbf{T}$ onto $[f]$; the track of a simple closed path is homeomorphic to the unit circle.

In order to illustrate the nature of paths, let us establish some approximation results that we shall need later.

Theorem 16.2.1 *Suppose that $\gamma : [0, 1] \to U$ is a path from a to b in an open subset U of a normed space $(E, \|.\|_E)$, and that $\delta > 0$. Then there is a polygonal path $\beta : [0, 1] \to U$ from a to b with $|\beta(t) - \gamma(t)| < \delta$ for $t \in [0, 1]$.*

Proof Since $[\gamma]$ is compact, we can suppose, by taking a smaller value of δ if necessary, that $N_\delta([\gamma]) = \cup\{N_\delta(z) : z \in [\gamma]\} \subseteq U$. Since γ is uniformly continuous on $[0, 1]$ there exist $0 = t_0 < t_1 < \cdots < t_k = 1$ such that $|\gamma(t) - \gamma(t_j)| < \delta/2$ for $t \in [t_{j-1}, t_j]$, for $1 \leq j \leq k$. We consider the polygonal path with vertices $t_0, \ldots, t_k$. Let

$$\beta = \sigma(\gamma(t_0), \gamma(t_1)) \vee \sigma(\gamma(t_1), \gamma(t_2)) \vee \ldots \sigma(\gamma(t_{k-1}), \gamma(t_k)),$$

parametrized so that $\beta(t_j) = \gamma(t_j)$ for $0 \leq j \leq k$. Then $[\beta] \subseteq U$, since $[\gamma(t_{j-1}), \gamma(t_j)] \subseteq N_\delta(t_j) \subseteq U$. If $t_{j-1} \leq t \leq t_j$ then

$$\|\beta(t) - \gamma(t)\|_E \leq \|\beta(t) - \beta(t_j)\|_E + \|\gamma(t_j) - \gamma(t)\|_E < \delta.$$

$\square$

Corollary 16.2.2 *If $E = \mathbf{R}^d$, then β can be chosen as a rectilinear path.*

Proof Replace each of the paths $\sigma(\gamma(t_{j-1}), \gamma(t_j))$ by the juxtaposition of finitely many linear paths, each parallel to an axis. $\square$

Corollary 16.2.3 *Suppose that* $\gamma : [0,1] \to U$ *is a path in an open subset* U *of* $\mathbf{R}^d$ *and that* $\delta > 0$. *Then there is a dyadic rectilinear path* $\beta : [0,1] \to U$ *with* $|\beta(t) - \gamma(t)| < \delta$ *for* $t \in [0,1]$.

Proof Approximate the vertices of the path of the previous corollary by vertices each of whose coordinates is a dyadic rational number. $\qquad\square$

16.3 Path-connectedness

A topological space (X, τ) is *path-connected* if for each $x, y \in X$ there is a path from x to y.

Proposition 16.3.1 *If* (X, τ) *is a path-connected topological space then* (X, τ) *is connected.*

Proof Suppose that F is a non-empty open and closed subset of X, and that $x \in F$. If $y \in X$ there exists a path from x to y. Let T be its track. Then $x \in T$ and T is connected, and so $T \subseteq F$. But $y \in T$, and so $y \in F$. This holds for all $y \in X$, and so $F = X$; X is connected. $\qquad\square$

Example 16.3.2 A connected compact subset of the plane $\mathbf{R}^2$ which is not path-connected.

Let

$$I = \{(0,y) : -1 \le y \le 1\}, \quad J = \{(x, \sin(1/x)) : 0 < x \le 1\}, \quad K = I \cup J.$$

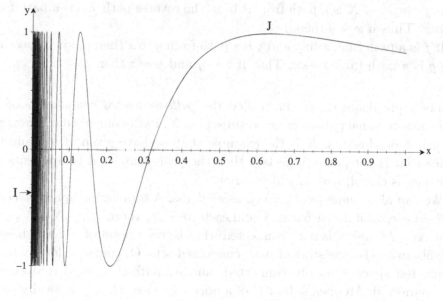

Figure 16.3. A connected set which is not path-connected.

The mapping $(x, \sin(1/x)) \to x$ is a homeomorphism of J onto $(0, 1]$, so that J is connected. K is the closure of J, so that K is compact and connected. We show that K is not path-connected. Suppose that $f : [a, b] \to K$ is a path from $(0, 0)$ to $(1, \sin 1)$ and set $f(t) = (g(t), h(t))$. Let $S = \sup\{s \in [a, b] : g(s) = 0\}$. By continuity, $g(s) = 0$, and so $a \le s < b$. Using the intermediate value theorem inductively, we can find a decreasing sequence $(t_n)_{n=0}^{\infty}$ in $[s, b]$ such that $g(t_n) = 2/\pi(2n+1)$. Let $T = \lim_{n\to\infty} t_n$. Then $h(t_n) = \sin((2n + 1)\pi/2) = (-1)^n$, so that $h(t_n)$ does not converge to $h(T)$ as $n \to \infty$, contradicting the continuity of h. Note also that J is path-connected, but that $\overline{J} = K$ is not.

Nevertheless, as we shall see, path connectedness provides a valuable test for connectedness. For example, a normed space $(E, \|.\|)$ is path-connected, and therefore connected: if $x, y \in E$, let $f(t) = (1 - t)x + ty$, for $t \in [0, 1]$. Here the track is the line segment $[x, y]$.

We can also partition a topological space into path-connected components. For this, we need the following.

Proposition 16.3.3   *Suppose that (X, τ) is a topological space. Define a relation $\sim$ on X by setting $x \sim y$ if there exists a path in X from x to y. Then $\sim$ is an equivalence relation on X.*

Proof   If $x \in X$, let $f(t) = x$ for $x \in [0, 1]$. Then f is a path from x to x, and so $x \sim x$.

If $f : [a, b] \to X$ is a path from x to y, the reverse path $f^{\leftarrow}$ is a path from y to x. Thus if $x \sim y$ then $y \sim x$.

If f is a path from x to y and g is a path from y to z then the juxtaposition $f \vee g$ is a path from $x \to z$. Thus if $x \sim y$ and $y \sim z$ then $x \sim z$. $\square$

The equivalence classes are called the *path-connected components* of X. They are maximal path-connected subsets of X. Path-connected components need not be closed. If K is the example that we have given of a connected space that is not path-connected, then the path-connected components are I, which is closed, and J, which is not.

We can also define *local path connectedness*. A topological space is *locally path-connected* if for each $x \in X$ and each $M \in \mathcal{N}_x$ there exists $N \in \mathcal{N}_x$ such that $N \subseteq M$ and N is path connected; that is, each point of X has a base of neighbourhoods consisting of path-connected sets. Of course, a locally path-connected space is locally connected, and its path-connected components are connected. An open subset U of a normed space $(E, \|, \|)$ is locally path-connected; if $x \in U$ and $N_\epsilon(x) \subseteq U$ then $N_\epsilon(x)$ is path-connected.

Proposition 16.3.4   *If (X, τ) is a locally path-connected topological space, then the path-connected components are open and closed.*

Proof   Suppose that E is a path-connected component of X and that $x \in E$. Let N be a path-connected neighbourhood of X. Then $x \in N^\circ \subseteq N \subseteq E$, and so E is open. The complement of E is the union of path-connected components, and therefore it is open. Thus E is closed. $\square$

Corollary 16.3.5   *A locally path-connected space is connected if and only if it is path-connected.*

In particular an open subset of a normed space, or of $\mathbf{C}$, is connected if and only if it is path-connected.

Results concerning the path-connectedness of product spaces are easier to prove than the corresponding ones for connectedness.

Theorem 16.3.6   *If $(X, \tau) = \prod_{\alpha \in A}(X_\alpha, \tau_\alpha)$ is the product of path-connected topological spaces (X_α, τ_α) then (X, τ) is path-connected.*

Proof   Suppose that $x, y \in X$. Then for each $\alpha \in A$ there exists a path $f_\alpha : [0, 1] \to X$ from x_α to y_α. Let $f(t) = (f_\alpha(t))_{\alpha \in A}$. Then f is a path from x to y. $\square$

Corollary 16.3.7   *If (X, τ) is a finite product of locally path-connected topological spaces then (X, τ) is locally path-connected.*

The example (the infinite product of two-point sets) that we gave for local connectedness shows that this result does not extend to infinite products.

Exercises

16.3.1 Suppose that (X, d) is a path-connected metric space. Show that the set $F(X)$ of finite non-empty subsets of X, with the Hausdorff metric, is path-connected.

16.3.2 Suppose that (X, d) is a path-connected locally path-connected compact metric space. Show that the configuration space $C(X)$, with the Hausdorff metric, is path-connected.

16.4 *Hilbert's path*

(This section can be omitted on a first reading.)

The definition of a path, and its track, are very straightforward and natural. Paths are however not at all straightfoward. In 1890, Peano gave an example of a path in the plane whose track is the unit square $[0, 1] \times [0, 1]$. We shall construct a path, *Hilbert's path*, with the same track; this was described by Hilbert in 1891.

We start with the unit square $[0, 1] \times [0, 1]$, which we list as $S_1^{(0)}$. It can be divided into a set Q_1 of four squares with side-length $1/2$, namely $[0, 1/2] \times [0, 1/2]$, $[0, 1/2] \times [1/2, 1]$, $[1/2, 1] \times [0, 1/2]$ and $[1/2, 1] \times [1/2, 1]$. We can divide each of these squares into four smaller squares, and iterate the procedure. Thus at the nth level we have a set Q_n of 4^n squares of side-length $1/2^n$. Note that there is a natural *parity* on Q_n; if $S = [(j-1)/2^n, j/2^n] \times [(k-1)/2^n, k/2^n]$, we say that S has *odd parity* if $j + k$ is odd, and *even parity* if $j + k$ is even. If we colour the squares with odd parity white, and those with even parity black, then we have a checker-board colouring. Note that if two squares in Q_n share a common side, then they have different parities.

We shall show that there is a unique listing $(S_j^{(n)})_{j=1}^{4^n}$ of Q_n for $n \in \mathbf{N}$ such that

(i) $x_0 = (0, 0) \in S_1^{(n)}$ and $x_1 = (1, 0) \in S_{4^n}^{(n)}$;

(ii) $S_j^{(n)}$ and $S_{j+1}^{(n)}$ share a common side, for $1 \leq j < 4^n$;

(iii) $S_j^{(n)} = S_{4j-3}^{(n+1)} \cup S_{4j-2}^{(n+1)} \cup S_{4j-1}^{(n+1)} \cup S_{4j}^{(n+1)}$ for $1 \leq j \leq 4^n$;

for all $n \in \mathbf{N}$. Note that if we have such a listing, then $S_j^{(n)}$ has odd parity if j is even, and even parity if j is odd.

To show this, note, by considering possible cases, that if we divide a square S into a set Q of four squares with equal side-length, and if $T \in Q$ and e is a side of S, then there is a unique listing $(T_j)_{j=1}^4$ of Q so that consecutive terms have a common side and so that $T_1 = T$ and $T_4 \cap e \neq \emptyset$.

We begin by listing Q_1: we set

$$S_1^{(1)} = [0, 1/2] \times [0, 1/2],$$

$$S_2^{(1)} = [0, 1/2] \times [1/2, 1],$$

$$S_3^{(1)} = [1/2, 1] \times [1/2, 1],$$

$$S_4^{(1)} = [1/2, 1] \times [0, 1/2].$$

This satisfies the conditions, and is the only listing that does so.

Suppose that Q_n has been listed. Let $e_j^{(n)}$ be the common side of $S_j^{(n)}$ and $S_{j+1}^{(n)}$, for $1 \leq j < 4^n$. We list the four elements of Q_{n+1} contained in $S_1^{(n)}$ so that consecutive terms have a common side and so that $x_0 \in$

$S_1^{(n+1)}$ and $S_4^{(n+1)} \cap e_1^{(n)} \neq \emptyset$; this listing is unique. We then set $S_5^{(n+1)}$ to be the element of $S_2^{(n)}$ which has a side in common with $S_4^{(n+1)}$, and iterate. We continue in this way until we reach $S_{4^{n+1}-3}^{(n+1)}$, which is contained in $S_{4^n}^{(n)}$; up to here the listing is unique. Now $S_{4^{n+1}-3}^{(n+1)}$ has even parity, and $[1 - 1/2^{n+1}, 1] \times [0, 1/2^{n+1}]$, the element of $\mathcal{Q}_{n+1}$ to which x_1 belongs, has odd parity, and so we can complete the listing to satisfy the conditions, and in a unique way.

We now use these listings to define approximations h_n to Hilbert's path. Let $x_0^{(n)} = x_0$, let $x_j^{(n)}$ be the centre of the square $S_j^{(n)}$, for $1 \leq j \leq 4^n$ and let $x_{4^n+1}^{(n)} = x_1$. Let

$$t_0 = 0, \text{ let } t_j = (j - 1/2)/4^n \text{ for } 1 \leq j \leq 4^n, \text{ and let } t_{4^n+1} = 1.$$

Then set

$$h_n((1 - \lambda)t_j + \lambda t_{j+1}) = (1 - \lambda)x_j^{(n)} + \lambda x_{j+1}^{(n)}$$

for $0 \leq \lambda \leq 1$ and $0 \leq j \leq 4^n$. Then h_n is a simple path from x_0 to x_1 which spends equal time $1/4^n$ in each of the squares of $\mathcal{Q}_n$ in turn. Because of condition (iii), if $(j - 1)/4^n \leq t \leq j/4^n$ then $h_n(t) \in S_j^{(n)}$ and also $h_m(t) \in S_j^{(n)}$ for all $m \geq n$.

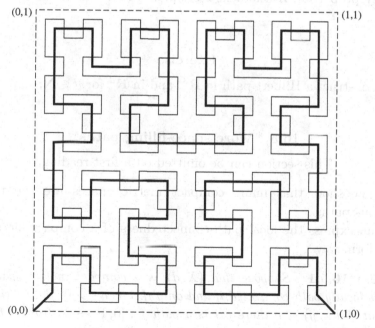

Figure 16.4. The paths H_3 and H_4.

Thus $\|h_n(t) - h_m(t)\|_2 \leq \sqrt{2}/2^n$. Since this holds for all $t \in [0,1]$, $\|h_n - h_m\|_\infty \leq \sqrt{2}/2^n$, and $(h_n)_{n=1}^\infty$ is a Cauchy sequence in $C_{\mathbf{R}^2}([0,1])$. Since $C_{\mathbf{R}^2}([0,1])$ is a Banach space, the uniform limit h is a path from x_0 to x_1. We must show that $[h] = [0,1] \times [0,1]$. If not, then, since $[h]$ is a closed subset of $[0,1] \times [0,1]$, its complement is a non-empty open subset of $[0,1] \times [0,1]$, and this contains a closed square $S_j^{(n)}$, for some j and n. But if $(j-1)/4^n \leq t \leq j/4^n$ then $h_m(t) \in S_j^{(n)}$ for $m \geq n$. As $h_m(t) \to h(t)$ as $m \to \infty$, it follows that $h(t) \in S_j^{(n)}$, giving a contradiction.

Hilbert's path has a great deal of self-similarity. For example, the mapping $c : (x,y) \to (y/2, x/2)$ maps $[0,1] \times [0,1]$ onto $[0,1/2] \times [0,1/2]$, and $c(f(t)) = f(t/4)$ for $0 \leq t \leq 1$. Similarly if $b(x,y) = (x/2, (y+1)/2)$ then $b(f(t)) = f((t+1)/4)$.

The examples of Peano and Hilbert overturned many intuitions about dimension: a two-dimensional object can be the continuous image of a one-dimensional one. But the consequence of this was the development of a rich theory of dimension. One other conclusion that should be drawn from this example is that the notion of continuity is not a straightforward one; the sketches that we make when we consider continuous functions are quite untypical of what can happen.

Hilbert's path is interesting because of its explicit construction and its self-similarity properties. In the next section, we shall establish a more general result.

Exercise

16.4.1 Construct a Hilbert path in $\mathbf{R}^3$, and in $\mathbf{R}^d$ for $d \in \mathbf{N}$.

16.5 *More space-filling paths*

(This section can be omitted on a first reading.)

We now show that many compact metric spaces are the tracks of continuous paths.

We can express the local path-connectedness of a compact metric space in a uniform way.

Theorem 16.5.1 *Suppose that (X, d) is a compact metric space. Then (X, d) is locally path-connected if and only if, given $\epsilon > 0$, there exists $\eta > 0$ such that if $d(x,y) < \eta$ there exists a path $f : [0,1] \to X$ from x to y such that $d(f(s), f(t)) < \epsilon$ for $0 \leq s < t \leq 1$.*

Proof The proof is a standard compactness proof. Suppose that (X, d) is locally path-connected and that $\epsilon > 0$. For each $a \in X$ there exists $\delta(a) > 0$ and a path-connected set C_a such that $N_{\delta(a)}(a) \subseteq C_a \subseteq N_{\epsilon/2}(x)$. Thus if $b, c \in N_{\delta(a)}(a)$ then there is a path $f : [0,1] \to X$ from b to c with $d(f(s), f(t)) < \epsilon$ for $0 \le s < t \le 1$. The sets $N_{\delta(a)/2}(a)$ form an open cover of X, and so there is a finite subcover $\{N_{\delta(a)/2}(a) : a \in F\}$. Let $\eta = \min\{\delta(a)/2 : a \in F\}$. If $d(x, y) < \eta$ then there exists $a \in F$ such that $x \in N_{\delta(a)/2}(a)$, and so $y \in N_{\delta(a)}(a)$. Thus there is a path from x to y with the required properties.

Conversely, suppose that the condition is satisfied, and that $N_\epsilon(x)$ is a neighbourhood of x in X, and let η satisfy the condition. Let $C_\epsilon(x)$ be the set of points y in X for which there is a path $f : [0,1] \to X$ from x to y for which $d(f(s), f(t)) < \epsilon$ for $0 \le s < t \le 1$. Any such path is contained in $N_\epsilon(x)$ (take $s = 0$), so that $C_\epsilon(x) \subseteq N_\epsilon(x)$, and the condition implies that $N_\eta(x) \subseteq C_\epsilon(x)$. Since $C_\epsilon(x)$ is clearly path connected, the sets $\{C_\epsilon(x) : \epsilon > 0\}$ form a base of path-connected neighbourhoods of x. $\square$

Corollary 16.5.2 *If (X, d) is a locally path-connected compact metric space there exists an increasing real-valued function h on $(0, \operatorname{diam} X]$, for which $h(u) \to 0$ as $u \searrow 0$, such that if $x, y \in X$ there exists a path $f : [0,1] \to X$ from x to y for which $d(f(s), f(t)) \le h(d(x, y))$ for $0 \le s < t \le 1$.*

Proof If $x, y \in X$, let $P(x, y)$ be the set of paths from x to y. Let $r(x, y) = \inf\{\operatorname{diam}([p]) : p \in P(x, y)\}$; then $d(x, y) \le r(x, y) \le \operatorname{diam}(X)$. If $0 < u < \operatorname{diam}(X)$ let $h(u) = 2\sup\{r(x, y) : d(x, y) \le u\}$. Then the theorem implies that $h(u) \to 0$ as $u \searrow 0$, and the construction ensures that the other requirements of the corollary are satisfied. $\square$

Theorem 16.5.3 *Suppose that (X, d) is a locally path-connected compact metric space. Then there exists a path in X whose track is X.*

Proof Let h be a function satisfying the conditions of Corollary 16.5.2. By Corollary 15.4.9, there exists a continuous surjective mapping f from the Cantor set C onto X. We extend this to a continuous mapping $\widetilde{f} : [0,1] \to X$. Suppose that (c, d) is a connected component of $[0,1] \setminus C$. There is a path $\gamma : [c, d] \to X$ from $f(c)$ to $f(d)$ such that

$$d(\gamma(s), \gamma(t)) \le h(d(f(c), f(d))) \text{ for } c \le s < t \le d.$$

We define $\widetilde{f}(s) = \gamma(s)$ for $s \in (c, d)$.

The function $\widetilde{f}$ maps $[0,1]$ onto X; it remains to show that $\widetilde{f}$ is continuous. If $s \in [0,1] \setminus C$ then $\widetilde{f}$ is continuous at s, since the path γ is continuous at s.

We must show that $\widetilde{f}$ is continuous at each point t of C. This takes just a little care. It is enough to show that $\widetilde{f}$ is continuous on the right on $[0,1)$ and continuous on the left on $(0,1]$, and it is clearly sufficient to show the former. If t is the left-hand end-point of a connected component of $[0,1]\setminus C$, then $\widetilde{f}$ is continuous on the right at t. Otherwise, suppose that $\epsilon > 0$. Then there exists $\delta > 0$ such that $h(u) < \epsilon/2$ for $0 < u < \delta$. Since f is continuous on C, there exists $0 < \eta \le \delta$ such that $t + \eta < 1$ and $|\widetilde{f}(r) - \widetilde{f}(t)| = |f(r) - f(t)| < \epsilon/2$ for $r \in (t, t+\eta) \cap C$. Since t is not the left-hand end-point of a connected component of $[0,1]\setminus C$, there exists $t' \in (t, t+\eta) \cap C$. If $t < s < t'$ then either $s \in C$, in which case $|\widetilde{f}(s) - \widetilde{f}(t)| < \epsilon/2$, or $s \in [0,1]\setminus C$. In the latter case, there is a connected component (c,d) of $[0,1]\setminus C$ for which $s \in (c,d)$. But then $(c,d) \subseteq (t,t')$, so that $d - c < \delta$, and $|\widetilde{f}(s) - \widetilde{f}(c)| < \epsilon/2$. Thus $|\widetilde{f}(s) - \widetilde{f}(t)| < \epsilon$ for $t < s < t'$; f is continuous on the right at t. $\qquad\square$

Corollary 16.5.4 *Suppose that K is a compact convex subset of a normed space $(E, \|.\|)$. Then there exists a path in K whose track is K.*

Proof If $k_1, k_2 \in K$ then the line segment $[k_1, k_2]$ is contained in K, and so K is locally path-connected. $\qquad\square$

16.6 Rectifiable paths

In Part Five, we turn to the problem of integrating a function along a path. We have however seen that paths like Hilbert's path can behave very badly. We must therefore consider a more restricted class of paths.

The trouble with Hilbert's path and other space-filling paths is that they have infinite length. Let us make this explicit. If $\gamma : [a,b] \to (X,d)$ is a path, its *length* $l(\gamma) = l_{[a,b]}(\gamma)$ is defined as

$$l(\gamma) = \sup\left\{ \sum_{j=1}^n d(\gamma(t_{j-1}), \gamma(t_j)) : a = t_0 < t_1 < \ldots t_n = b, \ n \in \mathbf{N} \right\}.$$

It is possible that $l_{[a,b]}(\gamma) = \infty$; as a simple example, if $\gamma(0) = (0,0)$ and $\gamma(t) = (t, t\sin\pi/t)$ for $0 < t \le 1$ then $\gamma : [0,1] \to \mathbf{R}^2$ is a simple path from $(0,0)$ to $(1,0)$ in $\mathbf{R}^2$ of infinite length. Hilbert's path h has infinite length, since if $n \in \mathbf{N}$ then the approximation h_n has length $2^n + (\sqrt{2}-1)/2^n$, and $l(h) \ge l(h_n)$ for all n.

A path of finite length is called a *rectifiable path*.

Proposition 16.6.1 *Suppose that $\gamma : [a,b] \to (X,d)$ is a path.*
   (i) $l_{[a,b]}(\gamma) \ge d(a,b)$.

(ii) $l_{[a,b]}(\gamma^{\leftarrow}) = l_{[a,b]}(\gamma)$.

(iii) If $a < c < b$ then the restrictions of γ to $[a, c]$ and to $[c, b]$ are both paths, and $l_{[a,c]}(\gamma) + l_{[c,b]}(\gamma) = l_{[a,b]}(\gamma)$.

(iv) If $\beta : [c, e] \to (X, d)$ is a path similar to γ then $l_{[c,e]}(\beta) = l_{[a,b]}(\gamma)$.

Proof   All these results follow immediately from the definitions. $\square$

As a cautionary example, consider the norm $\|(x_1, x_2)\|_{\infty} = \max(|x_1|, |x_2|)$ on $\mathbf{R}^2$. Let $a = (0, 0)$, $b = (1, 0)$ and $c = (2, 1)$, and let γ be the path $\sigma(a, b) \vee \sigma(b, c)$. Then $l_{[a,c]}(\gamma) = 2 = \|c - a\|$, although γ is not a linear path. This phenomenon cannot occur in Euclidean spaces, or indeed in an inner product space.

Proposition 16.6.2   *Suppose that $\gamma : [a, b] \to V$ is a simple path in a real inner product space $(V, \langle ., . \rangle)$ for which $l_{[a,b]}(\gamma) = \|\gamma(b) - \gamma(a)\|$. Then γ is similar to the linear path $\sigma(\gamma(a), \gamma(b))$.*

"A straight line is the shortest distance between two points" *(Thomas Carlyle).*

Proof   If the result is not true then there exists $a < c < b$ for which $c - a$ and $b - c$ are not linearly dependent. Then, using the Cauchy–Schwarz inequality,

$$\|\gamma(b) - \gamma(a)\|^2 = \|(\gamma(c) - \gamma(a)) + (\gamma(b) - \gamma(c))\|^2$$
$$= \|\gamma(c) - \gamma(a)\|^2 + \|\gamma(b) - \gamma(c)\|^2 + 2\langle\gamma(c) - \gamma(a), \gamma(b) - \gamma(c)\rangle$$
$$< \|\gamma(c) - \gamma(a)\|^2 + \|\gamma(b) - \gamma(c)\|^2 + 2\|\gamma(c) - \gamma(a)\| \|\gamma(b) - \gamma(c)\|$$
$$= (\|\gamma(c) - \gamma(a)\| + \|\gamma(b) - \gamma(c)\|)^2 \leq (l_{[a,b]}(\gamma))^2,$$

giving a contradiction. $\square$

It is convenient to use path length to parametrize a rectifiable path.

Proposition 16.6.3   *Suppose that $\gamma : [a, b] \to (X, d)$ is a rectifiable path. Let $l(a) = 0$, and let $l(t) = l_{[a,t]}(\gamma)$, for $t \in (a, b]$. Then l is a continuous increasing function on $[a, b]$.*

Proof   If $a < s < t \leq b$ then $l(t) = l(s) + l_{[s,t]}(\gamma) \geq l(s)$, so that l is an increasing function on $[a, b]$.

We shall show that if $a \leq t < b$ then l is continuous on the right at t; the proof of left continuity is exactly similar. Suppose that $\epsilon > 0$. Recall that $l(t^+) = \inf\{l(s) : s > t\}$. There exists $\delta > 0$ with $t + \delta \leq b$ such that $d(\gamma(s), \gamma(t)) < \epsilon/3$ and $l(s) < l(t^+) + \epsilon/3$ for $t < s < t + \delta$. Thus if

$t < s < r < t + \delta$ then $l_{[s,r]}(\gamma) = l(r) - l(s) < \epsilon/3$. Suppose that $t < s < t + \delta$. There exist $t = t_0 < t_1 < \ldots t_n = s$ such that

$$\sum_{j=1}^{n} d(\gamma(t_{j-1}), \gamma(t_j)) > l_{[t,s]} - \epsilon/3 = l(s) - l(t) - \epsilon/3.$$

Then

$$l(s) - l(t) < d(\gamma(t_0), \gamma(t_1)) + \sum_{j=2}^{n} d(\gamma(t_{j-1}), \gamma(t_j)) + \epsilon/3$$

$$\leq d(\gamma(t_0), \gamma(t_1)) + l_{[t_1,s]}(\gamma) + \epsilon/3$$

$$\leq \epsilon/3 + (l(s) - l(t_1)) + \epsilon/3 \leq \epsilon.$$

$\square$

Proposition 16.6.4 *Suppose that $\gamma : [a, b] \to (X, d)$ is a rectifiable path. Then there exists an equivalent path $\beta : [0, l_{[a,b]}(\gamma)] \to [\gamma]$ such that $l_{[0,s]}(\beta) = s$, for $0 < s \leq l_{[a,b]}(\gamma)$.*

Proof Let $l(a) = 0$ and let $l(t) = l_{[a,t]}(\gamma)$, for $t \in (a, b]$. Then l is a continuous increasing function on $[a, b]$, and $l([a, b]) = [0, l(b)]$. Suppose that $0 \leq t \leq l(b)$. Let $I_t = l^{-1}(\{t\})$. Then I_t is a (possibly degenerate) closed interval containing t. Suppose that I_t is not degenerate, and that $r, s \in I_t$ with $r < s$. Then $l(s) = l(r) + l_{[r,s]}(\gamma) = l(r)$, so that $l_{[r,s]}(\gamma) = 0$. Since $d(\gamma(r), \gamma(s)) \leq l_{[r,s]}(\gamma)$, it follows that $\gamma(r) = \gamma(s)$. Thus if we set $\beta(t) = \gamma(s)$ for some $s \in I_t$, then β is properly defined, and $\gamma = \beta \circ l$. Since γ and l are continuous, it follows from Proposition 15.1.6 that β is continuous. Clearly $[\beta] = [\gamma]$, and $l_{[0,t]}(\beta) = t$, for $t \in [0, l(b)]$. $\square$

This path is called the *path-length parametrization* of γ. Its use simplifies many problems involving paths. Let $\alpha(t) = \beta(t/l(\gamma))$ for $0 \leq t \leq 1$; the path $\alpha : [0, 1] \to X$ is the *normalized path-length parametrization*.

Note that if γ is a simple rectifiable path, then the function $l(t) = l_{[a,t]}(\gamma)$ is strictly increasing on $(a, b]$, and the mapping l is a homeomorphism of $[a, b]$ onto $[0, l(b)]$. In this case, the proof is much easier; simply set $\beta = \gamma \circ l^{-1}$.

Exercise

16.6.1 Show that the function $\rho(x, y) = \sqrt{|x - y|}$ on $[0, 1]$ is a metric on $[0, 1]$ which is uniformly equivalent to the usual metric. Let $\gamma(t) = t$ for $t \in [0, 1]$. Is γ a rectifiable path in $([0, 1], \rho)$?

Part Four

Functions of a vector variable

17

Differentiating functions of a vector variable

17.1 Differentiating functions of a vector variable

In Part Two, we considered continuity and limiting properties of real-valued functions of a real variable – functions defined on a subset of $\mathbf{R}$. In Part Three we extended these ideas to functions between metric spaces, or between topological spaces. In particular, these results apply to functions of several real variables – functions defined on a subset of $\mathbf{R}^d$.

We now turn to differentiation. This involves linearity: we therefore consider functions defined on a subset U of a real normed space $(E, \|.\|_E)$ taking values in a real normed space $(F, \|.\|_F)$. In fact, our principal concern will be with functions of several real variables (functions defined on an open subset of $\mathbf{R}^d$), but it is worth proceeding in a more general way. First, this illustrates more clearly the basic ideas that lie behind the theory. Secondly, even in the case where we consider functions defined on a finite-dimensional Euclidean space, there are advantages in proceeding in a coordinate free way; not only is the notation simpler, but also the results are seen to be independent of any particular choice of coordinates.

Recall (Volume I, Section 7.1) that a real-valued function f defined on an open interval I is differentiable at a point a of I if and only if there exists a real number $f'(a)$ such that if

$$r(h) = f(a+h) - f(a) - f'(a)h$$

for all non-zero h in $I - a = \{x \in \mathbf{R} : x + a \in I\}$, then $r(h)/|h| \to 0$ as $h \to 0$; that is, $r(h) = o(|h|)$. Let us set $Df_a(x) = f'(a)x$, for $x \in \mathbf{R}$. Then Df_a is a linear mapping from $\mathbf{R}$ into $\mathbf{R}$ and

$$f(a+h) = f(a) + Df_a(h) + r(h)$$

for all $h \in I - a = \{x \in \mathbf{R} : x + a \in I\}$. Thus f is differentiable at a if and only if we can write f as the sum of a constant (the value at a), a linear term $Df_a(h)$, and a small order term $r(h)$. From this point of view, differentiation is a matter of linear approximation.

These ideas extend naturally to vector-valued functions of a vector variable. Suppose that f is a function defined on an open subset U of a real normed space $(E, \|.\|_E)$, taking values in a real normed space $(F, \|.\|_F)$, and that $a \in U$. We say that f is *differentiable* at a, with *derivative* Df_a, if there is a continuous linear operator $Df_a \in L(E, F)$ such that if

$$r(h) = f(a + h) - f(a) - Df_a(h)$$

for all non-zero $h \in U - a = \{x \in E : x + a \in U\}$, then $r(h)/\|h\| \to 0$ as $h \to 0$. Again, we express f as the the sum of a constant (the value at a), a linear term $Df_a(h)$, and a small order term $r(h)$.

Note that we require Df_a to be a *continuous* linear mapping; this condition is automatically satisfied if E is finite-dimensional, since any linear operator from a finite-dimensional normed space into a normed space is continuous (Corollary 15.9.3). Note also that the conditions remain the same if we replace the norm on E and the norm on F by equivalent norms. In particular, when E is finite-dimensional then we can use any norm on E (and similarly for F), since any two norms on a finite-dimensional space are equivalent (Theorem 15.9.2).

Let us consider three special cases. First, when $E = \mathbf{R}$, we set $f'(a) = Df_a(1)$, so that

$$f(a + h) = f(a) + hf'(a) + r(h) \text{ for all } h \in U - a = \{x \in E : x + a \in U\};$$

$f'(a)$ is an element of F, while Df_a is a linear mapping from $\mathbf{R}$ into F. Note that if $\|Df_a\|$ is the operator norm of Df_a, then

$$\|Df_a\| = \sup\{\|Df_a(h)\|_F : |h| \le 1\} = \|f'(a)\|_F .$$

Secondly, suppose that H is a real Hilbert space, and that $F = \mathbf{R}$. In this case, Df_a is a continuous linear functional on H. By the Fréchet-Riesz representation theorem (Theorem 14.3.7), linear functionals can be expressed in terms of the inner product; there exists an element ∇f_a of H such that $Df_a(h) = \langle \nabla f_a, h \rangle$. The vector ∇f_a is called the *gradient* of f at a. The symbol ∇ was introduced by Hamilton, and named *'nabla'* by Maxwell; 'nabla' is the Greek word for a Hebrew harp. Nowadays, it is usually more prosaically called *'grad'*.

Thirdly, suppose that $f = (f_1, \ldots, f_n)$ is a mapping from an open subset U of a normed space $(E, \|.\|_E)$ into a finite product $(F_1, \|.\|_{F_1}) \times \cdots \times (F_n, \|.\|_{F_n})$ of normed spaces. Then f is differentiable at a point a of E if and only if f_j is differentiable at a for each $1 \leq j \leq n$; if so, then $Df_a = ((Df_1)_a, \ldots, (Df_n)_a)$.

If f is differentiable at every point of U, we say that f is *differentiable on U*. If so, then $a \to Df_a$ is a mapping from U to the normed space $(L(E, F), \|.\|)$ of continuous linear mappings from E into F. We say that f is *continuously differentiable at a* if this mapping is continuous at a, and that f is *continuously differentiable on U* if it is continuously differentiable at each point of U.

As a first example, if $T \in L(E, F)$ then $T(a + h) = T(a) + T(h)$ for all a and h in E, so that T is differentiable at every point of E, and $DT_a = T$.

We now have the following elementary results.

Proposition 17.1.1 *Suppose that f and g are functions defined on an open subset U of a normed space $(E, \|.\|_E)$, taking values in a normed space $(F, \|.\|_F)$, that $a \in U$, and that f and g are differentiable at a.*

(i) Df_a is uniquely determined.

(ii) If $\epsilon > 0$, there exists $\delta > 0$ such that $N_\delta(a) \subseteq U$ and such that $\|f(a + h) - f(a)\|_F \leq (\|Df_a\| + \epsilon) \|h\|_E$ for $\|h\|_E < \delta$.

(iii) f is continuous at a.

(iv) If $\lambda, \mu \in \mathbf{R}$ then $\lambda f + \mu g$ is differentiable at a, with derivative $\lambda Df_a + \mu Dg_a$.

Proof   (i) Suppose that $T_1, T_2 \in L(E, F)$ and that

$$f(a + h) = f(a) + T_1(h) + s_1(h) = f(a) + T_2(h) + s_2(h) \text{ for } h \in U - a,$$

where $s_1(h)/\|h\|_E \to 0$ and $s_2(h)/\|h\|_E \to 0$ as $h \to 0$. Suppose that x is a non-zero element of E. Let $y = T_1(x) - T_2(x)$. Since

$$\frac{y}{\|x\|_E} = \frac{T_1(\lambda x) - T_2(\lambda x)}{\|\lambda x\|_E} = -\frac{s_1(\lambda x) - s_2(\lambda x)}{\|\lambda x\|_E} \to 0$$

as $\lambda \to 0$, $y = 0$. Since this holds for all non-zero x in E, $T_1 = T_2$.

(ii) Let $f(x + h) = f(x) + Df_a(h) + r(h)$. There exists $\delta > 0$ such that $N_\delta(a) \subseteq U$ and such that $\|r(h)\|_F \leq \epsilon\|h\|_E$ for $\|h\|_E < \delta$. Then

$$\|f(a + h) - f(a)\|_F = \|Df_a(h) + r(h)\|_F \leq \|Df_a(h)\|_F + \|r(h)\|_F$$

$$\leq \|Df_a\|\|h\|_E + \epsilon\|h\|_E = (\|Df_a\| + \epsilon)\|h\|_E.$$

(iii) Suppose that $\epsilon > 0$. Let δ satisfy the conclusions of (ii), and let $\eta = \delta\epsilon/(\delta+1)(\|Df_a\|+\epsilon)$. If $\|h\|_E < \eta$ then $\|h\|_E < \delta$, so that

$$\|f(a+h) - f(a)\|_F \le (\|Df_a\|+\epsilon)\|h\|_E < \epsilon.$$

(iv) As easy as in the real scalar case. □

Theorem 17.1.2 (The chain rule) *Suppose that U is an open subset of a normed space $(E, \|.\|_E)$, that f is a function defined on U, taking values in a normed space $(F, \|.\|_F)$, and that f is differentiable at a point a of U. Suppose that V is an open set of F containing $f(U)$ and that k is a function defined on V, taking values in a normed space $(G, \|.\|_G)$, and differentiable at $f(a)$. Then the function $k \circ f$ is differentiable at a, with derivative $Dk_{f(a)} \circ Df_a$.*

Proof Let us set $b = f(a)$, and suppose that

$$f(a+h) = f(a) + Df_a(h) + r(h).$$

First we simplify the problem, by showing that we can replace k by a function j for which $Dj_b = 0$. Let $j(y) = k(y) - Dk_b(y)$, for $y \in V$. By Proposition 17.1.1, j is differentiable at b, and $Dj_b(y) = Dk_b(y) - Dk_b(y) = 0$.

Since Dk_b is linear,

$$Dk_b(f(a+h)) = Dk_b(f(a)) + Dk_b(Df_a(h)) + Dk_b(r(h)).$$

But $\|Dk_b(r(h))\|_G \le \|Dk_b\|\,\|r(h)\|_F$, so that $Dk_b(r(h))/\|h\|_E \to 0$ as $h \to 0$. Thus $Dk_b \circ f$ is differentiable at a, with derivative $Dk_b \circ Df_a$. Since $(k \circ f)(x) = (j \circ f)(x) + (Dk_b \circ f)(x)$, it is therefore sufficient to show that $j \circ f$ is differentiable at a, with derivative 0. In other words, we must show that $\|j(f(a+h)) - j(f(a))\|_G / \|h\|_E \to 0$ as $h \to 0$.

Suppose that $\epsilon > 0$. Let $L = \|Df_a\| + \epsilon$. By Proposition 17.1.1 (ii), there exist $\delta > 0$ such that $N_\delta(a) \subseteq U$ and $\|f(a+h) - f(a)\|_F \le L\|h\|_E$ for $\|h\|_E < \delta$. Since j is differentiable at b, there exists $\eta > 0$ such that $N_\eta(b) \subseteq V$ and

$$\|j(b+l) - j(b)\|_G < \epsilon\|l\|_F/L, \text{ for } \|l\|_F < \eta.$$

If $\|h\|_E < \min(\delta, \eta/L)$ then $\|f(a+h) - f(a)\|_F < \eta$. Set $l = f(a+h) - f(a)$; then $b + l = f(a+h)$, so that

$$\|j(f(a+h)) - j(f(a))\|_G < \epsilon\|f(a+h) - f(a)\|_F/L < \epsilon\|h\|_E.$$

Since this holds for all $\epsilon > 0$, the result follows. □

Corollary 17.1.3 *If f is continuously differentiable on U and k is continuously differentiable on V then $k \circ f$ is continuously differentiable on U.*

Proof For the functions $x \to Dk_{f(x)}$ and $x \to Df_x$ are continuous, and the composition of two continuous functions is continuous. □

Let us give some examples.

Example 17.1.4 The derivative of a bilinear mapping.

Suppose that $(E, \|.\|_E)$, $(F, \|.\|_F)$ and $(G, \|.\|_G)$ are normed spaces, and that $\|.\|$ is a product norm on $E \times F$. Suppose that B is a continuous bilinear mapping from the product $E \times F$ into G. Then

$$B((x, y) + (h, k)) = B(x, y) + B(h, y) + B(x, k) + B(h, k),$$

and $\|B(h, k)\|_G \leq \|B\|_\infty \|h\|_E \|k\|_F \leq \|B\|_\infty \cdot \|(h, k)\|^2$, where $\|B\|_\infty = \sup\{\|B(x, y)\|_G : \|x\|_E \leq 1, \|y\|_F \leq 1\}$ (see Exercises 14.3.1 and 14.3.2), so that $\|B(h, k)\|_G / \|(h, k)\| \to 0$ as $(h, k) \to 0$. Thus B is differentiable at each point of $E \times F$, and $DB_{(x,y)}(h, k) = B(h, y) + B(x, k)$.

Example 17.1.5 The derivative of the norm of a real inner-product space.

Suppose that E is a real inner-product space. Let $N(x) = \|x\|$. Can we differentiate N on E? If $x \in E \setminus \{0\}$ let $l_x(\lambda) = \lambda x$. Then $\|l_x(\lambda)\| = |\lambda| \|x\|$, so that the mapping $N \circ l_x$ is not differentiable at 0. Consequently N is not differentiable at 0.

Suppose on the other hand that $x \neq 0$. We write N as a product of mappings, consider each factor separately, and use the chain rule. We write $N(x) = (S \circ P \circ J)(x)$, where $J : E \setminus \{0\} \to E \times E$ is defined as $J(x) = (x, x)$, $P : E \times E \to \mathbf{R}$ is the inner product map $P(x, y) = \langle x, y \rangle$ and $S : (0, \infty) \to (0, \infty)$ is the square root map $S(x) = \sqrt{x}$. Then $S(P(J(x))) = \|x\|$, and

$$DJ_x = J, \text{ since } J \text{ is linear,}$$
$$DP_{(x,y)}(h, k) = \langle h, y \rangle + \langle x, k \rangle, \text{ by the previous example,}$$
$$\text{and } DS_x(h) = h/2\sqrt{x}.$$

By the chain rule, N is differentiable at x and

$$DN_x(h) = DS_{P(J(x))} DP_{J(x)} DJ_x(h) = DS_{P(J(x))} DP_{J(x)}(h, h)$$
$$= DS_{P(J(x))}(2\langle x, h \rangle) = \frac{\langle x, h \rangle}{\|x\|}.$$

Thus N is differentiable at each point of $E \setminus \{0\}$.

Example 17.1.6 The derivative of the mapping $J : U \to U^{-1}$ from $GL(E)$ to itself.

Suppose that U is in the general linear group $GL(E)$ of a Banach space E. Since $GL(E)$ is an open subset of $L(E)$, there exists $\delta > 0$ such that $N_\delta(U) \subseteq GL(E)$. Suppose that $0 < \|T\| < \delta$. Since $T = (U + T) - U$, it follows that $(U + T)^{-1}TU^{-1} = U^{-1} - (U + T)^{-1}$, and so

$$(U + T)^{-1} = U^{-1} - U^{-1}TU^{-1} + r(T),$$

where $r(T) = (U^{-1} - (U + T)^{-1})TU^{-1}$. Thus

$$\|r(T)\| \le \|U^{-1} - (U + T)^{-1}\| \cdot \|T\| \cdot \|U^{-1}\|.$$

Since $(U + T)^{-1} \to U^{-1}$ as $T \to 0$, it follows that $r(T)/\|T\| \to 0$ as $T \to 0$. Thus J is differentiable at U, and $DJ_U(T) = -U^{-1}TU^{-1}$. Consequently, J is continuously differentiable on $GL(E)$.

Exercises

17.1.1 Find the points of $\mathbf{R}^d$ at which the norms $\|x\|_1 = \sum_{j=1}^{d} |x_j|$ and $\|x\|_\infty = \max\{|x_j| : 1 \le j \le d\}$ are differentiable, and determine the derivatives at these points.

17.1.2 Suppose that E is a real inner product space. Let $\rho(x) = x/\|x\|$, for $x \in E \setminus \{0\}$. Show that ρ is differentiable and that

$$D\rho_x(h) = \frac{h}{\|x\|} - \frac{\langle x, h \rangle \, x}{\|x\|^3}.$$

Verify that $\langle D\rho_x(h), x \rangle = 0$. Explain the geometric reason for this.

17.1.3 Suppose that $(E, \|.\|_E)$ is a normed space. If $T \in L(E)$, let $s(T) = T^2$. Show that s is a differentiable mapping $L(F) \to L(F)$, and that $Ds_T(S) = ST + TS$.

Show that if $n \in \mathbf{N}$ then the mapping $p^{(n)} : L(E) \to L(E)$ defined by $p^{(n)}(T) = T^n$ is differentiable, and determine its derivative.

17.1.4 Let $M_d(\mathbf{R})$ be the vector space of $d \times d$ real matrices. Show that the mapping $T \to \det T : M_d(\mathbf{R}) \to \mathbf{R}$ is differentiable. Show that if I is the identity matrix then $D\det_I(S) = \mathrm{tr}(S)$ (where $\mathrm{tr}(S) = \sum_{j=1}^{d} s_{jj}$ is the *trace* of S). Show that if T is invertible then $D\det_T(S) = \det T(\mathrm{tr}(T^{-1}S))$.

17.1.5 Suppose that $f : U \to F$ is a mapping from an open subset U of a normed space $(E, \|.\|_E)$ into a normed space $(F, \|.\|_F)$ which is differentiable at a point a of U, and suppose that there exists $T \in L(F, E)$ such that $x = TDf_a(x)$, for all $x \in E$. Show that there exists $\delta > 0$ such that $N_\delta(a) \subseteq U$, and such that if $x \in N_\delta(a)$ then $f(x) \neq f(a)$.

Does there necessarily exist $\delta > 0$ such that $N_\delta(a) \subseteq U$, and such that if $x, y \in N_\delta(a)$ then $f(x) \neq f(y)$?

17.2 The mean-value inequality

The mean-value theorem is a powerful result for real-valued functions on a closed interval in $\mathbf{R}$. We cannot hope for an equivalent result for vector-valued functions. For example, let $f : [0, 2\pi] \to \mathbf{R}^2$ be defined by setting $f(t) = (\cos t, \sin t)$ for $t \in [0, 2\pi]$. Then $f(0) = f(2\pi) = (1, 0)$, while $f'(t) = (-\sin t, \cos t) \neq (0, 0)$ for any t, so that there exists no t in $[0, 2\pi]$ for which $f(2\pi) - f(0) = 2\pi f'(t)$. We can however prove an inequality, known as the mean-value inequality, which is extremely useful. First we consider functions in a closed interval.

Theorem 17.2.1 *Suppose that $f : I \to F$ is a path from a closed interval $[a, b]$ in $\mathbf{R}$ into a normed space $(F, \|.\|_F)$ which is differentiable at each point of (a, b). Then*

$$\|f(b) - f(a)\|_F \leq (b - a) \sup\{\|f'(c)\|_F : c \in [a, b]\}.$$

Proof Let $M = \sup\{\|f'(c)\|_F : c \in [a, b]\}$. If $M = \infty$ there is nothing to prove. Otherwise, suppose that $a < a' < b' < b$ and that $\epsilon > 0$. We shall show that $\|f(b') - f(a')\|_F \leq (b' - a')(M + \epsilon)$. Then since ϵ is arbitrary, $\|f(b') - f(a')\|_F \leq (b - a)M$. Since f is continuous at a and b, and since the mapping $x \to \|x\|$ is continuous, it follows that $\|f(b) - f(a)\|_F \leq (b - a)M$.

Let $B = \{t \in [a', b'] : \|f(t) - f(a')\|_F \leq (t - a')(M + \epsilon)\}$. Since the function $t \to \|f(t) - f(a')\|_F - (t - a')(M + \epsilon)$ is continuous on $[a', b']$, B is a non-empty closed subset of $[a', b']$. Let $c = \sup B$. If $c < b'$, it follows from Proposition 17.1.1 (ii) that there exists $c < d < b'$ such that

$$\|f(d) - f(c)\|_F \leq (d - c)(\|f'(c)\|_F + \epsilon) \leq (d - c)(M + \epsilon).$$

But then

$$\|f(d) - f(a')\|_F \leq \|f(d) - f(c)\|_F + \|f(c) - f(a')\|_F$$
$$\leq (d - c)(M + \epsilon) + (c - a')(M + \epsilon) = (d - a')(M + \epsilon).$$

Thus $d \in B$, giving a contradiction. Thus $c = b'$. But B is closed, and so $b' \in B$; thus $\|f(b') - f(a')\|_F \leq (b' - a')(M + \epsilon)$. $\qquad\square$

We now extend this result to functions of a vector variable.

Theorem 17.2.2 (The mean-value inequality) *Suppose that $f : U \to F$ is a continuous function from an open subset U of a normed space $(E, \|.\|_E)$ into a normed space $(F, \|.\|_F)$. Suppose that the closed segment $[a, b]$ is contained in U, and that f is differentiable at each point of the open segment (a, b). Then*

$$\|f(b) - f(a)\|_F \leq \|b - a\|_E \sup\{\|Df_c\| : c \in (a, b)\}.$$

Proof   This follows from the chain rule. Let $l(t) = (1-t)a + tb$. Then $f \circ l$ is continuous on $[0, 1]$ and differentiable on $(0, 1)$, and $(f \circ l)'(t) = Df_{l(t)}(b-a)$, by the chain rule. Thus

$$\|f(b) - f(a)\|_F \leq \sup\{\|(f \circ l)'(t)\|_F : t \in (0, 1)\}$$
$$= \sup\{\|Df_c(b - a)\|_F : c \in (a, b)\}$$
$$\leq \|b - a\|_E \sup\{\|Df_c\| : c \in (a, b)\}.$$

$\qquad\square$

Corollary 17.2.3   *Suppose that $f : U \to F$ is a continuous function from a non-empty open connected subset U of a normed space $(E, \|.\|_E)$ into a normed space $(F, \|.\|_F)$, that f is differentiable at each point of U and that $Df_a = 0$ for all $a \in U$. Then f is a constant.*

Proof   Let x_0 be an arbitrary element of U, and let $C = \{x \in U : f(x) = f(x_0)\}$. On the one hand, C is a closed subset of U, since f is continuous. On the other hand, if $c \in C$, there exists $\delta > 0$ such that $N_\delta(c) \subseteq U$. If $d \in N_\delta(c)$ the closed segment $[c, d]$ is contained in U, and so $\|f(d) - f(c)\| = 0$, by the theorem. Thus $d \in C$, and so $N_\delta(c) \subseteq C$. Hence C is open; since U is connected, $C = U$. $\qquad\square$

Corollary 17.2.4   *Suppose that $G : U \to L(E, F)$ is a continuous function from a non-empty connected open subset U of a normed space $(E, \|.\|_E)$ into $L(E, F)$, where F is a normed space $(F, \|.\|_F)$. If f_1 and f_2 are any two solutions of the partial differential equation $Df_x = G(x)$, for $x \in U$, then $f_1 - f_2$ is a constant function.*

Proof   For $D(f_1 - f_2) = 0$. $\qquad\square$

Corollary 17.2.5 *Suppose that $f : U \to F$ is a continuous function from a non-empty convex open subset U of a normed space $(E, \|.\|_E)$ into a normed space $(F, \|.\|_F)$, that f is differentiable at each point of U and that $\|Df_a\| \leq M$ for all $a \in U$. Then f is a Lipschitz function on U, with constant M.*

Proof For if $a, b \in U$ then $[a, b] \subseteq U$, so that $\|f(b) - f(a)\| \leq M \|b - a\|$. □

The following form of the mean-value inequality is also useful.

Corollary 17.2.6 *Suppose that $f : U \to F$ is a continuous function from an open subset U of a normed space $(E, \|.\|_E)$ into a normed space $(F, \|.\|_F)$. Suppose that the closed segment $[a, b]$ is contained in U, and that f is differentiable at each point of the open segment (a, b). If $T \in L(E, F)$ then*

$$\|f(b) - f(a) - T(b - a)\| \leq \|b - a\| \sup\{\|Df_c - T\| : c \in (a, b)\}.$$

Proof Let $g(x) = f(x) - T(x)$. Then g is differentiable, with derivative $Df_x - T$, for $x \in (a, b)$. Apply the mean-value value inequality to g. □

The mean-value inequality allows us to obtain a more general version of Theorem 12.1.8.

Theorem 17.2.7 *Suppose that $(f_n)_{n=1}^\infty$ is a sequence of differentiable real-valued functions on a bounded open convex subset U of a normed space $(E, \|.\|_E)$, taking values in a Banach space $(F, \|.\|_F)$. Suppose that*
  (i) there exists $c \in U$ such that $f_n(c)$ converges, to $f(c)$ say, as $n \to \infty$, and
  (ii) the sequence $(Df_n)_{n=1}^\infty$ of derivatives converges in the operator norm uniformly on U to a function g from U to $L(E, F)$.
  Then there exists a function $f : U \to F$ such that $f_n \to f$ uniformly on U. Further, f is differentiable on U, and $Df(x) = g(x)$ for all $x \in U$.

Proof This follows by making straightforward changes to the proof of Theorem 12.1.8; the details are left as a worthwhile exercise for the reader. □

Important examples of rectifiable paths are given by piecewise continuously differentiable paths. Suppose that $(E, \|.\|)$ is a Banach space. A path $\gamma : [a, b] \to E$ is *continuously differentiable* if it is differentiable on $[a, b]$ (with one-sided derivatives at a and b), with derivative γ' continuous on $[a, b]$. A *piecewise continuously differentiable path*, is a juxtaposition of finitely many continuously differentiable paths.

Theorem 17.2.8 *If $(E, \|.\|)$ is a Banach space and γ is a piecewise continuously differentiable path in E then γ is rectifiable, and $l_{[a,b]}(\gamma) = \int_a^b \|\gamma'(t)\| \, dt$.*

Proof   It is clearly sufficient to consider the case where γ is continuously differentiable. Then γ' is bounded on $[a, b]$; let $M = \sup_{t \in [a,b]} \|\gamma'(t)\|$. Suppose that $\epsilon > 0$. Let $\eta = \epsilon/4(b - a)$. Since γ' is uniformly continuous on $[a, b]$, there exists $\delta > 0$ such that if $s, t \in [a, b]$ and $|s - t| < \delta$ then $\|\gamma'(s) - \gamma'(t)\| < \eta$.

Suppose that

$$a = t_0 < t_1 < \cdots < t_n = b, \text{ with } t_j - t_{j-1} < \delta \text{ for } 1 \le j \le n.$$

By Corollary 17.2.6

$$\|\gamma(t_j) - \gamma(t_{j-1}) - (t_j - t_{j-1})\gamma'(t_{j-1})\| \le \eta(t_j - t_{j-1}),$$

for $1 \le j \le n$. Then $\|\gamma(t_j) - \gamma(t_{j-1})\| \le (M + \eta)(t_j - t_{j-1})$, so that

$$\sum_{j=1}^{n} \|\gamma(t_j) - \gamma(t_{j-1})\| \le (M + \eta)(b - a),$$

and γ is rectifiable.

Now $\|\gamma'(t) - \gamma'(t_{j-1})\| < \eta$ for $t \in [t_{j-1}, t_j]$, so that

$$\left| \int_{t_{j-1}}^{t_j} \|\gamma'(t)\| \, dt - (t_j - t_{j-1}) \|\gamma'(t_{j-1})\| \right| \le \eta(t_j - t_{j-1}),$$

and so

$$\left| \int_{t_{j-1}}^{t_j} \|\gamma'(t)\| \, dt - \|\gamma(t_j) - \gamma(t_{j-1})\| \right| \le 2\eta(t_j - t_{j-1}),$$

Adding, we see that

$$\left| \int_a^b \|\gamma'(t)\| \, dt - \sum_{j-1}^{n} \|\gamma(t_j) - \gamma(t_{j-1})\| \right| \le 2\eta(b - a) = \epsilon/2$$

and so

$$\left| \int_a^b \|\gamma'(t)\| \, dt - l_{[a,b]}(\gamma) \right| < \epsilon.$$

Since ϵ is arbitrary, the result follows. $\square$

Corollary 17.2.9 *If $l(t) = l_{[a,t]}(\gamma)$ then l is piecewise differentiable on $[a, b]$ and $l'(t) = \|\gamma'(t)\|$.*

Corollary 17.2.10 *If β is the path-length parametrization of γ then β is a piecewise continuously differentiable path, and $\|\beta'(t)\| = 1$ (suitably interpreted at points of juxtaposition).*

Proof For $\beta = \gamma \circ l^{-1}$. $\square$

Corollary 17.2.11 *If $\delta : [0, d] \to E$ is a continuously differentiable parametrization of γ and $\|\delta'(t)\| = 1$ for $t \in [0, d]$, then δ is the path-length parametrization of γ.*

Proof For $l(t) = \int_0^t \|\delta'(s)\| \, ds = t$. $\square$

As an example, the circular path $\kappa_r(w)$ in $\mathbf{R}^2$ is differentiable, and $(\kappa_r(w))'(t) = (-r\sin t, r\cos t)$, so that $\|(\kappa_r(w))'(t)\| = r$, and $l_{[0,2\pi]}(\kappa_r(w)) = 2\pi r$.

Recall that a path is *piecewise-linear*, or *polygonal*, if it is the juxtaposition of finitely many linear paths. We can approximate a rectifiable path in a Banach space by a piecewise-linear path, without increasing path-length.

Proposition 17.2.12 *Suppose that $\gamma : [a, b] \to U$ is a rectifiable path in an open subset U of a Banach space $(E, \|.\|)$, and that $\epsilon > 0$. Then there exists a piecewise-linear path $\delta : [a, b] \to U$ such that $\|\delta(t) - \gamma(t)\| \le \epsilon$ for $t \in [a, b]$, and $l_{[a,b]}(\delta) \le l_{[a,b]}(\gamma)$.*

Proof Since $[\gamma]$ is compact and since γ is uniformly continuous on $[a, b]$ there exists $\eta > 0$ such that if $s, t \in [a, b]$ and $|s - t| < \eta$ then $N_\epsilon(\gamma(t)) \subseteq U$ and $\|\gamma(s) - \gamma(t)\| < \epsilon/2$.

Let $a = t_0 < t_1 < \cdots < t_k = b$ be a dissection of $[a, b]$ with $t_j - t_{j-1} < \eta$ for $1 \le j \le k$.

If $t = (1 - \lambda)t_{j-1} + \lambda t_j \in [t_{j-1}, t_j]$, let $\delta(t) = (1 - \lambda)\gamma(t_{j-1}) + \lambda\gamma(t_j)$.

Since $\delta(t) = \gamma(t_{j-1}) + \lambda(\gamma(t_j) - \gamma(t_{j-1}))$, $\delta(t) \in U$, and

$$\|\delta(t) - \gamma(t)\| \le \|\delta(t) - \gamma(t_{j-1})\| + \|\gamma(t) - \gamma(t_{j-1})\| < \epsilon.$$

Also, $l_{[a,b]}(\delta) = \sum_{j=1}^k \|\gamma(t_j) - \gamma(t_{j-1})\| \le l_{[a,b]}(\gamma)$. $\square$

Exercises

17.2.1 We have used a connectedness argument to prove the mean-value inequality. It can also be proved using a compactness argument. With the notation introduced in Theorem 17.2.1, show that there exist

$a' = t_0 < t_1 < \cdots < t_k = b'$ such that

$$\|f(t_j) - f(t_{j-1})\|_F \leq (t_j - t_{j-1})(M + \epsilon) \text{ for } 1 \leq j \leq k,$$

and deduce the mean-value inequality.

17.2.2 Give the details of the proof of Theorem 17.2.7.

17.2.3 *[The Newton–Raphson method]* Suppose that f is a differentiable mapping from an open neighbourhood $N_t(x_0)$ of a point x_0 of a Banach space $(E, \|.\|_E)$ into a normed space $(F, \|.\|_F)$, and that there exists $s > 0$ such that

- $\|f(x_0)\|_F \leq t/2s$;
- if $x, y \in N_t(x_0)$ then $\|Df_x - Df_y\| \leq 1/2s$; and
- if $x \in N_t(x_0)$ then there exists $J_x \in L(F, E)$ with $\|J_x\| \leq s$ such that $J_x \circ Df_x = Df_x \circ J_x = I$, where I is the identity mapping on E.

Define (x_n) by setting $x_n = x_{n-1} - J_x(f(x_{n-1}))$, for $n \in \mathbf{N}$. Use Corollary 17.2.6 to show that

$$\|x_n - x_{n-1}\|_E \leq t/2^n \text{ and } \|f(x_n)\|_F \leq t/2^{n+1}s.$$

Show that (x_n) converges to a point x_∞ of $N_t(x_0)$, that $f(x_\infty) = 0$ and that x_∞ is the only point in $N_t(x_0)$ with this property.

17.2.4 Suppose that γ is a rectifiable path in $\mathbf{R}^2$. Show that, given $\epsilon > 0$ there is a finite set of closed rectangles whose union contains the track $[\gamma]$, and has area less than ϵ. Deduce that the interior of $[\gamma]$ is empty. Deduce that a space-filling path in $\mathbf{R}^2$ is not rectifiable.

17.2.5 Suppose that f is a function defined on a connected open subset U of a normed space $(E, \|.\|_E)$ taking values in a normed space $(F, \|.\|_F)$, and that $\|f(x) - f(y)\|_F \leq K \|x - y\|_E^\alpha$, for $x, y \in U$, where $K > 0$ and $\alpha > 1$. Show that f is constant.

17.3 Partial and directional derivatives

The derivative of a vector-valued function of a vector variable is a linear operator. It is desirable, where possible, to express it in simpler terms. Suppose first that $(E, \|.\|_E) = \prod_{j=1}^n (E_j, \|.\|_j)$ is the product of normed spaces $(E_j, \|.\|_j)$, and that $f : U \to F$ is a function from an open subset U of E into a normed space $(F, \|.\|_F)$. We can vary each variable separately. If $h_j \in E_j$, let $i_j(h_j) = \widetilde{h}_j = (0, \ldots, 0, h_j, 0, \ldots, 0)$, where h_j occurs in the jth place. If

$a \in U$ let

$$k_{a,j}(h_j) = a + i_j(h_j) = (a_1, \ldots, a_{j-1}, a_j + h_j, a_{j+1}, \ldots a_n).$$

Then $k_{a,j}$ is differentiable at every point of E_j, and $Dk_{a,j} = i_j$. Also $(k_{a,j})^{-1}(U)$ is an open subset of E_j, containing 0. If the mapping $f \circ k_{a,j} : (k_{a,j})^{-1}(U) \to F$ is differentiable at 0, we denote its derivative by $D_j f_a$; this is the jth *partial derivative of* f *at* a, and is an element of $L(E_j, F)$.

If $E = \mathbf{R}^d$, we set

$$D_j f_a(1) = (\partial f / \partial x_j)(a) = \frac{\partial f}{\partial x_j}(a).$$

Then

$$\frac{\partial f}{\partial x_j}(a) \in F \text{ and } D_j f_a(\lambda) = \lambda \frac{\partial f}{\partial x_j}(a).$$

Suppose that f is differentiable at a. Then

$$f(k_{a,j}(h_j)) = f(a + \tilde{h}_j) = f(a) + Df_a(\tilde{h}_j) + r(h_j)$$
$$= f(k_{a,j}(0)) + Df_a(\tilde{h}_j) + r(h_j),$$

so that the jth partial derivative $D_j f_a$ exists, for $1 \le j \le d$, and $D_j f_a(h_j) = Df_a(\tilde{h}_j)$, for $h_j \in E_j$. Further, if $h = (h_j)_{j=1}^d$ then

$$Df_a(h) = \sum_{j=1}^{d} Df_a(\tilde{h}_j) = \sum_{j=1}^{d} D_j f_a(h_j).$$

In particular, if $E = \mathbf{R}^d$ and $F = \mathbf{R}^k$, and $f = (f_1, \ldots, f_k)$ then

$$f_i(a + h) = f_i(a) + \sum_{j=1}^{d} h_j \frac{\partial f_i}{\partial x_j}(a) + r_i(h).$$

where $(\partial f_i / \partial x_j)(a) \in \mathbf{R}$: the derivative $Df_a \in L(\mathbf{R}^d, \mathbf{R}^k)$ is represented by the $k \times d$ real matrix $(\partial f_i / \partial x_j(a))$.

When $E = \mathbf{R}^d$ and $F = \mathbf{R}$, then

$$\nabla f_a = \left(\frac{\partial f}{\partial x_1}(a), \ldots, \frac{\partial f}{\partial x_d}(a) \right).$$

In the special case where $d = k$, we shall need to know when the linear operator $Df_a : \mathbf{R}^d \to \mathbf{R}^d$ is invertible. This is the case if and only if the determinant of the matrix $(\partial f_i / \partial x_j(a))_{i=1, j=1}^{d, d}$ is non-zero (Appendix B,

Corollary B.3.2). This determinant is called the *Jacobian* of f at a, and is denoted by

$$J_f(a) \text{ or } \frac{\partial(f_1, \ldots, f_d)}{\partial(x_1, \ldots, x_d)}(a).$$

We can also consider directional derivatives. Suppose that f is a mapping from an open subset U of a normed space $(E, \|.\|_E)$ into a normed space $(F, \|.\|_F)$. Suppose that $y \in E$ and that $y \neq 0$. There exists $\epsilon > 0$ such that the interval $[a, a + \epsilon y)$ is contained in U. We say that f has a *directional derivative in the direction y* if there exists an element $f'_y(a)$ of F such that if

$$r_y(\lambda) = f(a + \lambda y) - f(a) - \lambda f'_y(a), \text{ for } 0 < \lambda < \epsilon,$$

then $r_y(\lambda)/\lambda \to 0$ as $\lambda \to 0$. The vector $f'_y(a)$ is then the *directional derivative in the direction y*.

If f is differentiable at a, then it has directional derivatives in all directions, and $f'_y(a) = Df_a(y)$, for $y \neq 0$. In the appropriate case, it also has partial derivatives. The converse statements are *not* true, as the following simple example shows. Let

$$f(x, y) = \frac{x^3 y}{x^4 + y^2} \text{ for } (x, y) \neq (0, 0) \text{ and let } f(0, 0) = 0.$$

Then the reader should verify the following statements:

- f is a continuous real-valued function on $\mathbf{R}^2$;
- f is differentiable at every $(a, b) \neq (0, 0)$, and the mapping $(a, b) \to Df_{(a,b)}$ is continuous and bounded on $\mathbf{R}^2 \setminus \{(0, 0)\}$ (use the chain rule, rather than elaborate calculations);
- f has partial derivatives at $(0, 0)$ and

$$\frac{\partial f}{\partial x_1}(0, 0) = \frac{\partial f}{\partial x_2}(0, 0) = 0;$$

- f has directional derivatives in all directions at $(0, 0)$, all equal to 0.

On the other hand, f is not differentiable at 0. If it were, the derivative would have to be 0, and so $f(x, y)/\|(x, y)\|$ would tend to 0 as $(x, y) \to 0$. But $f(t, t^2) = t/2$ for all $t \in \mathbf{R} \setminus \{0\}$, and so this is not the case.

This is inconvenient, to say the least. For example, if we are investigating the differentiability of a function defined on an open subset of $\mathbf{R}^d$, the first step will be to find out whether or not partial derivatives exist. If they do, can we use them to tell whether the function is differentiable or not? The next theorem gives an extremely useful test.

Theorem 17.3.1 *Suppose that $(E, \|.\|_E) = E_1 \times E_2$ is the product of normed spaces $(E_1, \|.\|_1)$ and $(E_2, \|.\|_2)$ and that $f : U \to F$ is a function from an open subset U of E into a normed space $(F, \|.\|_F)$. Suppose that $D_1 f_a$ exists at a and that $D_2 f_x$ exists for all x in a neighbourhood $N_\theta(a)$ of a, and is continuous at a. Then f is differentiable at a.*

Proof   If $a + h \in U$, let

$$r(h_1, h_2) = f(a + h) - f(a) - D_1 f_a(h_1) - D_2 f_a(h_2).$$

We must show that $r(h)/\|h\|_E \to 0$ as $h \to 0$. Now $D_1 r_0 = 0$, $D_2 r_0 = 0$ and $D_2 r$ is continuous at 0. Suppose that $\epsilon > 0$. There exists $0 < \delta < \theta$ such that

$$\|r(h_1, 0)\|_F \leq \epsilon \|h\|_E /2 \text{ and } \|D_2 r_h\| = \|D_2 f_{a+h} - D_2 f_a\| < \epsilon/2 \text{ for } \|h\|_E < \delta.$$

By the mean-value inequality, if $\|h\|_E < \delta$ then

$$\|r(h_1, h_2) - r(h_1, 0)\|_F \leq \|h_2\|_2 \sup\{\|D_2 r_{(h_1, \lambda h_2)}\| : 0 \leq \lambda \leq 1\}$$
$$< \epsilon \|h\|_E /2,$$

and so

$$\|r(h)\|_F \leq \|r(h_1, 0)\|_F + \|r(h_1, h_2) - r(h_1, 0)\|_F < \epsilon \|h\|_E.$$

$\square$

Corollary 17.3.2 *Suppose that $(E, \|.\|_E) = \prod_{j=1}^n (E_j, \|.\|_j)$ is a product of normed spaces $(E_j, \|.\|_j)$ and that $f : U \to F$ is a function from an open subset U of E into a normed space $(F, \|.\|_F)$. Suppose that all the partial derivatives $(\partial f/\partial x_j)(b)$ exist, for all b in a neighbourhood of a, and are continuous at a. Then f is differentiable at a.*

Proof   A simple inductive argument. $\square$

Exercises

17.3.1 Let g be a real-valued function on the unit sphere S^{d-1}. Let $k(0) = 0$ and let $k(x) = \|x\|^2 .g(x/\|x\|)$, for $x \neq 0$. Show that k has directional derivatives in all directions at 0. Give examples to show that k need not be continuous, and to show that k can be continuous, but not differentiable at 0.

17.3.2 Suppose that $(E, \|.\|_E)$ is a normed space and that $f : E \times E \to$ $(F, \|.\|_F)$ is a differentiable mapping. If $c \in E$, let $g_c(x) = f(x, c - x)$. Show that $g_c : E \to F$ is differentiable, and determine its derivative. Suppose that $D_1 f = -D_2 f$. Show that there is a differentiable function $k : E \to F$ such that $f(x, y) = k(x - y)$.

17.4 The inverse mapping theorem

We have seen in Volume I, Propositions 6.4.4 and 6.4.5 that a continuous function f on an open interval (a, b) of $\mathbf{R}$ is injective and has a continuous inverse if and only if it is strictly monotonic. A sufficient condition for this is that f is differentiable on (a, b), and that $f'(x) \neq 0$ for all $x \in (a, b)$ (Volume I, Corollary 7.3.3). Thus if $f'(a) \neq 0$ and f' is continuous at a then there is a neighbourhood $N_\epsilon(a)$ such that f is a homeomorphism of $N_\epsilon(a)$ onto $f(N_\epsilon(a))$. (The condition that f' is continuous cannot be dropped: see Exercise 7.5.9. in Volume I.)

We now prove a corresponding result for vector valued functions. If $f : W \to F$ is a mapping from an open subset W of a Banach space $(E, \|.\|_E)$ into a Banach space $(F, \|.\|_F)$ which is differentiable at a point a of W, we say that Df_a is *invertible* if Df_a is a bijection of E onto F. By the isomorphism theorem (Corollary 14.7.9), Df_a^{-1} is a continuous linear mapping, so that Df_a is an isomorphism of $(E, \|.\|_E)$ onto $(F, \|.\|_F)$. (Df_a^{-1} is trivially continuous when E and F are finite-dimensional.)

Theorem 17.4.1 (The differentiable inverse mapping theorem) *Suppose that $f : W \to F$ is a differentiable mapping from an open subset W of a Banach space $(E, \|.\|_E)$ into a Banach space $(F, \|.\|_F)$, and that $a \in W$. Suppose that the derivative Df_x is continuous at a and that Df_a is invertible. Then there is a neighbourhood $N_\theta(a)$ such that $f(N_\theta(a))$ is open in F, $f : N_\theta(a) \to f(N_\theta(a))$ is a homeomorphism, and the inverse mapping f^{-1} is differentiable at $f(a)$, with derivative $(Df_a)^{-1}$.*

Proof   The proof uses the Lipschitz inverse function theorem (Theorem 14.6.6). The first step is to simplify the problem. Let $V = W - a$, and let $g(x) = f(x + a) - f(a)$. Then $g(0) = 0$ and $Dg_x = Df_{x+a}$; the mapping g from V to E is differentiable, and the mapping $Dg : V \to L(E, F)$ is continuous at 0. Now let $k = Df_a^{-1} \circ g = Dg_0^{-1} \circ g$. Then

- $k(0) = 0$,
- $k : V \to E$ is differentiable,

- the mapping $x \to Dk_x$ is continuous at 0, and
- $Dk_0 = I$.

We prove the result for the function k. Then since $f(x) = g(x-a) + f(a) = Df_a(k(x-a)) + f(a)$, the result follows for f.

We denote the open ball $\{x : \|x\|_E < \alpha\}$ with radius α by U_α. Let $j(x) = k(x) - x$, so that $Dj_0 = 0$. Since the mapping $x \to Dj_x$ is continuous at 0, there exists a ball U_θ such that $\|Dj_x\| < 1/2$ for $x \in U_\theta$. If $x, y \in U_\theta$ then $[x, y] \subseteq U_\theta$, so that

$$M = \sup \left\{ \|Dj_{(1-t)x+ty}\| : 0 \le t \le 1 \right\} \le 1/2.$$

Hence, by the mean-value inequality,

$$\|j(x) - j(y)\|_E \le M \|x - y\|_E \le \|x - y\|_E /2;$$

thus j is a Lipschitz mapping on $N_\theta(a)$, with constant $1/2$. It therefore follows from the Lipschitz inverse function theorem that $k(U_\theta)$ is open and that k is a homeomorphism of U_θ onto $k(U_\theta)$.

It remains to show that $k^{-1} : k(U_\theta) \to U_\theta$ is differentiable at 0, with derivative I. Suppose that $0 < \epsilon \le 1$. There exists $0 < \delta \le \theta$ such that if $h \in U_\delta$ then $\|j(h)\|_E < \epsilon \|h\|_E /2$. Since $k^{-1} : k(U_\theta) \to U_\theta$ is continuous, and since $k(U_\theta)$ is an open neighbourhood of 0, there exists $\eta > 0$ such that $U_\eta \subseteq k(U_\theta)$ and $k^{-1}(U_\eta) \subseteq U_\delta$.

Suppose that $\|y\| \le \eta$. Let $k^{-1}(y) = y + s(y)$. We shall show that $\|s(y)\|_E < \epsilon \|y\|_E$, so that k^{-1} is differentiable at 0, with derivative I. First, $\|y + s(y)\|_E = \|k^{-1}(y)\|_E < \delta$, so that

$$\|j(y + s(y))\|_E < \epsilon \|y + s(y)\|_E /2.$$

Next,

$$y = k(k^{-1}(y)) = k(y + s(y)) = y + s(y) + j(y + s(y)),$$

so that $s(y) = -j(y + s(y))$. Thus

$$\|s(y)\|_E = \|j(y + s(y))\|_E \le \epsilon \|y + s(y)\|_E /2 \le \epsilon \|y\|_E /2 + \epsilon \|s(y)\|_E /2;$$

since $\epsilon < 1$, $\|s(y)\|_E \le \epsilon \|y\|_E$. $\quad\square$

Note that if the conditions of the theorem are satisfied then Df_a is a linear isomorphism of $(E, \|.\|_E)$ onto $(F, \|.\|_F)$. In practice, the theorem is usually applied when $(E, \|.\|_E) = (F, \|.\|_F)$.

Corollary 17.4.2 *Suppose that f is a continuously differentiable function on W and that Df_x is invertible, for each $x \in W$. If V is an open subset of W then $f(V)$ is open in F.*

Proof   This follows immediately from the theorem.   □

Corollary 17.4.3 *Suppose that f is a continuously differentiable function on W and that Df_x is invertible, for each $x \in W$. If f is injective then f is a homeomorphism of W onto $f(W)$, f^{-1} is continuously differentiable on $f(W)$ and $Df^{-1}_{f(x)} = (Df_x)^{-1}$, for $x \in W$.*

Proof   The mapping f^{-1} is differentiable at each point of $f(W)$: we must show that Df^{-1} is continuous on $f(W)$. But the mapping $y \to Df^{-1}_y$ is the composition of the mapping $y \to f^{-1}(y) : f(W) \to W$, the mapping $w \to Df_w : W \to GL(E)$ and the mapping $J : U \to U^{-1} : GL(E) \to GL(E)$, each of which is continuous.   □

A mapping f which satisfies the conditions of this corollary is called a *diffeomorphism* of W onto $f(W)$.

17.5 The implicit function theorem

We have an implicit function theorem for differentiable functions.

Theorem 17.5.1 (The implicit function theorem) *Suppose that $(E_1, \|.\|_1)$, $(E_2, \|.\|_2)$ and $(F, \|.\|_F)$ are normed spaces, that $(E_2, \|.\|_2)$ and $(F, \|.\|_F)$ are complete and that f is a differentiable mapping from an open subset U of $E_1 \times E_2$ into F. Suppose that $a = (a_1, a_2) \in U$, that the partial derivative $D_2 f_a : E_2 \to F$ is invertible, and that the mapping $x \to Df_x$ is continuous at a. Then there is a neighbourhood N of a_1 in E_1 and a unique mapping $\phi : N \to E_2$ such that*

1. *ϕ is continuous;*
2. *the cross-section $N \times \{a_2\}$ is contained in U;*
3. *$(x, \phi(x)) \in U$ for $x \in N$;*
4. *$f(x, \phi(x)) = f(a_1, a_2)$ for $x \in N$;*
5. *ϕ is differentiable at a_1, and $D\phi_{a_1} = -(D_2 f_a)^{-1} D_1 f_a$.*

Thus the theorem says that there is a neighbourhood of a_1 on which there is a unique solution to the equation $f(x, y) = f(a_1, a_2)$, that the solution is continuous in the neighbourhood, and that the solution is differentiable at a_1.

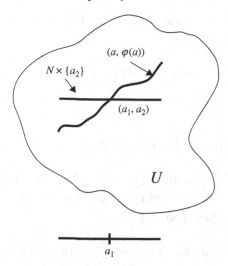

Figure 17.5. The implicit function theorem.

Proof   As with the inverse function theorem, we simplify the problem. Let $V = U - a$, and let $g(x_1, x_2) = f(x_1 + a_1, x_2 + a_2) - f(a_1, a_2)$, for $(x_1, x_2) \in V$. Let $k = (D_2 f_a^{-1}) \circ g$. Then k maps V into E_2, and

- $k(0) = 0$,
- $k : V \to E_2$ is differentiable,
- the mapping $x \to Dk_x$ is continuous at 0, and
- $D_2 k_0 = I$.

We set $T = D_1 k_0 = (D_2 f_a^{-1}) \circ D_1 f_a$, and set

$$j(x_1, x_2) = k(x_1, x_2) - x_2, \quad r(x_1, x_2) = k(x_1, x_2) - T(x_1) - x_2.$$

We give $E_1 \times E_2$ the norm $\|(x_1, x_2)\| = \|x_1\|_1 + \|x_2\|_2$. Let $K = 2\|T\| + 1$. Since r is continuously differentiable at 0, and $Dr_0 = 0$, there exists $\delta > 0$ such that $N_\delta(0) \subseteq V$ and $\|Dr_x\| \le 1/2$ for $x \in N_\delta(0)$. By Corollary 17.2.5, $\|r(x) - r(y)\|_2 \le \frac{1}{2}\|x - y\|$ for $x, y \in N_\delta(0)$. Let $\eta = \delta/K$.

Let $X_1 = \{x_1 \in E_1 : \|x_1\|_1 < \eta\}$ and let $X_2 = \{x_2 \in E_2 : \|x_2\|_2 < \delta\}$. It follows from the inequality above that if $x_1 \in X_1$ and $x_2, x_2' \in X_2$ then

$$\|j(x_1, x_2) - j(x_1, x_2')\|_2 = \|r(x_1, x_2) - r(x_1, x_2')\|_2 \le \frac{1}{2}\|x_2 - x_2'\|_2.$$

It now follows from the Lipschitz implicit function theorem (Theorem 14.6.4) that there exists a unique continuous function $\psi : X_1 \to Y$ such that $j(x_1, \psi(x_1)) = \psi(x_1)$ for $x_1 \in X_1$. Thus $k(x_1, \psi(x_1)) = 0$ for $x_1 \in X_1$, and ψ is the unique function with this property.

Next we show that ψ is differentiable at 0, with derivative $-T$. If $x_1 \in X_1$, let $s(x_1) = r(x_1, \psi(x_1))$. Then

$$\psi(x_1) = -T(x_1) - s(x_1).$$

Thus

$$\|\psi(x_1)\|_2 \leq \|T\| \, \|x_1\|_1 + \tfrac{1}{2}(\|x_1\|_1 + \|\psi(x_1)\|_2) = (K/2) \, \|x_1\|_1 + \tfrac{1}{2} \|\psi(x_1)\|_2 \, ;$$

hence $\|\psi(x_1)\|_2 \leq K \, \|x_1\|_1$, and $\|(x_1, \psi(x_1))\| \leq (K+1) \, \|x_1\|_1$. Consequently $s(x_1)/ \|x_1\|_1 \to 0$ as $x_1 \to 0$, and ψ is differentiable at 0, with derivative $-T$.

Finally, it follows that if we set

$$N = X_1 + a_1 \text{ and } \phi(x) = a_2 + \psi(x - a_1) \text{ for } x \in N,$$

then N and ϕ satisfy the requirements of the theorem. $\qquad\square$

17.6 Higher derivatives

Suppose that $f : U \to F$ is a mapping from an open subset U of a normed space $(E, \|.\|_E)$ into a normed space $(F, \|.\|_F)$ which is differentiable on U. Then Df is a mapping from U into $L(E, F)$. We consider the case where the mapping Df is differentiable at a point a of U. If Df is differentiable at $a \in U$ then we denote its derivative by $D(Df)_a$, and say that f is *twice differentiable* at a. The linear operator $D(Df)_a$ is an element of $L(E, L(E, F))$; if $h \in E$ then $D(Df)_a(h) \in L(E, F)$. Thus if $k \in E$ then $(D(Df)_a(h))(k) \in F$. We have seen (Exercise 14.3.2) that there is a natural isometric isomorphism j of $L(E, L(E, F))$ onto the normed space $B(E, E; F)$ of continuous bilinear mappings from $E \times E$ into F. We denote the bilinear mapping $j(D(Df)_a)$ by $D^2 f_a$. Thus $D^2 f_a(h, k) = (D(Df)_a(h))(k)$. The mapping $D^2 f_a$ is the *second derivative* of f at a.

Example 17.6.1 The second derivative of the mapping $J : U \to U^{-1}$ from $GL(E)$ to itself.

Suppose that U is in the general linear group $GL(E)$ of a Banach space E. Since $GL(E)$ is an open subset of $L(E)$, there exists $\delta > 0$ such that $N_\delta(U) \subseteq GL(E)$. Suppose that $0 < \|T\| < \delta$, and suppose that $S \in L(E)$. Then

$$DJ_{U+T}(S) - DJ_U(S) = -(U + T)^{-1} S (U + T)^{-1} + U^{-1} S U^{-1}.$$

Since $(U+T)^{-1} = U^{-1} - U^{-1}TU^{-1} + r(T)$, where $r(T) = o(\|T\|)$,

$$DJ_{U+T}(S) - DJ_U(S) =$$
$$-(U^{-1} - U^{-1}TU^{-1})S(U^{-1} - U^{-1}TU^{-1}) + U^{-1}SU^{-1} + s_S(T),$$

where $s_S(T) = -r(T)S(U+T)^{-1} - (U+T)^{-1}Sr(T) = o(\|T\|)$. Now

$$-(U^{-1} - U^{-1}TU^{-1})S(U^{-1} - U^{-1}TU^{-1}) + U^{-1}SU^{-1} =$$
$$U^{-1}TU^{-1}SU^{-1} + U^{-1}SU^{-1}TU^{-1} - U^{-1}TU^{-1}SU^{-1}TU^{-1}.$$

But $U^{-1}TU^{-1}SU^{-1}TU^{-1} = o(\|T\|)$, so that J is twice differentiable, and $D^2J(T,S) = U^{-1}TU^{-1}SU^{-1} + U^{-1}SU^{-1}TU^{-1}$.

This second derivative is symmetric in S and T. This is an important general property.

Theorem 17.6.2 *Suppose that $f : U \to F$ is a mapping from an open subset U of a normed space $(E, \|.\|_E)$ into a normed space $(F, \|.\|_F)$ which is differentiable on U, and twice differentiable at $a \in U$. Then $D^2f_a(h,k) = D^2f_a(k,h)$ for all $(h,k) \in E \times E$.*

Proof   Before beginning the proof, let us see why this is a result that we should expect. For small h, the difference $\Delta f_a(h) = f(a+h) - f(a)$ is a good approximation to $Df_a(h)$, and so for small h and k the second difference $\Delta^2 f_a(h,k) = \Delta(\Delta f_a(h))(k)$ is a good approximation to $D^2f_a(h,k)$. But

$$\Delta^2 f_a(h,k) = f(a+h+k) - f(a+h) - f(a+k) + f(a)$$

is symmetric in h and k, and so we can expect $D^2f_a(h,k)$ to have the same property. As we shall see, the proof is quite complicated.

Suppose that $\epsilon > 0$. There exists $\delta > 0$ such that $N_\delta(a) \subseteq U$ and

$$\|Df_{a+x} - Df_a - (D(Df_a))(x)\| \leq \epsilon \|x\|_E, \text{ for } \|x\|_E < \delta.$$

That is,

$$\|Df_{a+x}(y) - Df_a(y) - (D^2 f_a)(x,y))\|_F \leq \epsilon \|x\|_E \|y\|_E,$$

for $\|x\|_E < \delta$ and $y \in E$. First suppose that $\|h\|_E < \delta/4$ and $\|k\|_E < \delta/4$. Let

$$g(t) = \Delta f_{a+th}(k) = f(a+th+k) - f(a+th), \text{ for } t \in (-2,2),$$

so that g is a differentiable mapping from $(-2,2)$ to F. By the chain rule, $g'(t) = Df_{a+th+k}(h) - Df_{a+th}(h)$. Since

$$\left\|Df_{a+th+k}(h) - Df_a(h) - D^2f_a(th+k,h)\right\|_F \leq \epsilon \|th+k\|_E \|h\|_E$$

$$\text{and } \left\|Df_{a+th}(h) - Df_a(h) - D^2f_a(th,h)\right\|_F \leq \epsilon \|th\|_E \|h\|_E,$$

it follows from the triangle inequality that if $-1 \leq t \leq 1$ then

$$\left\|Df_{a+th+k} - Df_{a+th} - D^2f_a(k,h)\right\|_F \leq 2\epsilon(\|h\|_E + \|k\|_E) \|h\|_E;$$

that is,

$$\left\|g'(t) - D^2f_a(k,h)\right\|_F \leq 2\epsilon(\|h\|_E + \|k\|_E) \|h\|_E. \qquad (*)$$

Thus

$$\left\|g'(t) - g'(0)\right\|_F \leq \left\|g'(t) - D^2f_a(k,h)\right\|_F + \left\|g'(0) - D^2f_a(k,h)\right\|_F$$
$$\leq 4\epsilon(\|h\|_E + \|k\|_E) \|h\|_E.$$

Applying the mean-value inequality of Corollary 17.2.6,

$$\left\|g(1) - g(0) - g'(0)\right\|_F \leq \sup\{\left\|g'(t) - g'(0)\right\|_F : 0 \leq t \leq 1\}$$
$$\leq 4\epsilon(\|h\|_E + \|k\|_E) \|h\|_E,$$

and so

$$\left\|g(1) - g(0) - D^2f_a(k,h)\right\|_F$$
$$\leq \left\|g(1) - g(0) - g'(0)\right\|_F + \left\|g'(0) - D^2f_a(k,h)\right\|_F$$
$$\leq 6\epsilon(\|h\|_E + \|k\|_E) \|h\|_E,$$

by $(*)$. But

$$g(1) - g(0) = f(a+h+k) - f(a+h) - f(a+k) + f(a) = \Delta^2 f_a(h,k),$$

and this is symmetric in h and k. Exchanging h and k, we see that

$$\left\|g(1) - g(0) - D^2f_a(h,k)\right\|_F \leq \sup\{\left\|g'(t) - g'(0)\right\| : 0 \leq t \leq 1\}$$
$$\leq 6\epsilon(\|h\|_E + \|k\|_E) \|k\|_E,$$

so that $\left\|D^2f_a(k,h) - D^2f_a(h,k)\right\|_F \leq 6\epsilon(\|h\|_E + \|k\|_E)^2$.

So far, we have only proved this inequality for small h and k. The following simple scaling argument shows that it holds in general. Suppose that h and

k are arbitrary members of E. There exists $\lambda > 0$ such that $\|\lambda h\|_E < \delta/4$ and $\|\lambda k\|_E < \delta/4$. Then

$$\left\|D^2 f_a(k, h) - D^2 f_a(h, k)\right\|_F = \left\|D^2 f_a(\lambda k, \lambda h) - D^2 f_a(\lambda h, \lambda k)\right\|_F / \lambda^2$$

$$\leq 6\epsilon (\|\lambda h\|_E + \|\lambda k\|_E)^2 / \lambda^2$$

$$= 6\epsilon (\|h\|_E + \|k\|_E)^2.$$

But ϵ is arbitrary, and so $D^2 f_a(h, k) = D^2 f_a(k, h)$. $\qquad\square$

Note that if f is twice differentiable at a then $D^2 f(h, k) = D_h(D_k f)_a$, where D_h and D_k are directional derivatives in the directions h and k respectively. Thus $D_h(D_k f)_a = D_k(D_h f_a)$. In particular, if $E = \mathbf{R}^d$ and f is twice differentiable at a then

$$D^2 f_a(e_i, e_j) = \frac{\partial}{\partial x_i} \left(\frac{\partial f}{\partial x_j} \right)(a) = \frac{\partial^2 f}{\partial x_i \partial x_j}(a),$$

so that

$$D^2 f_a(h, k) = \sum_{i=1}^{d} \sum_{j=1}^{d} h_i k_j \frac{\partial^2 f}{\partial x_i \partial x_j}(a),$$

where

$$\frac{\partial^2 f}{\partial x_i \partial x_j}(a) = \frac{\partial^2 f}{\partial x_j \partial x_i}(a) \text{ for } 1 \leq i, j \leq d.$$

The results that we have established depend in an essential way upon the fact that f is twice differentiable at a. The existence of second directional and partial derivatives does *not* imply the symmetry result of the theorem. Let

$$f(0,0) = 0, \text{ and } f(x, y) = \frac{xy(x^2 - y^2)}{x^2 + y^2} \text{ for } (x, y) \neq (0, 0).$$

The reader should verify the following:

- f is continuous and differentiable at every point of $\mathbf{R}^2$;
- Df is continuous and differentiable at every point of $\mathbf{R}^2 \setminus \{(0,0)\}$;
- $$\frac{\partial^2 f}{\partial x_1 \partial x_2}(0, 0) = -\frac{\partial^2 f}{\partial x_2 \partial x_1}(0, 0) = -1.$$

There are further examples of bad behaviour, and rather specialized positive results, some of which are included in the exercises, but we shall not investigate this further.

We can also consider higher derivatives. Suppose that f is a mapping from an open subset U of a normed space $(E, \|.\|_E)$ into a normed space $(F, \|.\|_F)$ which is $(k-1)$-times differentiable on U and is k-times differentiable at *a.* Let $M_k(E, F)$ denote the space of continuous k-linear mappings from E^k into F. Then we can consider $D^k f_a$ as a k-linear mapping from E^k into F: $D(D^{k-1}f)_a \in L(E, M_{k-1}(E, F))$, and $D(D^{k-1}f_a)(h_1)(h_2, \ldots, h_k) = D^k f_a(h_1, \ldots, h_k)$. A function which is k-times continuously differentiable is called a $C^{(k)}$-function. A function which is infinitely differentiable (a $C^{(k)}$-function, for each $k \in \mathbf{N}$) is called a *smooth* function.

Theorem 17.6.3 *Suppose that U is an open subset of a normed space $(E, \|.\|_E)$, that f is a $C^{(k)}$-function defined on U, taking values in a normed space $(F, \|.\|_F)$, that V is an open set of F containing $f(U)$ and that g is a $C^{(k)}$-function defined on V, taking values in a normed space $(G, \|.\|_G)$. Then the function $g \circ f$ is a $C^{(k)}$-function on U.*

Proof   The proof is by induction on k. The result is true for $k = 1$, by Corollary 17.1.3. Suppose that it holds for $k - 1$ and that f and g are $C^{(k)}$-functions. The function $x \to Df_x$ is a $C^{(k-1)}$-function on U, and, by the inductive hypothesis, the function $x \to Dg_{f(x)}$ is a $C^{(k-1)}$-function on U. By the inductive hypothesis again, the function $x \to Dg_{f(x)} \circ Df_x$ is a $C^{(k-1)}$-function on U: that is to say, $g \circ f$ is a $C^{(k)}$-function on U. $\square$

Corollary 17.6.4 *If f and g are smooth, then so is $g \circ f$.*

Corollary 17.6.5 *The inversion mapping $J : GL(E) \to GL(E)$ is a smooth mapping.*

Proof   We need a preliminary lemma.

Lemma 17.6.6 *Suppose that $(E, \|.\|_E)$ is a normed space. Let $B : L(E) \to L(E)$ be defined as $B(S)(T) = -STS$. Then B is a smooth function.*

Proof   For $(DB_S(H))(T) = -HTS - STH$, and

$$(D^2 B_S(H, K))(T) = -HTK - KTH,$$

so that $D^3 B = 0$. $\square$

We now prove the corollary. The proof is by induction on k. The mapping J is continuously differentiable. Suppose that it is a $C^{(k-1)}$-function. Since $DJ = B \circ J$, the derivative DJ is a $C^{(k-1)}$-function; that is, J is a $C^{(k)}$-function. $\square$

We also have the following result.

Theorem 17.6.7 *Suppose that $f : W \to F$ is a diffeomorphism from an open subset W of a Banach space $(E, \|.\|_E)$ onto a subset $f(W)$ of a*

Banach space $(F, \|.\|_F)$, *and that* f *is a* $C^{(k)}$*-function. Then* f^{-1} *is also a* $C^{(k)}$*-function.*

Proof Since $(E, \|.\|_E)$ and $(F, \|.\|_F)$ are isomorphic Banach spaces, we can suppose that $E = F$. The proof is again by induction on k. Suppose that the result holds for $k - 1$, and that f is a $C^{(k)}$-function. Then the mapping $f^{-1} : f(W) \to W$ is a $C^{(k-1)}$-function, by hypothesis, the function $Df : U \to GL(E)$ is a $C^{(k-1)}$-function, and the inversion function $J : GL(E) \to GL(E)$ is a smooth function, and so, applying Theorem 17.6.3, it follows that the mapping $y \to (Df_{f^{-1}(y)})^{-1}$ is a $C^{(k-1)}$-function. Thus f^{-1} is a $C^{(k)}$-function. □

A diffeomorphism which is a $C^{(k)}$-function is called a $C^{(k)}$-*diffeomorphism*, and a diffeomorphism which is a smooth function is called a *smooth diffeomorphism*.

We have the following symmetry result.

Theorem 17.6.8 *Suppose that* f *is a mapping from an open subset* U *of a normed space* $(E, \|.\|_E)$ *into a normed space* $(F, \|.\|_F)$ *which is* $(k-1)$*-times differentiable on* U *and is* k*-times differentiable at* a. *If* σ *is a permutation of* $\{1, \dots, k\}$, *then* $D^k f_a(h_1, \dots, h_k) = D^k(h_{\sigma(1)}, \dots, h_{\sigma(k)})$.

Proof The proof is by induction on k. It is trivially true if $k = 1$, and it is true when $k = 2$, by Theorem 17.6.2. Suppose that it is true for $j < k$, and that f is $(k - 1)$-times differentiable on U and is k-times differentiable at a. Let G be the set of permutations of $\{1, \dots, k\}$ for which equality holds. Then G is a subgroup of the group Σ_k of permutations of $\{1, \dots, k\}$. Let $H = \{\sigma \in \Sigma_k : \sigma(1) = 1\}$, and let $\tau_{i,j}$ denote the permutation which transposes i and j. If $\sigma \in H$ then by the inductive hypothesis

$$\begin{aligned} D^k f_a(h_1, \dots, h_k) &= D(D^{k-1} f_a(h_2, \dots, h_k))(h_1) \\ &= D(D^{k-1} f_a(h_{\sigma(2)}, \dots, h_{\sigma(k)}))(h_{\sigma(1)}) \\ &= D^k(h_{\sigma(1)}, \dots, h_{\sigma(k)}), \end{aligned}$$

so that $H \subseteq G$. In particular, $\tau_{i,j} \in G$ if neither i nor j is equal to 1. On the other hand, by Theorem 17.6.2,

$$\begin{aligned} D^k f_a(h_1, \dots, h_k) &= D^2(D^{k-2} f_a(h_3, \dots, h_k))(h_1, h_2) \\ &= D^2(D^{k-2} f_a(h_3, \dots, h_k))(h_2, h_1) \\ &= D^k f_a(h_2, h_1, h_3, \dots h_k), \end{aligned}$$

so that $\tau_{1,2} \in G$. Thus $\tau_{1,j} = \tau_{2,j}\tau_{1,2}\tau_{2,j} \in G$. Since any permutation can be written as a product of such transpositions (Exercise 11.7.1), it follows that $G = \Sigma_k$. □

If $h \in E$, let $h^j = (h, \ldots, h) \in E^j$. Since the directional derivative of f in the direction h is $Df_a(h)$, it follows that the jth directional derivative in the direction h is $D^j f_a(h^j)$.

We can also establish a version of Taylor's theorem. We prove it for Hilbert spaces. A corresponding result holds for Banach spaces, but this needs the Hahn–Banach theorem, whose proof is beyond the scope of this book

Theorem 17.6.9 *Suppose that f is a k-times differentiable mapping from an open subset U of a normed space $(E, \|.\|_E)$ into a Hilbert space $(F, \|.\|_F)$, and suppose that the segment $[a, a + h]$ is contained in U. Suppose that $\sup\{\|D^k f_{a+th}\| : 0 \le t \le 1\} = M < \infty$. Then*

$$f(a + h) = f(a) + \sum_{j=1}^{k-1} \frac{D^j f_a(h^j)}{j!} + r_k(h),$$

where $\|r_k(h)\|_F \le M \|h\|_E^k / k!$.

Proof   We reduce this to the scalar result. If $r_k(h) = 0$, there is nothing to prove. Otherwise, there exists an open interval I in $\mathbf{R}$ containing $[0, 1]$ such that $a + th \in U$ for $t \in I$. Let $\phi = r_k(h) / \|r_k(h)\|_F$, so that $\|\phi\|_F = 1$ and $\langle r_k(h), \phi \rangle = \|r_k(h)\|_F$.

Then the mapping $g : I \to \mathbf{R}$ defined by $g(t) = \langle f(a + th), \phi \rangle$ is k-times differentiable. By the chain rule, $(d^j g/dx^j)(t) = \langle D^j f_{a+th}(h^j), \phi \rangle$, and so by Taylor's theorem

$$\langle f(a + h), \phi \rangle = g(1) = \langle f(a), \phi \rangle + \sum_{j=1}^{k-1} \frac{\langle D^j f_a(h^j), \phi \rangle}{j!} + \frac{1}{k!}\frac{d^k g}{dx^k}(c)$$

for some $0 < c < 1$. Thus

$$\|r_k(h)\|_F = \langle r_k(h), \phi \rangle = \left\langle D^k f_{a+ch}(h^k)/k!, \phi \right\rangle$$
$$\le \frac{\|D^k f_{a+ch}\| \|h\|_E^k}{k!} \le \frac{M \|h\|_E^k}{k!}.$$

□

Let us consider the case where $E = \mathbf{R}^d$. Then the jth term in the Taylor expansion is

$$\frac{1}{j!} D^j f_a(h^j) = \frac{1}{j!} \sum_{i_1=1}^{d} \cdots \sum_{i_j=1}^{d} h_{i_1} \dots h_{i_j} \frac{\partial}{\partial x_{i_1}} \cdots \frac{\partial}{\partial x_{i_j}} f(a).$$

Using the symmetry established in Theorem 17.6.9 and gathering terms together, we see that

$$\frac{1}{j!} D^j f_a(h^j) = \sum_{j_1+\cdots+j_d=j} \frac{h_1^{j_1} \dots h_d^{j_d}}{j_1! \dots j_d!} \frac{\partial^{j_1}}{\partial x_1^{j_1}} \cdots \frac{\partial^{j_d}}{\partial x_d^{j_d}} f_a.$$

Exercises

17.6.1 There exists a continuous real-valued function g on $\mathbf{R}$ which is not differentiable at any point. Let $h(x) = \int_0^x g(t)\,dt$. Use h to obtain a continuous function f on $\mathbf{R}^2$ such that
(a) f is continuously differentiable at every point of $\mathbf{R}^2$;
(b) $\partial f/\partial x_2$ is continuously differentiable at every point of $\mathbf{R}^2$;
(c) $\partial^2 f/\partial x_1 \partial x_2$ exists at every point of $\mathbf{R}^2$, and is continuous on $\mathbf{R}^2$;
(d) $\partial^2 f/\partial x_2 \partial x_1$ does not exist at any point of $\mathbf{R}^2$.

17.6.2 Suppose that f is a real-valued function defined on an open subset U of $\mathbf{R}^2$, and that $\partial f/\partial x_1$ and $\partial^2 f/\partial x_2 \partial x_1$ exist at every point of U. Suppose that the closed rectangle $R = [a, a+h] \times [b, b+k]$ is contained in U. By considering the function $g(t) = \Delta f_{(a+th,b)}((0,k))$ and applying the mean-value theorem twice, show that there exists (u,v) in the interior of R for which

$$hk \frac{\partial^2 f}{\partial x_2 \partial x_1}(u,v) = \Delta^2 f_a((h,0),(0,k))$$

$$= f(a+h, b+k) - f(a+h, b) - f(a, b+k) + f(a,b).$$

17.6.3 Suppose that f is a real-valued function on an open subset U of $\mathbf{R}^2$, and that $\partial f/\partial x_1$, $\partial f/\partial x_2$ and $\partial^2 f/\partial x_2 \partial x_1$ exist at every point of U, and that $\partial^2 f/\partial x_2 \partial x_1$ is continuous at (a,b). Suppose that $\epsilon > 0$. Show the following.
(a) There exists $\delta > 0$ such that $N_\delta(a,b) \subseteq U$ and

$$\left| \frac{\Delta^2 f_{(a,b)}((h,0),(0,k))}{hk} - \frac{\partial^2 f}{\partial x_2 \partial x_1}(a,b) \right| < \epsilon,$$

for $0 < |h| < \delta/2$, $0 < |k| < \delta/2$.

(b) If $0 < |h| < \delta$ then

$$\left| \frac{1}{h} \left(\frac{\partial f}{\partial x_2}(a + h, b) - \frac{\partial f}{\partial x_2}(a, b) \right) - \frac{\partial^2 f}{\partial x_2 \partial x_1}(a, b) \right| \leq \epsilon.$$

(c) $\partial^2 f / \partial x_1 \partial x_2(a, b)$ exists, and is equal to $\partial^2 f / \partial x_2 \partial x_1(a, b)$.

17.6.4 Suppose that $(E, \|.\|_E)$ is a Banach space. Show that if J is the mapping $U \to U^{-1}$ of $GL(E)$ to $GL(E)$, then

$$D^k J_U(T_1, \ldots T_k) = (-1)^k \sum_{\sigma \in \Sigma_k} U^{-1} T_{\sigma(1)} U^{-1} \ldots U^{-1} T_{\sigma(k)} U^{-1}.$$

18

Integrating functions of several variables

18.1 Elementary vector-valued integrals

We now consider the problem of integrating vector-valued functions of several variables. We begin by considering dissections, step functions and elementary integrals, as in the case of real-valued functions of a single variable. A *cell* C in $\mathbf{R}^d$ is a subset of $\mathbf{R}^d$ of the form $I_1 \times \cdots \times I_d$, where $I_1, \ldots, I_d$ are intervals (open, closed, or neither) in $\mathbf{R}$. Thus a one-dimensional cell is an interval and a two-dimensional cell is a rectangle. The d-dimensional *volume* or *content* $v_d(C)$ is defined to be $v_d(C) = \prod_{j=1}^d l(I_j)$.

Suppose that $C = I_1 \times \cdots \times I_d$ is a compact cell, and suppose that $D_j = \{a_j = x_{j,0} < x_{j,1} < \cdots < x_{j,k_j} = b_j\}$ is a dissection of I_j, for $1 \leq j \leq d$, with constituent intervals $I_{j,1}, \ldots, I_{j,k_j}$. Then $D = D_1 \times \cdots \times D_d$ is a *dissection* of C. The collection of cells

$$\{I_{1,i_1} \times \cdots \times I_{d,i_d} : 1 \leq i_j \leq k_j, 1 \leq j \leq d\}$$

is then the set of *constituent cells* of the dissection D. We list them as $C_1, \ldots, C_k$, where $k = k_1 \ldots k_d$, and we denote the indicator function of C_j by χ_j. The *mesh size* $\delta(D)$ of a dissection D is the maximum diameter of a constituent cell.

We order the dissections of C by inclusion: we say that D' *refines* D, and write $D \leq D'$, if D'_j refines D_j, for $1 \leq j \leq d$, This is a partial order on the set Δ of all dissections of C, and Δ is a lattice:

$$D \vee D' = (D_1 \cup D'_1) \times \cdots \times (D_d \cup D'_d) \text{ and } D \wedge D' = D \cap D'.$$

Δ has a least element, with one cell $\{C\}$, but has no greatest element.

Suppose now that C is a compact cell in $\mathbf{R}^d$, that D is a dissection of C, with constituent cells $C_1, \ldots, C_k$, and that $(F, \|.\|)$ is a Banach space. We

513

denote by $S_F(C, D)$ the set of all F-valued functions of the form

$$f(x) = \sum_{j=1}^{k} f_j \chi_j(x), \text{ where } f_j \in F \text{ for } 1 \leq j \leq k.$$

The elements of $S_F(C, D)$ are F-valued *step functions* on C.

We define the *elementary integral* $\int_C f(x)\, dx$ of a step function $f = \sum_{j=1}^{k} f_j \chi_j$ in $S_F(C, D)$ to be $\sum_{j=1}^{k} f_j v_d(C_j)$. As in the real-valued case it is necessary, and straightforward, to show that this is well-defined.

Proposition 18.1.1 *Suppose that f and g are step functions and that c is a scalar. Then $f + g$ and cf are step functions, and*

(i) $\int_C (f(x) + g(x))\, dx = \int_C f(x)\, dx + \int_C g(x)\, dx$

(ii) $\int_C cf(x)\, dx = c \int_C f(x)\, dx$.

Proof The proofs are the same as the proofs of the corresponding results in the real-valued one-dimensional case.   □

Thus the set $S_F(C)$ of F-valued step functions on C is a linear subspace of the space $B_F(C)$ of all bounded F-valued functions on C.

If f is an F-valued step functions on C then $\|f\|$ is a real-valued step function on C.

Proposition 18.1.2 *Suppose that $f \in S_F(C)$. Then*

$$\left\| \int_C f(x)\, dx \right\| \leq \int_C \|f(x)\|\, dx.$$

The function $\int_a^b \|f(x)\|\, dx$ is a norm on $S_F(C)$.

Proof By the triangle inequality,

$$\left\| \int_C f(x)\, dx \right\| \leq \sum_{j=1}^{k} \|f_j\|\, v_d(C_j) = \int_C \|f(x)\|\, dx.$$

Clearly $\int_C \|cf(x)\|\, dx = |c| \int_C \|f(x)\|\, dx$, and $\int_C \|f(x)\|\, dx = 0$ if and only if $f = 0$. If $f, g \in S_F(C)$ there exists a dissection D such that

$$f = \sum_{j=1}^{k} f_j \chi_j \text{ and } g = \sum_{j=1}^{k} g_j \chi_j.$$

Then

$$\int_C \|f(x) + g(x)\| \, dx = \sum_{j=1}^{k} \|f_j + g_j\| \, v_d(C_j) \leq \sum_{j=1}^{k} (\|f_j\| + \|g_j\|) v_d(C_j)$$

$$= \int_C \|f(x)\| \, dx + \int_C \|g(x)\| \, dx.$$

$\square$

Exercise

18.1.1 Show that the elementary integral of a step function is well-defined.

18.2 Integrating functions of several variables

We now consider the Riemann integral of a real-valued function of several variables. We follow the procedure for integrating functions of a single variable very closely, and we therefore omit many of the details.

Suppose that f is a bounded real-valued function defined on a compact cell C in $\mathbf{R}^d$. We define the upper and lower integrals of f:

$$\overline{\int}_C f(x) \, dx = \inf \left\{ \int_C g(x) \, dx : g \text{ a step function, } g \geq f \right\}$$

$$\underline{\int}_C f(x) \, dx = \sup \left\{ \int_C h(x) \, dx : h \text{ a step function, } h \leq f \right\}.$$

The function f is *Riemann integrable* if the upper and lower integrals are equal; if so, the common value is the *Riemann integral* of f.

Recall that if f is a function on a set S taking values in a metric space (X, d), and A is a subset of X then the oscillation $\Omega(f, A)$ of f on A is defined to be $\sup\{d(f(a), f(b)) : a, b \in A\}$; when f is real-valued then $\Omega(f, A) = \sup\{|f(a) - f(b)| : a, b \in A\}$, and when f takes values in a Banach space $(F, \|.\|)$ then $\Omega(f, A) = \sup\{\|f(a) - f(b)\| : a, b \in A\}$.

Theorem 18.2.1 *Suppose that f is a bounded real-valued function on a compact cell C in $\mathbf{R}^d$. The following are equivalent.*

(i) f is Riemann integrable.

(ii) Whenever $\epsilon > 0$ there exists a dissection D of C with constituent cells $C_1, \ldots, C_k$ such that

$$\sum_{j=1}^{k} \Omega(f, C_j) v_d(C_j) < \epsilon.$$

Proof   Suppose that f is Riemann integrable and that $\epsilon > 0$. There exist step functions g and h on C such that $h \leq f \leq g$ and

$$\int_C g(x)dx - \int_C f(x)\,dx < \epsilon/2 \text{ and } \int_C f(x)dx - \int_C h(x)\,dx < \epsilon/2,$$

so that $\int_C(g(x) - h(x))\,dx < \epsilon$. Let D be a dissection of C, with constituent cells $C_1, \ldots, C_k$, such that g and h are constant on each C_j. Then $\Omega(f, C_j) \leq g(x) - h(x)$ for $x \in C_j$ for $1 \leq j \leq k$, and so

$$\sum_{j=1}^{k} \Omega(f, C_j)v_d(C_j) \leq \int_C (g(x) - h(x))\,dx < \epsilon.$$

Thus (i) implies (ii).

Conversely, if D is a dissection of C with constituent cells $C_1, \ldots, C_k$ for which $\sum_{j=1}^{k} \Omega(f, C_j)v_d(C_j) < \epsilon$, let

$$g = \sum_{j=1}^{k} \sup\{f(x) : x \in C_j\}\chi_j \text{ and } h = \sum_{j=1}^{k} \inf\{f(x) : x \in C_j\}\chi_j.$$

Then g and h are step functions with $h \leq f \leq g$, and

$$\int_C g(x)\,dx - \int_C h(x)\,dx = \sum_{j=1}^{k} \Omega(f, C_j)v_d(C_j) < \epsilon,$$

so that (ii) implies (i).   □

Corollary 18.2.2   *If f is a continuous real-valued function on a compact cell C in $\mathbf{R}^d$, then f is Riemann integrable.*

Proof   Suppose that $\epsilon > 0$. Since f is continuous and C is compact, f is uniformly continuous on C, and so there exists $\delta > 0$ such that if $\|x - y\| < \delta$ then $|f(x) - f(y)| < \epsilon/v_d(C)$. Let D be any dissection of C with mesh-size less than δ. If c_j is any constituent cell, then $\Omega(f, C_j) < \epsilon/v_d(C)$, and so

$$\sum_{j=1}^{k} \Omega(f, C_j)v_d(C_j) < \frac{\epsilon}{v_d(C)} \sum_{j=1}^{k} v_d(C_j) = \epsilon.$$

□

Similarly, we have the following elementary results. The proofs are the same as the proofs for real-valued functions of a single variable, and details are left to the reader.

Theorem 18.2.3 *Suppose that f and g are Riemann integrable functions on a compact cell C in $\mathbf{R}^d$.*

(i) If $c \in \mathbf{R}$ then $f + g$ and cf are Riemann integrable, and

$$\int_C (f(x) + g(x))\, dx = \int_C f(x)\, dx + \int_C g(x)\, dx,$$

$$\int_C cf(x)\, dx = c \int_C f(x)\, dx.$$

(ii) If $f(x) \le g(x)$ for all $x \in C$ then $\int_C f(x)\, dx \le \int_C g(x)\, dx$.

(iii) If f takes values in $[-R, R]$ and ϕ is a continuous real-valued function on $[-R, R]$ then $\phi \circ f$ is Riemann integrable.

(iv) The functions f^+, f^-, $|f|$, f^2 and fg are Riemann integrable.

(v) $\left| \int_C f(x)\, dx \right| \le \int_C |f(x)|\, dx$.

Exercise

18.2.1 Suppose that f is a real-valued function on $[0, 1] \times [0, 1]$, that the mappings $x \to f(x, y)$ are increasing for each $y \in [0, 1]$, and that the mappings $y \to f(x, y)$ are increasing for each $x \in [0, 1]$. Show that f is Riemann integrable. Extend this result to functions on compact cells in $\mathbf{R}^d$.

18.3 Integrating vector-valued functions

We now consider the problem of integrating vector-valued functions defined on a compact cell in $\mathbf{R}^d$. Suppose that $f : C \to (F, \|.\|)$ is a function on a compact cell C in $\mathbf{R}^d$ taking values in a bounded subset of a Banach space $(F, \|.\|)$. Since in general there is no order on F, we cannot use upper and lower integrals to determine when f is Riemann integrable, and to define the Riemann integral. Instead, we start with the characterization of Riemann integrability given in Theorem 18.2.1 (i). We say that f is *Riemann integrable* if, whenever $\epsilon > 0$, there exists a dissection D of C, with constituent cells $C_1, \ldots, C_k$, such that $\sum_{j=1}^k v_d(C_j)\Omega(f, C_j) < \epsilon$.

Before defining the integral, we need to establish some fundamental properties of Riemann integrable functions.

Theorem 18.3.1 *Suppose that f is a function on a compact cell C in $\mathbf{R}^d$ taking values in a bounded subset of a Banach space F. The following are equivalent.*

(i) *The function f is Riemann integrable.*

(ii) *Whenever $\epsilon > 0$ there exists a dissection D of C with constituent cells $C_1, \ldots, C_k$ and a partition $G \cup B$ of $\{1, \ldots, k\}$ such that*

$$\Omega(f, C_j) < \epsilon \text{ for } j \in G, \text{ and } \sum_{j \in B} v_d(C_j) < \epsilon.$$

(iii) *The real-valued function $\|f - g\|$ is Riemann integrable, for each step function. If $\epsilon > 0$, then there exists a step function g on C for which*

$$\inf \left\{ \int_C \|f(x) - g(x)\| \, dx : g \text{ a step function} \right\} = 0.$$

Proof   Suppose that (i) holds and that $\epsilon > 0$. There exists a dissection D of C with constituent cells $C_1, \ldots, C_k$ such that

$$\sum_{j=1}^{k} \Omega(f, C_j) v_d(C_j) < \epsilon^2.$$

Let $G = \{j : \Omega(f, C_j) < \epsilon\}$ and let $B = \{j : \Omega(f, C_j) \geq \epsilon\}$. Then

$$\epsilon \left(\sum_{j \in B} v_d(C_j) \right) \leq \sum_{j=1}^{k} \Omega(f, C_j) v_d(C_j) < \epsilon^2,$$

so that (ii) holds.

Suppose that (ii) is satisfied, and that $\epsilon > 0$. Let

$$\eta = \epsilon / (v_d(C) + \Omega(f, C)).$$

There exists a dissection D of C with constituent cells $C_1, \ldots, C_k$ such that the condition holds, for η. Then

$$\sum_{j=1}^{k} \Omega(f, C_j) v_d(C_j)$$

$$= \sum_{j \in G} \Omega(f, C_j) v_d(C_j) + \sum_{j \in B} \Omega(f, C_j) v_d(C_j)$$

$$\leq (\sup_{j \in G} \Omega(f, C_j)) \sum_{j \in G} v_d(C_j) + \Omega(f, C) \sum_{j \in B} v_d(C_j)$$

$$< \eta v_d(C) + \eta \Omega(f, C) = \epsilon.$$

Thus (ii) implies (i).

Suppose that f is Riemann integrable, that g is a step function and that $\epsilon > 0$. There exists a dissection D of C, with constituent cells $C_1, \ldots, C_k$, for which $\sum_{j=1}^{k} v_d(C_j)\Omega(f, C_j) < \epsilon$, and for which g is constant on each C_j. If $x, y \in C_j$ then

$$| \|f(x) - g(x)\| - \|f(y) - g(y)\| | \leq \|f(x) - f(y)\|,$$

so that $\Omega(\|f - g\|, C_j) \leq \Omega(f, C_j)$. Hence $\sum_{j=1}^{k} v_d(C_j)\Omega(\|f - g\|, C_j) < \epsilon$, and so $\|f - g\|$ is Riemann integrable, by Theorem 18.2.1 (i). Now choose $y_j \in C_j$, for $1 \leq j \leq k$, and let g be the step function $g = \sum_{j=1}^{k} f(y_j)\chi_j$. Then $\|f(x) - g(x)\| \leq \Omega(f, C_j)$ for $x \in C_j$, so that

$$\int_C \|f(x) - g(x)\| \, dx = \sum_{j=1}^{k} \left(\int_{C_j} \|f(x) - g(x)\| \, dx \right)$$

$$\leq \sum_{j=1}^{k} v_d(C_j)\Omega(f, C_j) < \epsilon.$$

Thus (i) implies (iii).

Conversely, suppose that (iii) holds, that $\epsilon > 0$ and that g is a step function for which $\int_C \|f(x) - g(x)\| < \epsilon/2$. There exists a dissection D of C, with constituent cells $C_1, \ldots, C_k$, for which

$$\sum_{j=1}^{k} v_d(C_j) \left(\sup_{x \in C_j} \|f(x) - g(x)\| \right) < \epsilon/2,$$

and for which g is constant on each C_j. If $y, z \in C_j$ then

$$\|f(y) - f(z)\| \leq 2 \left(\sup_{x \in C_j} \|f(x) - g(x)\| \right),$$

so that $\sum_{j=1}^{k} v_d(C_j)\Omega(f, C_j) < \epsilon$. Thus f is Riemann integrable, and (iii) implies (i). □

We are now ready to define the Riemann integral of a Riemann integrable function. Suppose that f is a Riemann integrable function on a compact cell C. For each $\epsilon > 0$, let

$$A_\epsilon(f) = \left\{ g : g \text{ a step function on } C, \int_C \|f(x) - g(x)\| \, dx < \epsilon \right\},$$

and let $J_\epsilon = \{\int_C g(x)\,dx : g \in A_\epsilon(f)\}$. Then $A_\epsilon(f)$ is non-empty, and J_ϵ is a non-empty subset of F. If $g, h \in A_\epsilon(f)$ then

$$\left\| \int_C g(x)\,dx - \int_C h(x)\,dx \right\|$$

$$\leq \int_C \|g(x) - h(x)\|\,dx$$

$$\leq \int_C \|f(x) - g(x)\|\,dx + \int_C \|f(x) - h(x)\|\,dx < 2\epsilon.$$

Thus J_ϵ has diameter at most 2ϵ, and so therefore has its closure $\overline{J}_\epsilon$. It now follows from Corollary 14.1.12 that the intersection $\cap\{\overline{J}_\epsilon : \epsilon > 0\}$ is a singleton set $\{I\}$. We define I to be the *Riemann integral* $\int_C f(x)\,dx$ of f. Note that if $g \in A_\epsilon(f)$, then

$$\left\| I - \int_C g(x)\,dx \right\| \leq \operatorname{diam}(J_\epsilon) \leq 2\epsilon.$$

Corollary 18.3.2 *If f is a Riemann integrable function on C then there exists a sequence $(f_n)_{n=1}^\infty$ of step functions on C for which $\int_C \|f(x) - f_n(x)\|\,dx \to 0$ as $n \to \infty$, and, for any such sequence, $\int_C f_n(x)\,dx \to \int_C f(x)\,dx$ as $n \to \infty$.*

Proof   Pick $f_n \in A_{1/n}$.   □

We have the following fundamental inequality.

Theorem 18.3.3 (The mean-value inequality for integrals)   *Suppose that f is a Riemann integrable function on a compact cell taking values in a Banach space $(F, \|.\|)$. Then*

$$\left\| \int_C f(x)\,dx \right\| \leq \int_C \|f(x)\|\,dx.$$

Proof   Suppose that $\epsilon > 0$. Then there exists a step function g such that $\int_C \|f(x) - g(x)\|\,dx < \epsilon/3$. By the remark above,

$$\left\| \int_C f(x)\,dx - \int_C g(x)\,dx \right\| \leq 2\epsilon/3.$$

Thus, applying Proposition 18.1.2,

$$\left\| \int_C f(x)\,dx \right\| \leq \left\| \int_C g(x)\,dx \right\| + 2\epsilon/3$$

$$\leq \int_C \|g(x)\|\,dx + 2\epsilon/3$$

$$\leq \int_C \|f(x)\|\,dx + \int_C \|f(x) - g(x)\|\,dx + 2\epsilon/3$$

$$\leq \int_C \|f(x)\|\,dx + \epsilon.$$

Since ϵ is arbitrary, the result follows. □

Corollary 18.3.2 enables us to establish standard properties of the Riemann integral.

Proposition 18.3.4 *Suppose that f and g are Riemann integrable functions on a compact cell C taking values in a Banach space $(F, \|.\|)$, that h is a real-valued Riemann integrable function on C and that $\alpha \in \mathbf{R}$.*
(i) The functions $f + g$ and αf are Riemann integrable and

$$\int_C f(x) + g(x)\,dx = \int_C f(x)\,dx + \int_C g(x)\,dx, \quad \int_C \alpha f(x) = \alpha \int_C f(x)\,dx.$$

(ii) The function hf is Riemann integrable.
(iii) Suppose that $\phi : f(C) \to (G, \|.\|_G)$ is a uniformly continuous mapping from the image $f(C)$ of C into a Banach space G. Then $\phi \circ f$ is Riemann integrable.

Proof That $f + g$ and hf are Riemann integrable follows from the facts (which the reader should verify) that

$$\Omega(f + g, A) \leq \Omega(f, A) + \Omega(g, A)$$
$$\text{and } \Omega(hf, A) \leq \Omega(h, A)\|f\|_\infty + \|h\|_\infty \Omega(f, A),$$

and the definition.

There exist sequences $(f_n)_{n=1}^\infty$ and $(g_n)_{n=1}^\infty$ of step functions such that $\int_C \|f(x) - f_n(x)\|\,dx \to 0$ and $\int_C \|g(x) - g_n(x)\|\,dx \to 0$ as $n \to \infty$. Then

$$\int_C \|(f(x) + g(x)) - (f_n(x) + g_n(x))\|\,dx \leq$$

$$\int_C \|f(x) - f_n(x)\|\,dx + \int_C \|g(x) - g_n(x)\|\,dx,$$

so that

$$\int_C \|(f(x) + g(x)) - (f_n(x) + g_n(x))\| \, dx \to 0 \text{ as } n \to \infty,$$

and so

$$\int_C f(x) + g(x) \, dx = \int_C f(x) \, dx + \int_C g(x) \, dx.$$

The proof of the result for scalar multiplication is even easier.

(iii) Suppose that $\epsilon > 0$. There exists $\delta > 0$ such that if $\|f(x) - f(y)\|_F < \delta$, then $\|\phi(f(x)) - \phi(f(y))\|_G < \epsilon$. By Theorem 18.3.1, there exists a dissection D of C with constituent cells $C_1, \ldots, C_k$ and a partition $G \cup B$ of $\{1, \ldots, k\}$ such that

$$\Omega(f, C_j) < \delta \text{ for } j \in G, \text{ and } \sum_{j \in B} v_d(C_j) < \epsilon.$$

Then $\Omega(\phi \circ f, C_j) < \epsilon$ for $j \in G$, so that, by Theorem 18.3.1, $\phi \circ f$ is Riemann integrable. $\qquad\square$

The uniform limit of Riemann integrable functions is Riemann integrable.

Theorem 18.3.5 *Suppose that $(f_n)_{n=1}^\infty$ is a sequence of Riemann integrable functions on a compact cell C, taking values in a Banach space $(F, \|.\|)$, which converges uniformly to f. Then f is Riemann integrable, and*

$$\int_C f_n(x) \, dx \to \int_C f(x) \, dx \text{ as } n \to \infty.$$

Proof   Suppose that $\epsilon > 0$. There exists N such that

$$\|f - f_n\|_\infty = \sup\{\|f(x) - f_n(x)\| : x \in C\} < \epsilon/4v_d(C)$$

for $n \geq N$, and there exists a dissection D of C with constituent cells $C_1, \ldots, C_k$ such that $\sum_{j=1}^k v_d(C_j)\Omega(f_N, C_j) < \epsilon/2$. Then $\Omega(f, C_j) \leq \Omega(f_N, C_j) + 2\|f - f_N\|_\infty$, so that

$$\sum_{j=1}^k v_d(C_j)\Omega(f, C_j) \leq \sum_{j=1}^k v_d(C_j)\left(\Omega(f_N, C_j) + 2\|f - f_N\|_\infty\right)$$

$$< \epsilon/2 + 2v_d(C)\|f - f_N\|_\infty < \epsilon.$$

Thus f is Riemann integrable.

Suppose now that $n \geq N$. Then

$$\left\|\int_C f(x) \, dx - \int_C f_n(x) \, dx\right\| \leq \int_C \|f(x) - f_n(x)\| \, dx \leq \epsilon/4,$$

so that

$$\int_C f_n(x)\,dx \to \int_C f(x)\,dx \text{ as } n \to \infty.$$

□

Riemann integrability can also be characterized, and the integral calculated, using dissections with decreasing mesh size. See Exercise 1.

So far, then, everything appears to be very straightforward. In fact, there are very real technical difficulties. These arise in the following circumstances.

(i) We frequently wish to integrate functions over more general bounded subsets of $\mathbf{R}^d$ than cells.

(ii) We would like to evaluate integrals by repeatedly calculating one-dimensional integrals. For example, if f is a Riemann integrable function on $C_0 = [0,1] \times [0,1]$ can we calculate

$$\int_{C_0} f(x_1, x_2)\,dx \text{ as a repeated integral } \int_0^1 \left(\int_0^1 f(x_1, x_2)\,dx_2 \right) dx_1?$$

(iii) Can we establish a 'change of variables' formula of general applicability? If U and V are bounded open sets, and $\phi : U \to V$ is a continuously differentiable homeomorphism, can we show that

$$\int_V f(x)\,dx = \int_U f(\phi(y))|J_\phi(y)|\,dy \quad ?$$

As far as (i) is concerned, a bounded subset A of a compact cell C is said to be *Jordan measurable* if its characteristic function I_A is Riemann integrable. If so, the Riemann integral of I_A is called the *Jordan content* or *volume* of A, and is denoted by $v_d(A)$, or $v(A)$. Clearly a cell is Jordan measurable, and the definition of its content as an integral agrees with the definition at the beginning of this section. On the other hand, we have seen that the indicator function of a fat Cantor set is not Riemann integrable, and so a fat Cantor set is not Jordan measurable.

If A is a Jordan measurable subset of a compact cell C, with indicator function I_A, and f is a Riemann integrable function on C, then it follows from Proposition 18.3.4 (ii) that fI_A is Riemann integrable; we define

$$\int_A f(x)\,dx = \int_C f(x)I_A(x)\,dx.$$

As for (ii), let

$$A = \{(x,y) \in [0,1] \times [0,1] : y \text{ is rational if } x = 1/2\}.$$

Then A is a Jordan measurable subset of $[0,1] \times [0,1]$, but $\int_0^1 I_A(1/2, y)\, dy$ is not defined.

As for (iii), we shall see in Section 18.4 that it can happen that U is Jordan measurable, but V is not.

These difficulties suggest that we need a more sophisticated theory of integration, and the Lebesgue integral, which is studied in Volume III, provides such a theory. The Riemann integral is however adequate for the integration of continuous functions. Further, the change of variables results that we obtain in Section 18.7 are essential for corresponding change of variable results in the Lebesgue integral setting.

As with functions of a scalar variable, we can define improper integrals, but some care is needed; for example, if f is defined on $\mathbf{R}^d$, it is natural to consider limits such as

$$\lim_{R \to \infty} \int_{\|x\|_2 \leq R} f(x)\, dx \quad \text{and} \quad \lim_{R \to \infty} \int_{\|x\|_\infty \leq R} f(x)\, dx,$$

and it is relatively easy to give examples where the limits exist and are different. Similar remarks apply to Cauchy principal value integrals. In each case, it is necessary to make explicit the limiting procedure that is used.

Exercises

18.3.1 Suppose that f is a function on a compact cell C in $\mathbf{R}^d$ taking values in a bounded subset of a Banach space $(F, \|.\|)$. Suppose that $(D_r)_{r=1}^\infty$ is a sequence of dissections of C whose mesh-sizes tend to 0, and that $C_{r,1}, \ldots, C_{r,k_r}$ are the constituent cells of D_r. Show that f is Riemann integrable if and only if there exists $J \in F$ such that if $y_{r,j} \in C_{r,j}$ for $1 \leq j \leq k_r$ and $r \in \mathbf{N}$ then

$$\sum_{j=1}^{k_r} f(y_{r,j}) v_d(C_{r,j}) \to J \text{ as } r \to \infty.$$

[Hint: If D' is a dissection of C, with constituent cells $C_1', \ldots, C_k'$, let T_r be the set of constituent cells of D_r which are not contained in one of the cells C_j'. Show that

$$\sum \{v_d(C_{r,j}) : C_{r,j} \in T_r\} \to 0 \text{ as } r \to \infty.]$$

18.3.2 [**The fundamental theorem of calculus for vector-valued functions**] Suppose that f is a Riemann integrable function on $[a, b]$, taking values in a Banach space $(E, \|.\|)$.

(i) Show that if $a < c < b$ then f is Riemann integrable on $[a, b]$ if and only if it is Riemann integrable on $[a, c]$ and $[c, b]$, and if so then $\int_a^b f(x)\, dx = \int_a^c f(x)\, dx + \int_c^b f(x)\, dx$.

(ii) Set $F(t) = \int_a^t f(x)\, dx$, for $a \leq t \leq b$. Show that F is continuous on $[a, b]$.

(iii) Show that If f is continuous at t then F is differentiable at t, and $F'(t) = f(t)$. (If $t = a$ or b, then F has a one-sided derivative.)

(iv) Suppose that f is differentiable on $[a, b]$ (with one-sided derivatives at a and b). Show that if f' is Riemann integrable then $f(x) = f(a) + \int_a^x f'(t)\, dt$ for $a \leq x \leq b$.

18.3.3 Suppose that f and g are Riemann integrable functions on a cell C in $\mathbf{R}^d$, taking values in $L(E)$, where $(E, \|.\|)$ is a Banach space and $L(E)$ is given the operator norm. Show that $f \circ g$ and $g \circ f$ are Riemann integrable. Is $\int_C f(x) \circ g(x)\, dx = \int_C g(x) \circ f(x)\, dx$?

18.3.4 Construct a continuous real valued function f on $\mathbf{R}^2$ for which

$$\lim_{R \to \infty} \int_{\|x\|_2 \leq R} f(x)\, dx = 0$$

and for which

$$\lim_{R \to \infty} \int_{\|x\|_\infty \leq R} f(x)\, dx \text{ does not exist.}$$

18.4 Repeated integration

To begin with, let us consider a continuous function f defined on a compact cell $C = I_1 \times \cdots \times I_d$ in $\mathbf{R}^d$, taking values in a Banach space $(F, \|.\|)$. Let $\widetilde{C} = I_1 \times \cdots \times I_{d-1}$, so that $C = \widetilde{C} \times I_d$. Denote a point $x = (x_1, \ldots, x_d)$ in $\mathbf{R}^d$ by $(\widetilde{x}, t)$, where $\widetilde{x} = (x_1, \ldots, x_{d-1})$ and $t = x_d$.

Theorem 18.4.1 *Let f be a continuous function from a compact cell $C = I_1 \times \cdots \times I_d$ in $\mathbf{R}^d$ into a Banach space $(F, \|.\|)$. With the notation above, if $\widetilde{x} \in C_{d-1}$ let $\phi(\widetilde{x}) = \int_{I_d} f(\widetilde{x}, t)\, dt$. Then ϕ is a continuous function on $\widetilde{C}$, and*

$$\int_C f(x)\, dx = \int_{\widetilde{C}} \phi(\widetilde{x})\, d\widetilde{x} = \int_{\widetilde{C}} \left(\int_{I_d} f(\widetilde{x}, t)\, dt \right) d\widetilde{x}.$$

Proof We use the fact that f is uniformly continuous on C. Given $\epsilon > 0$ there exists $\delta > 0$ such that if $x, y \in C$ and $d(x, y) < \delta$ then $\|f(x) - f(y)\| <$

$\epsilon/l(I_d)$. Thus if $d(\widetilde{x}, \widetilde{y}) < \delta$ then $\|f(\widetilde{x}, t) - f(\widetilde{y}, t)\| < \epsilon/l(I_d)$, and so

$$\|\phi(\widetilde{x}) - \phi(\widetilde{y})\| = \left\| \int_{I_d} f(\widetilde{x}, t) - f(\widetilde{y}, t) \, dt \right\|$$

$$\leq \int_{I_d} \|f(\widetilde{x}, t) - f(\widetilde{y}, t)\| \, dt \leq \epsilon.$$

Thus ϕ is continuous on $\widetilde{C}$.

Further, there exists a step function g on C such that $\|f(x) - g(x)\| \leq \epsilon/2v_d(C)$ for $x \in C$. Let $\psi(\widetilde{x}) = \int_{I_d} g(\widetilde{x}, t) \, dt$, for $\widetilde{x} \in \widetilde{C}$. Then

$$\left\| \int_C f(x) \, dx - \int_C g(x) \, dx \right\| \leq \int_C \|f(x) - g(x)\| \, dx \leq \epsilon/2,$$

and

$$\left\| \phi(\widetilde{x}) - \int_{I_d} g(\widetilde{x}, t) \, dt \right\| \leq \int_{I_d} \|f(\widetilde{x}, t) - g(\widetilde{x}, t)\| \, dt \leq \frac{\epsilon l(I_d)}{2v_d(C)},$$

so that

$$\left\| \int_{\widetilde{C}} \phi(\widetilde{x}) \, d\widetilde{x} - \int_{\widetilde{C}} \psi(\widetilde{x}) \, d\widetilde{x} \right\| \leq \epsilon/2.$$

But $\int_C g(x) \, dx = \int_{\widetilde{C}} \psi(\widetilde{x}) \, d\widetilde{x}$ and so

$$\left\| \int_C f(x) \, dx - \int_{\widetilde{C}} \phi(\widetilde{x}) \, d\widetilde{x} \right\| \leq \epsilon.$$

Since ϵ is arbitrary, the result follows. $\qquad\qquad\qquad\qquad\square$

Corollary 18.4.2

$$\int_C f(x) \, dx = \int_{I_1} \left(\int_{I_2} \cdots \left(\int_{I_d} f(x_1, \ldots, x_d) \, dx_d \right) \cdots dx_2 \right) dx_1.$$

Thus we can evaluate the integral by repeatedly evaluating one dimensional integrals. Further, if $\sigma \in \Sigma_n$ is a permutation of $\{1, \ldots, d\}$ then we could integrate with respect first to $x_{\sigma(d)}$, then $x_{\sigma(d-1)}$, and so on, and obtain the same result.

Corollary 18.4.3   *If $\sigma \in \Sigma_d$ then*

$$\int_{I_1} \left(\int_{I_2} \cdots \left(\int_{I_d} f(x_1, \ldots, x_d) \, dx_d \right) \cdots dx_2 \right) dx_1 =$$

$$\int_{I_{\sigma(1)}} \left(\int_{I_{\sigma(2)}} \cdots \left(\int_{I_{\sigma(d)}} f(x_1, \ldots, x_d) \, dx_{\sigma(d)} \right) \cdots dx_{\sigma(2)} \right) dx_{\sigma(1)}.$$

For example, if $B(x, y)$ is the beta function $\int_0^1 t^{x-1}(1 - t)^{y-1} \, dt$ then

$$\int_1^Y \left(\int_1^X B(x, y) \, dx \right) dy = \int_1^Y \left(\int_1^X \left(\int_0^1 t^{x-1}(1 - t)^{y-1} \, dt \right) dx \right) dy$$

$$= \int_0^1 \left(\int_1^X t^{x-1} \, dx \right) \left(\int_1^Y (1 - t)^{y-1} \, dy \right) dt$$

$$= \int_0^1 \frac{(1 - t^{X-1})(1 - (1 - t)^{Y-1})}{(\log t)(\log(1 - t))} \, dt.$$

We can also justify the 'quick and easy' proof of Theorem 9.4.4 of Volume I. The argument there now shows that if f and g are continuous 2π-periodic functions then the Fourier coefficients satisfy $(\widehat{f \star g})_n = \hat{f}_n \hat{g}_n$. This result extends easily to locally Riemann integrable 2π-periodic functions f and g. Let $M = \max(\|f\|_\infty, \|g\|_\infty)$. If $0 < \epsilon \le 1$, there exist continuous 2π-periodic functions f' and g' with

$$\int_0^{2\pi} |f(t) - f'(t)| \, dt < \epsilon/4(M + 1) \text{ and } \int_0^{2\pi} |f(t) - f'(t)| \, dt < \epsilon/4(M + 1).$$

Then

$$|(\widehat{f \star g})_n - (\widehat{f' \star g'})_n| < \epsilon/2 \text{ and } |\hat{f}_n \hat{g}_n - \hat{f}'_n \hat{g}'_n| < \epsilon/2,$$

so that $|(\widehat{f \star g})_n - \hat{f}_n \hat{g}_n| < \epsilon$. Since ϵ is arbitrary, the result follows.

These results can be applied to continuous functions on $\mathbf{R}^d$ of compact support: that is, functions which take the value 0 outside a bounded set. Frequently, though, we wish to consider improper integrals. As we have seen when we considered products of series, difficulties arise when convergence depends upon cancellation. For this reason, we restrict attention to the simplest case, where we integrate non-negative continuous real-valued functions.

Suppose then that $C = I_1 \times \cdots \times I_d$ is the product of open intervals (which may be semi-infinite, or infinite), and that f is a non-negative continuous real-valued function on C. We then define the improper Riemann integral as

$$\int_C f(x) \, dx = \sup \left\{ \int_K f(x) \, dx : K \text{ a compact cell, } K \subseteq C \right\}$$

The resulting integral can then be finite or infinite. We continue with the notation introduced at the beginning of the section.

Theorem 18.4.4 *Suppose that f is a non-negative continuous real-valued function on $C = I_1 \times \cdots \times I_d$. For $\tilde{x} \in \tilde{C}$, let $\phi(\tilde{x}) = \int_{I_d} f(\tilde{x}, t) \, dt$. Suppose* that $\phi(\widetilde{x}) < \infty$ for each $\widetilde{x} \in \widetilde{C}$, and that ϕ is a continuous function on $\widetilde{C}$. Then

$$\int_C f(x)\,dx = \int_{\widetilde{C}} \phi(\widetilde{x})\,d\widetilde{x} = \int_{\widetilde{C}} \left(\int_{I_d} f(\widetilde{x}, t)\,dt \right) d\widetilde{x}.$$

Proof   If $K = J_1 \times \cdots \times J_d$ is a compact cell contained in C then

$$\int_K f(x)\,dx = \int_{\widetilde{K}} \left(\int_{J_d} f(\widetilde{x}, t)\,dt \right) d\widetilde{x} \leq \int_{\widetilde{K}} \left(\int_{I_d} f(\widetilde{x}, t)\,dt \right) d\widetilde{x}$$

$$= \int_{\widetilde{K}} \phi(\widetilde{x})\,d\widetilde{x} \leq \int_{\widetilde{C}} \phi(\widetilde{x})\,d\widetilde{x}.$$

Consequently

$$\int_C f(x)\,dx \leq \int_{\widetilde{C}} \phi(\widetilde{x})\,d\widetilde{x}.$$

On the other hand, if $M < \int_{\widetilde{C}} \phi(\widetilde{x})\,d\widetilde{x}$, there exists a compact cell $\widetilde{K}$ contained in $\widetilde{C}$ such that

$$\int_{\widetilde{C}} \phi(\widetilde{x})\,d\widetilde{x} > M.$$

Let $(L_j)_{j=1}^{\infty}$ be an increasing sequence of compact intervals contained in I_d whose union is I_d, and let $\phi_j(\widetilde{x}) = \int_{L_j} f(\widetilde{x}, t)\,dt$. Then each ϕ_j is a continuous function on $\widetilde{K}$, and ϕ_j increases pointwise to the continuous function ϕ. It therefore follows from Dini's theorem (Theorem 15.2.12) that $\phi_j \to \phi$ uniformly on $\widetilde{K}$, as $j \to \infty$, and so

$$\int_{\widetilde{K}} \phi_j(\widetilde{x})\,d\widetilde{x} \to \int_{\widetilde{K}} \phi(\widetilde{x})\,d\widetilde{x} \text{ as } j \to \infty.$$

Thus there exists $j \in \mathbf{N}$ such that

$$\int_{\widetilde{K} \times L_j} f(x)\,dx = \int_{\widetilde{K}} \phi_j(\widetilde{x})\,d\widetilde{x} > M.$$

Hence $\int_C f(x)\,dx > M$. Since this holds for all $M < \int_{\widetilde{C}} \phi(\widetilde{x})\,d\widetilde{x}$, it follows that

$$\int_C f(x)\,dx \geq \int_{\widetilde{C}} \phi(\widetilde{x})\,d\widetilde{x}.$$

Thus we have equality. $\square$

Thus $\int_1^{\infty} (\int_1^{\infty} B(x,y)\,dx)\,dy = \int_0^1 1/(\log t \log(1-t))\,dt$.

With a little care, this result can be used in cases where f is not positive. Let us give an example.

Example 18.4.5 $\int_0^\infty \sin x / x \, dx = \pi/2$.

Proof Suppose that $K > 0$. If $x > 0$ then $1/x = \int_0^\infty e^{-xt} \, dt$, and so

$$\int_0^K \frac{\sin x}{x} \, dx = \int_0^K \sin x \left(\int_0^\infty e^{-xt} \, dt \right) dx.$$

Although sin is not a positive function, we can divide the interval $[0, K]$ into finitely many intervals, on each of which sin is either non-negative or non-positive, and apply the previous theorem to each of them. Thus

$$\int_0^K \sin x \left(\int_0^\infty e^{-xt} \, dt \right) dx = \int_0^\infty \left(\int_0^K e^{-xt} \sin x \, dx \right) dt.$$

Integrating the inner integral by parts twice,

$$\int_0^K e^{-xt} \sin x \, dx = [-e^{-xt} \cos x]_0^K - t \int_0^K e^{-xt} \cos x \, dx$$

$$= 1 - e^{-Kt} \cos K - t[e^{-xt} \sin x]_0^K - t^2 \int_0^K e^{-xt} \sin x \, dx$$

$$= 1 - e^{-Kt}(\cos K + t \sin K) - t^2 \int_0^K e^{-xt} \sin x \, dx.$$

so that

$$\int_0^K e^{-xt} \sin x \, dx = \frac{1}{1+t^2} + R_K(t),$$

where $R_K(t) = -e^{-Kt}(\cos K + t \sin K)/(1+t^2)$. Since

$$|\cos K + t \sin K| \leq 1 + t \leq 2(1+t^2),$$

$|R_K(t)| \leq 2e^{-Kt}$, and so $\int_0^\infty R_K(t) \, dt \to 0$ as $K \to \infty$. Consequently

$$\int_0^\infty \frac{\sin x}{x} \, dx = \lim_{K \to \infty} \int_0^K \frac{\sin x}{x} \, dx = \int_0^\infty \frac{dt}{1+t^2} = \frac{\pi}{2}.$$

$\square$

Exercises

18.4.1 Suppose that f is a real-valued Riemann integrable function on a compact cell C in $\mathbf{R}^d$ and that g is a Riemann integrable function on a compact cell D in $\mathbf{R}^e$. Show that the function $(x, y) \to f(x)g(y)$ is Riemann integrable on $C \times D$ and that, with the obvious notation,

$$\int_{C \times D} f(x)g(y) \, d(x, y) = \left(\int_C f(x) \, dx \right) \left(\int_D g(y) \, dy \right).$$

18.4.2 Give an example of a sequence of continuous real-valued functions on $[0, 1]$ which increases pointwise to a bounded function on $[0, 1]$ which is not Riemann integrable.

18.4.3 Let

$$P = \{(m/p, n/p) \in [0, 1] \times [0, 1] : p \text{ a prime}, m, n \in \mathbf{N}\}.$$

Show that P is dense in $[0, 1] \times [0, 1]$, and that

$$\int_0^1 \left(\int_0^1 I_P(x, y) \, dy \right) dx = \int_0^1 \left(\int_0^1 I_P(x, y) \, dx \right) dy.$$

Is I_P a Riemann integrable function on $[0, 1] \times [0, 1]$?

18.5 Jordan content

We now investigate some of the properties of Jordan measurable subsets of $\mathbf{R}^d$.

Suppose that A is a subset of a compact cell C and that D is a dissection of C, with constituent cells $C_1, \ldots, C_k$. We partition the cells of the partition into three: we set $D = J(A) \cup K(A) \cup L(A)$, where

$$J(A) = \{C_j : C_j \subseteq A\}$$
$$K(A) = \{C_j : C_j \cap A \neq \emptyset \text{ and } C_j \cap (C \setminus A) \neq \emptyset\}$$
$$L(A) = \{C_j : C_j \cap A = \emptyset\}.$$

Thus $J(A)$ is the set of cells contained in A, $K(A)$ is the set of cells which contain points of A and points not in A, and $L(A)$ is the set of cells disjoint from A.

Theorem 18.5.1 *A bounded subset A of a compact cell C in $\mathbf{R}^d$ is Jordan measurable if and only if given $\epsilon > 0$ there exists a dissection D of C, with constituent cells $C_1, \ldots, C_k$, such that*

$$\sum \{v_d(C_j) : C_j \in K(A)\} < \epsilon.$$

Let

$$\bar{v}_d(A) = \inf \left\{ \sum_{C_j \in J(A) \cup K(A)} v_d(C_j) : D \text{ a dissection of } C \right\},$$

$$\underline{v}_d(A) = \sup \left\{ \sum_{C_j \in J(A)} v_d(C) : D \text{ a dissection of } C \right\}.$$

Then A is Jordan measurable if and only if $\bar{v}_d(A) = \underline{v}_d(A)$. If so, then $v_d(A) = \bar{v}_d(A) = \underline{v}_d(A)$.

Proof   Let I_A be the indicator function of A. If C_j is a constituent cell in a dissection D, then $\Omega(I_A, C_j) = 0$ if $C_j \in J(A) \cup L(A)$, and $\Omega(I_A, C_j) = 1$ if $C_j \in K(A)$. Thus

$$\sum_{C_j \in D} v_d(C) \Omega(I_A, C_j) = \sum_{C_j \in K(A)} v_d(C_j),$$

so that A is Jordan measurable if and only if given $\epsilon > 0$ there exists a dissection D of C such that $\sum \{ v_d(C_j) : C_j \in K(A) \} < \epsilon$. Similarly $\bar{v}_d(A)$ is the upper integral of I_A, and $\underline{v}_d(A)$ is the lower integral of I_A; the remaining results follow from this. $\square$

Corollary 18.5.2   *A is Jordan measurable if (and only if) for each $\epsilon > 0$ there are Jordan measurable subsets B_1 and B_2 of C, with $B_1 \subseteq A \subseteq B_2$, such that $v_d(B_2) < v_d(B_1) + \epsilon$.*

The quantities $\bar{v}_d(A)$ and $\underline{v}_d(A)$ are called the *outer* and *inner Jordan contents* of A.

Thus a bounded set is Jordan measurable if it can be approximated from the outside and the inside by finite unions of cells, whose contents converge to a common value. This notion of approximating content, in two or three dimensions, by considering simple figures, goes back to the ancient Greeks. (The transition to Lebesgue measure is made by approximating by open sets (on the outside) and compact sets (on the inside).)

Since the sum and product of two real-valued Riemann integrable functions are Riemann integrable, it follows that the intersection and union of two Jordan measurable sets A and B are Jordan measurable, and that

$$v_d(A \cup B) + v_d(A \cap B) = v_d(A) + v_d(B).$$

Similarly $A \setminus B$ is Jordan measurable, and $v_d(A \setminus B) = v_d(A) - v_d(A \cap B)$. It is easy to see that if A is Jordan measurable, then so is a translate $A + x = \{ a + x : a \in A \}$, and that $v_d(A + x) = v_d(A)$. Similarly, if $\lambda > 0$ then the dilate $\lambda A = \{ \lambda a : a \in A \}$ is Jordan measurable, and $v_d(\lambda A) = \lambda^d v_d(A)$.

The notion of Jordan content makes the idea of the integral as the 'area under the curve' explicit.

Theorem 18.5.3   *Suppose that f is a bounded non-negative real-valued function defined on a compact cell C in $\mathbf{R}^d$. Let*

$$A_f = \{ (x, y) \in C \times \mathbf{R} : 0 \le y \le f(x) \}.$$

Then f is Riemann integrable if and only if A_f is a Jordan measurable subset of $\mathbf{R}^d \times \mathbf{R} = \mathbf{R}^{d+1}$. If so, then $\int_C f(x)\,dx = v_{d+1}(A_f)$.

Proof   The result is certainly true if f is a step function.

Suppose that f is Riemann integrable, and that $\epsilon > 0$. There exist step functions g and h with $0 \leq g \leq f \leq h$ such that $\int_C (h(x) - g(x))\,dx < \epsilon$. Then A_g and A_h are Jordan measurable, and

$$v_{d+1}(A_h) = \int_C h(x)\,dx \leq \int_C g(x)\,dx + \epsilon = v_{d+1}(A_g) + \epsilon.$$

Since $A_g \subseteq A_f \subseteq A_h$, it follows that A_f is Jordan measurable. Since $v_{d+1}(A_g) \leq v_{d+1}(A_f) \leq v_{d+1}(A_h)$ and $\int_C g(x)\,dx \leq \int_C f(x)\,dx \leq \int_C h(x)\,dx$, it also follows that $\int_C f(x)\,dx = v_{d+1}(A_f)$.

The converse is proved in a similar way. Suppose that D is a dissection of $C \times [0, M]$. If $x \in C$, let $I_x = \{(x, t) : 0 \leq t \leq M\}$, and let

$$J_x(A_f) = \{C_j \in J(A_f) : C_j \cap I_x \neq \emptyset\},$$

$$K_x(A_f) = \{C_j \in K(A_f) : C_j \cap I_x \neq \emptyset\}.$$

Let $m_D(x) = 0$ if $J_x(A_f) = \emptyset$, and let

$$m_D(x) = \sup\{t : (x, t) \in C_j, C_j \in J_x(A_f)\} \text{ otherwise;}$$

similarly, let $M_D(x) = 0$ if $J_x(A_f) \cup K_x(A_f) = \emptyset$, and let

$$M_D(x) = \sup\{t : (x, t) \in C_j, C_j \in J_x(A_f) \cup K_x(A_f)\} \text{ otherwise.}$$

Then $m_D \leq f \leq M_D$. It follows that if the condition is satisfied then f is Riemann integrable and $\int_C f(x)\,dx = v_{d+1}(A_f)$.

$\square$

A similar result holds if we consider the set

$$U_f = \{(x, y) \in C \times \mathbf{R} : 0 < y < f(x)\}.$$

Let us show that there is a useful class of Jordan measurable sets. A *convex body* in $\mathbf{R}^d$ is a convex subset with a non-empty interior.

Proposition 18.5.4   *A bounded convex body A in $\mathbf{R}^d$ is Jordan measurable.*

Proof   We need the following easy result about convex sets.

Lemma 18.5.5   *If A is a convex subset of a vector space V and $\lambda > 0$ then $(1 + \lambda)A = A + \lambda A$.*

Proof If A is any subset of V,

$$(1 + \lambda)A = \{a + \lambda a : a \in A\} \subseteq \{a + \lambda b : a, b \in A\} = A + \lambda A;$$

we need to establish the converse inclusion, when A is convex. If $x = a + \lambda b \in A + \lambda A$ then

$$y = \frac{1}{1 + \lambda}a + \frac{\lambda}{1 + \lambda}b \in A,$$

and so $x = (1 + \lambda)y \in (1 + \lambda)A$. □

Let us now prove Proposition 18.5.4. Without loss of generality, we can suppose that 0 is an interior point of A, so that there exists $\eta > 0$ such that $N_\eta(0) \subseteq A$. Suppose that C is a compact cell containing A and that $\epsilon > 0$. Let $\delta = \epsilon\eta$. Then

$$A + N_\delta(0) \subseteq A + \epsilon A = (1 + \epsilon)A,$$

so that if $x \notin (1 + \epsilon)A$ then $d(x, A) \geq \delta$. Thus if D is a dissection of C with mesh size less than δ, and $D = J(A) \cup K(A) \cup L(A)$, as above, then $\cup_{C_j \in K(A)} C_j \subseteq (1 + \epsilon)A$. Consequently

$$\overline{v}_d(A) \leq \underline{v}_d((1 + \epsilon)A) = (1 + \epsilon)^d \underline{v}_d(A).$$

Since ϵ is arbitrary, $\overline{v}_d(A) = \underline{v}_d(A)$, and A is Jordan measurable. □

Exercises

18.5.1 Suppose that f and g are Riemann integrable functions defined on a compact cell C in $\mathbf{R}^d$, taking values in $[a, b]$, and that $f \leq g$. Let $A = \{(x, y) \in C \times [a, b] : f(x) \leq y \leq g(x)\}$. Suppose that h is a continuous function on $C \times [a, b]$. Show that hI_A is Riemann integrable and that

$$\int_A f(w) \, dw = \int_C \left(\int_{f(x)}^{g(x)} h(x, y) \, dy \right) dx.$$

Does a similar result hold for Riemann integrable functions h?

18.5.2 Show that a bounded subset A of $\mathbf{R}^d$ is Jordan measurable if and only if its boundary ∂A has outer content 0.

18.5.3 Suppose that f and g are Riemann integrable real-valued functions on a cell C in $\mathbf{R}^d$, and that g is non-negative. Show that there exists $c \in \mathbf{R}$, with

$$\inf\{f(x) : x \in C\} \leq c \leq \sup\{f(x) : x \in C\},$$

such that $\int_C f(x)g(x)\,dx = c\int_C g(x)\,dx$.

18.5.4 Suppose that C is a compact convex body in $\mathbf{R}^d$. Show that C is homeomorphic to the closed unit ball in $\mathbf{R}^d$.

18.5.5 Use induction, and repeated integrals, to calculate the volume of the unit ball $B_d = \{x \in \mathbf{R}^d : \|x\|_2 \le 1\}$ in $\mathbf{R}^d$. How does $v_d(B_d)$ behave as $d \to \infty$?

18.5.6 Let $E_d = \{x \in B_d : |x_1| \le 1/\sqrt{d}\}$ be an equatorial strip in B_d. Show that $v_d(E_d)/v_d(B_d)$ converges to a non-zero limit as $d \to \infty$.

18.6 Linear change of variables

In this section and Section 18.8, we consider 'change of variables'. We begin by establishing a 'change of variables' formula for linear mappings. The problem here is that the notion of a cell depends upon the coordinates in $\mathbf{R}^d$.

We therefore need to appeal to some elementary plane geometry and to the structure of the general linear group GL_d, the group of invertible linear mappings of $\mathbf{R}^d$. We consider some simple elements of GL_d. A *scaling operator* D is an element of GL_d defined by an invertible diagonal matrix $\mathrm{diag}(\lambda_1, \ldots, \lambda_d)$, so that $D(x) = (\lambda_1 x_1, \ldots, \lambda_d x_d)$. An *elementary shear operator* R is an element of GL_d defined by an elementary shear matrix: a matrix T of the form $I + \alpha E_{ij}$, where $i \ne j$ and E_{ij} is the matrix with $(E_{ij})_{ij} = 1$, and with all other entries zero. Thus if $T(x) = y$ then $y_i = x_i + \alpha x_j$, and $y_k = x_k$ for all other indices.

Theorem 18.6.1 *If $T \in GL_d$ then T can be written as PDQ, where D is a scaling operator, and each of P and Q is the product of a finite number of elementary shear operators.*

This theorem is proved in Appendix B (Theorem B.2.3).

Proposition 18.6.2 *Suppose that T is an invertible linear mapping of $\mathbf{R}^d$ onto itself and that C is a cell in $\mathbf{R}^d$. Then $T(C)$ is Jordan measurable, and $v_d(T(C)) = |\det T|.v_d(C)$.*

Proof Since $T(C)$ is a convex body, it is Jordan measurable. Since $\det(ST) = \det S. \det T$ it is sufficient, by Theorem 18.6.1, to prove the result for scaling operators and for elementary shear operators.

If $T = \mathrm{diag}(\lambda_1, \ldots, \lambda_d)$ then $T(C)$ is a cell and

$$v_d(T(C)) = |\lambda_1 \ldots \lambda_d|.v_d(C) = |\det T|.v_d(C),$$

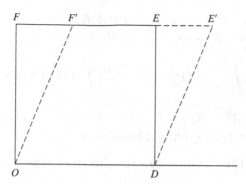

Figure 18.6. A cell and a sheared cell.

so that the result holds for scaling operators. It remains to show that the result also holds when $T = I + \alpha E_{ij}$ is an elementary shear operator. Since the translate of a cell is a cell with the same content, it is sufficient to consider the case where $C = [0,1]^d$.

We consider the case where $\alpha > 0$; the proof for $\alpha < 0$ is similar. Since $\det T = 1$, setting

$$e_i = D,\ e_i + e_j = E,\ e_j = F,\ T(e_i + e_j) = E' \text{ and } T(e_j) = F',$$
$$v_d(T(C)) = v_2(ODE'F') = v_2(ODEF') + v_2(DE'E)$$
$$= v_2(ODEF') + v_2(OF'F) = v_2(ODEF) = |\det T| v_d(C).$$

$\square$

Corollary 18.6.3 $T^{-1}(C)$ *is Jordan measurable, and*

$$v_d(C) = |\det T| . v_d(T^{-1}(C)).$$

Proof For $\det(T^{-1}) = 1/\det T$. $\square$

Corollary 18.6.4 *If A is a Jordan measurable subset of $\mathbf{R}^d$, then $T(A)$ is a Jordan measurable subset of $\mathbf{R}^d$, and $v_d(T(A)) = |\det T| v_d(A)$.*

Proof The result is true if A is the finite disjoint union of cells. If $\epsilon > 0$ there exist two such sets B_1 and B_2 with $B_1 \subseteq A \subseteq B_2$, and with $v_d(B_2) < v_d(B_1) + \epsilon/|\det T|$. Then $T(B_1) \subseteq T(A) \subseteq T(B_2)$ and $v_d(T(B_2)) < v_d(T(B_1)) + \epsilon$. Hence $T(A)$ is Jordan measurable, by Corollary 18.5.2, and it follows that $v_d(T(A)) = |\det T| v_d(A)$. $\square$

Corollary 18.6.5 *If f is a real-valued Riemann integrable function on $T(A)$, then $f \circ T$ is a Riemann integrable function on A, and*

$$\int_{T(A)} f(x)\, dx = |\det T| \int_A f(T(x))\, dx.$$

Proof If $(x, t) \in \mathbf{R}^d \times \mathbf{R}$, let $\widetilde{T}(x, t) = (T(x), t)$. Then $\det \widetilde{T} = \det T$, and so the result follows by applying Theorem 18.5.3. $\square$

We shall extend these results to a non-linear change of variables in Section 18.8.

18.7 Integrating functions on Euclidean space

The results of the previous section allow us to integrate functions defined on a subset of Euclidean space, when there is no coordinate system in place. We introduce coordinates, use them to define the integral, and then show that this does not depend on the choice of coordinates.

Suppose then that E is a d-dimensional Euclidean space, and that $(e_1, \ldots, e_d)$ and $(e_1', \ldots, e_d')$ are two orthonormal bases for E. If $x \in \mathbf{R}^d$, we set $L(x) = \sum_{j=1}^d x_j e_j$ and $L'(x) = \sum_{j=1}^d x_j e_j'$. L and L' are linear isometries of $\mathbf{R}^d$ onto E, $U = L^{-1} \circ L'$ is an orthogonal mapping on $\mathbf{R}^d$, and $|\det U| = 1$.

Suppose now that B is a bounded subset of E. We say that B is *Jordan measurable* if $L^{-1}(B)$ is a Jordan measurable subset of $\mathbf{R}^d$, and set $v_d(B) = v_d(L^{-1}(B))$. Since $L^{-1}(B) = U(L'^{-1}(B))$, it follows from Corollary 18.6.4 that these definitions do not depend upon the choice of basis. Similarly, if f is a real-valued function on B, we say that f is Riemann integrable if $f \circ L$ is Riemann integrable on $L^{-1}(B)$, and define the Riemann integral $\int_B f(x)\, dv_d(x)$

$$\int_B f(x)\, dv_d(x) = \int_{L^{-1}(B)} f(L(x))\, dx.$$

Again these definitions do not depend upon the choice of basis, this time by Corollary 18.6.5.

Finally, suppose that f is a function taking values in an e-dimensional Euclidean space F. Suppose that $(g_1, \ldots, g_e)$ is an orthonormal basis for F. We can then write $f(x) = \sum_{j=1}^e f_j(x) g_j$. We say that f is Riemann integrable if $f_1, \ldots, f_e$ are and we set

$$\int_B f(x)\, dv_d(x) = \sum_{j=1}^e \left(\int_B f_j(x)\, dv_d(x) \right) g_j.$$

This time, it is the linearity of the integral that ensures that this does not depend upon the choice of basis.

Exercise

18.7.1 What happens if we use bases of E which are not orthonormal? What happens if we use bases of F which are not orthonormal?

18.8 Change of variables

We now consider a more general change of variables.

Suppose that $C^{(\epsilon)}$ is a fat Cantor subset of $[0,1]$. If $x \in [0,1]$, let $f(x) = 2 - I_{C^{(\epsilon)}}(x)$, so that $f(x) = 1$ if $x \in C^{(\epsilon)}$ and $f(x) = 2$ otherwise. Let $U_f = \{(x,y) \in (0,1) \times (0,2) : 0 < y < f(x)\}$. Then U_f is an open subset of $\mathbf{R}^2$ which is not Jordan measurable. Further U_f is connected, and its complement is also connected. It therefore follows from the Riemann mapping theorem (which we shall prove in Volume III), that there is a smooth diffeomorphism ϕ of the open unit square $(0,1) \times (0,1)$ onto U_f. This clearly has bad consequences for 'change of variables' results. In fact, the bad behaviour results from bad behaviour of ϕ near the boundary of $(0,1) \times (0,1)$. If we avoid this possibility, then, as we shall see, we can obtain some positive results.

First, we consider what happens to compact cells.

Theorem 18.8.1 *Suppose that $\phi : U \to V$ is a diffeomorphism from an open subset U of $\mathbf{R}^d$ onto an open subset V of $\mathbf{R}^d$, and that $D\phi_x$ is invertible, for each $x \in U$. If C is a compact cell contained in U then $\phi(C)$ is Jordan measurable, and $v_d(\phi(C)) = \int_C |J_\phi(x)| \, dx$, where J_ϕ is the Jacobian of ϕ.*

Proof The idea of the proof is to find a fine enough dissection of C such that we can approximate ϕ linearly on the constituent cells, and to use the estimates in the Lipschitz inverse function theorem to obtain good approximations. Since we are working with cells, it is convenient to work with the supremum norm on $\mathbf{R}^d$: $\|x\|_\infty = \max\{|x_j| : 1 \leq j \leq d\}$, and with the corresponding operator norm on $L(\mathbf{R}^d)$.

Since $\phi(C)$ is a compact subset of the open set V, there exists $\theta > 0$ such that the compact set $K = \{y \in \mathbf{R}^d : d(y, \phi(C)) \leq \theta\}$ is contained in V.

Suppose that $0 < \epsilon < 1$. Choose $0 < \eta < \epsilon$ such that

$$1 - \epsilon < (1 - \eta)^d < (1 + \eta)^d < 1 + \epsilon.$$

The mappings J_ϕ and $D\phi$ are uniformly continuous on C and, by Corollary 14.6.9, $D\phi^{-1}$ is uniformly continuous on K. There therefore exist $1 \le M < \infty$ such that

$$\|D\phi(x)\| \le M, \text{ for } x \in C, \text{ and } \|D\phi^{-1}(y)\| \le M, \text{ for } y \in K,$$

and $0 < \delta < \theta$ such that

$$|J_\phi(x) - J_\phi(x')| < \eta/v_d(C), \text{ for } x, x' \in C, \|x - x'\| < \delta,$$
$$\|D\phi(x) - D\phi(x')\| < \eta/2M, \text{ for } x, x' \in C, \|x - x'\| < \delta,$$
$$\text{and } \|D\phi^{-1}(y) - D\phi^{-1}(y')\| < \eta/2M, \text{ for } y, y' \in K, \|y - y'\| < \delta.$$

Suppose now that D is a dissection with mesh size less than δ, with constituent cells $C_1, \ldots, C_k$, with midpoints $x_1, \ldots, x_k$. Let $S_j = D\phi_{x_j}$, for $1 \le j \le k$.

Let us consider a particular cell C_j. By translating U and V, we can suppose that $x_j = 0$ and that $\phi(x_j) = 0$. If $h \in C_j$ let $\psi(h) = S_j^{-1}(\phi(h))$. We show that

$$(1 - \eta)C_j \subseteq \psi(C_j) \subseteq (1 + \eta)C_j.$$

By Corollary 17.2.6, if $h, k \in C_j$, then

$$\|\phi(h) - \phi(k) - D\phi_0(h - k)\| \le \|h - k\| \sup\{\|D\phi_l - D\phi_0\| : l \in [h, k]\}$$
$$\le (\eta/M) \|h - k\|.$$

If $h \in C_j$, let $\chi(h) = \psi(h) - h$. Then

$$\|\chi(h) - \chi(k)\| = \|\psi(h) - \psi(k) - (h - k)\|$$
$$\le M \|\phi(h) - \phi(k) - D\phi_0(h - k)\| \le \delta \|h - k\|.$$

Thus χ is a Lipschitz function on C_j, with Lipschitz constant η. It therefore follows from the Lipschitz inverse function theorem (Theorem 14.6.6) that

$$(1 - \eta)C_j \subseteq \psi(C_j) \subseteq (1 + \eta)C_j.$$

Consequently,

$$(1 - \eta)S_j(C_j) \subseteq \phi(C_j) \subseteq (1 + \eta)S_j(C_j).$$

We use these inclusions to estimate the upper and lower Jordan contents of $\phi(C)$. First,

$$\overline{v}_d(\phi(C_j)) \le v_d((1 + \eta)S_j(C_j)) \le (1 + \epsilon)|J_\phi(x_j)|v_d(C_j),$$

and

$$\left| \int_{C_j} |J_\phi(x)| \, dx - |J_\phi(x_j)| v_d(C_j) \right| < \epsilon v_d(C_j)/v_d(C).$$

Thus

$$\bar{v}_d(\phi(C)) \leq \sum_{j=1}^{k} \bar{v}_d(\phi(C_j))$$

$$\leq \sum_{j=1}^{k} (1+\epsilon)|J_\phi(x_j)| v_d(C_j) \leq (1+\epsilon) \int_C |J_\phi(x)| \, dx + \epsilon.$$

Secondly,

$$\underline{v}_d(\phi(C_j)) \geq v_d(S_j((1-\eta)C_j)) \geq (1-\epsilon)|J_\phi(x_j)| v_d(C_j),$$

so that

$$\underline{v}_d(\phi(C)) \geq \sum_{j=1}^{k} v_d(S_j((1-\eta)C_j)) \geq (1-\epsilon) \sum_{j=1}^{k} |J_\phi(x_j)| v_d(C_j)$$

$$\geq (1-\epsilon) \int_C |J_\phi(x)| \, dx - \epsilon.$$

Since ϵ is arbitrary, it follows that $\phi(C)$ is Jordan measurable, and that $v_d(\phi(C)) = \int_C |J_\phi(x)| \, dx$. □

This result can extended to more general sets.

Corollary 18.8.2 *Suppose that B is a Jordan measurable subset of U contained in a compact subset K of U. Then $\phi(B)$ is Jordan measurable, and*

$$v_d(\phi(B)) = \int_B |J_\phi(x)| dx.$$

Proof Let $L = \sup\{|J_\phi(x)| : x \in K\}$. First suppose that B is a cell. Suppose that $\epsilon > 0$. There exists a compact cell C' contained in B such that $v_d(C') > v_d(B) - \epsilon/L = v_d(\overline{B}) - \epsilon/L$. Then

$$v_d(\phi(\overline{B})) - v_d(\phi(C')) = \int_B |J_\phi(x)| \, dx - \int_{C'} |J_\phi(x)| \, dx < \epsilon;$$

since ϵ is arbitrary, $\phi(B)$ is Jordan measurable, and $v_d(\phi(B)) = \int_B |J_\phi(x)| dx$.

In the general case, if $\epsilon > 0$ there exists a finite collection of disjoint cells $J(B) \cup K(B)$ such that

$$\cup\{C : C \in J(B)\} \subseteq B \subseteq \cup\{C : C \in J(B) \cup K(B)\},$$

for which $v_d(\cup\{C : C \in K(B)\}) < \epsilon/L$. Then

$$v_d(\phi(\cup\{C : C \in J(B)\})) > v_d(\phi(\cup\{C : C \in J(B) \cup K(B)\})) - \epsilon,$$

so that $\phi(B)$ is Jordan measurable, and it follows that $v_d(\phi(B)) = \int_B |J_\phi(x)|\,dx$. □

We now obtain a change of variables result for Riemann integrable functions.

Theorem 18.8.3 *Suppose that $\phi : U \to V$ is a diffeomorphism from an open subset U of $\mathbf{R}^d$ onto an open subset V of $\mathbf{R}^d$, and that $D\phi_x$ is invertible, for each $x \in U$. Suppose that K is a compact subset of U, that B is a Jordan measurable subset of K and that f is a Riemann integrable mapping from $\phi(B)$ into $\mathbf{R}^e$. Then $f \circ \phi$ is Riemann integrable, and*

$$\int_{\phi(B)} f(y)\,dy = \int_B f(\phi(x))|J_\phi(x)|\,dx.$$

Proof   We use Theorem 18.5.3. By considering the coordinates of f, we can suppose that f is real-valued, and by considering f^+ and f^-, we can suppose that f is non-negative. Let $M = \|f\|_\infty + 1$, let

$$\widetilde{U} = U \times (-M, M) \text{ and } \widetilde{V} = V \times (-M, M),$$

and let $\widetilde{\phi}(x, t) = (\phi(x), t)$ for $(x, t) \in \widetilde{U}$. Then $\widetilde{\phi} : \widetilde{U} \to \widetilde{V}$ is a diffeomorphism from $\widetilde{U}$ onto $\widetilde{V}$, and $D\widetilde{\phi}_{(x,t)}(h, s) = (D\phi_x(h), s)$, so that $D\widetilde{\phi}_x$ is invertible, for each $x \in U$. Further, $J_{\widetilde{\phi}}(x, t) = J_\phi(x)$, and $\widetilde{\phi}(A_{f \circ \phi}) = A_f$. Since f is Riemann integrable, the set A_f is Jordan measurable. Applying Corollary 18.8.2 to the mapping $\widetilde{\phi}^{-1}$, it follows that $A_{f \circ \phi}$ is Jordan measurable. Thus $f \circ \phi$ is Riemann integrable, and

$$\int_{\phi(B)} f(y)\,dy = v_{d+1}(A_f) = \int_{A_{f \circ \phi}} |J_{\widetilde{\phi}}(x, t)|\,d(x, t)$$

$$= \int_B \left(\int_0^{f(\phi(x))} dt \right) |J_\phi(x)|\,dx = \int_B f(\phi(x))|J_\phi(x)|\,dx.$$

□

In many cases, the integral can be extended to the whole of U, as the following example shows.

Proposition 18.8.4 $\int_0^\infty e^{-t^2/2}\,dt = \sqrt{\frac{\pi}{2}}.$

Proof   We use polar coordinates. Let $U = (0, \infty) \times (0, \pi/2)$, and let $\phi(r, \theta) = (r \cos \theta, r \sin \theta)$. Then ϕ is a continuously differentiable homeomorphism of U onto the quadrant $V = (0, \infty) \times (0, \infty)$ and

$$D\phi_{(r,\theta)} = \begin{bmatrix} \cos \theta & \sin \theta \\ -r \sin \theta & r \cos \theta \end{bmatrix}, \text{ so that } J_\phi(r, \theta) = r.$$

Set $f(y_1, y_2) = e^{-y_1^2/2} e^{-y_2^2/2}$ for $y = (y_1, y_2) \in V$, so that $f(\phi(r, \theta)) = e^{-r^2/2}$. Let $U_{\epsilon, R} = [\epsilon, R] \times [\epsilon, \pi/2 - \epsilon]$ for $0 < \epsilon < \pi/4 < R < \infty$. Then

$$\left(\int_0^\infty e^{-t^2/2} \, dt \right)^2 = \lim_{\epsilon \to 0, R \to \infty} \int_{\phi(U_{\epsilon, R})} f(y) \, dy$$

$$= \lim_{\epsilon \to 0, R \to \infty} \int_{U_{\epsilon, R}} e^{-r^2/2} r \, dr \, d\theta$$

$$= \lim_{\epsilon \to 0, R \to \infty} (\pi/2 - 2\epsilon)(e^{-\epsilon^2/2} - e^{-R^2/2}) = \pi/2.$$

$\square$

Exercises

18.8.1 Let $f(x, y) = (1/2\pi) e^{-(x^2 + y^2)/2}$. Let $D_r = \{x, y) \in \mathbf{R}^2 : x^2 + y^2 \le r^2\}$ and let $B_s = \{(x, y) \in \mathbf{R}^2; x \le sy\}$. Calculate

$$\int_{D_r} f(x, y) \, d(x, y), \quad \int_{B_s} f(x, y) \, d(x, y) \text{ and } \int_{D_r \cap B_s} f(x, y) \, d(x, y).$$

18.8.2 Let $Q = \{(x, y) \in \mathbf{R}^2 : x \ge 0, y \ge 0\}$. Let $f(x, y) = e^{-(x+y)}$, for $(x, y) \in Q$. Let

$$A_r = \{(x, y) \in Q : x + y \le r\}, \quad B_s = \{(x, y) \in \mathbf{R}^2; x \le sy\}.$$

Calculate

$$\int_{A_r} f(x, y) \, d(x, y), \quad \int_{B_s} f(x, y) \, d(x, y) \text{ and } \int_{A_r \cap B_s} f(x, y) \, d(x, y).$$

18.8.3 Explain (or find out) the probabilistic significance of these results.

18.8.4 Let $U = (0, \infty) \times (0, \pi) \times (-\pi, \pi)$. If $(r, \theta, \phi) \in U$, let

$$J_s(r, \theta, \phi) = (r \sin \theta \cos \phi, r \sin \theta \sin \phi, r \cos \theta).$$

Show that J_s is a homeomorphism of U onto $\mathbf{R}^3 \setminus H$, where H is the half-plane $\{(x, y, z) : x \le 0, y = 0\}$, and illustrate this with a sketch. Show that the Jacobian of J_s is

$$\frac{\partial(x, y, z)}{\partial(r, \theta, \phi)} = r^2 \sin \theta.$$

The coordinates (r, θ, ϕ) are the *spherical polar coordinates*; r is the *radius*, θ is the *inclination angle*, and ϕ is the *azimuth*.

18.8.5 Use spherical polar coordinates to calculate the volume of the set (described in spherical polar coordinates) $\{J_s(r, \theta, \phi) : r^3 \leq \sin \theta\}$.

18.8.6 Let $V = (0, \infty) \times (-\pi, \pi) \times \mathbf{R}$. If $(\rho, \phi, z) \in V$ let $J_c(\rho, \theta, z) = (\rho \cos \phi, \rho \sin \phi, z)$. Show that J_c is a homeomorphism of U onto $\mathbf{R}^3 \setminus H$, where H is the half-plane $\{(x, y, z) : x \leq 0, y = 0\}$, and illustrate this with a sketch. Show that the Jacobian of J_c is

$$\frac{\partial(x, y, z)}{\partial(\rho, \phi, z)} = \rho.$$

The coordinates (ρ, ϕ, z) are the *cylindrical polar coordinates*; ρ is the *radius* (notice that this not the same as the radius in spherical polar coordinates), ϕ is the *azimuth* and z is the *altitude*.

18.8.7 Use cylindrical polar coordinates to calculate the volume of the set (described in cylindrical polar coordinates)

$$J_c = \{(\rho, \phi, z) : \rho \leq z \cos \phi / 2, 0 \leq z \leq 1\}.$$

18.8.8 Let $P = (-1/2, 0)$ and $Q = (1/2, 0)$. Let

$$s = s(x, y) = d((x, y), P), \ t = t(x, y) = d((x, y), Q),$$

$$u = u(x, y) = s + t, \ v = v(x, y) = s - t,$$

for $(x, y) \in \mathbf{R}^2$. Show that the mapping $(x, y) \to (u(x, y), v(x, y))$ is a homeomorphism of the upper half space $H^+ = \{(x, y) : y > 0\}$ onto $(1, \infty) \times (0, 1)$. Show that

$$\frac{\partial(u, v)}{\partial(x, y)} = -\frac{2y}{st}.$$

Calculate

$$\int_{H^+} \frac{ye^{-(s+t)}}{st} \, d(x, y).$$

18.8.9 Let $P = (0, 0, -1/2)$ and $Q = (0, 0, 1/2)$. Let

$$s = s(x, y, z) = d((x, y, z), P), \ t = t(x, y, z) = d((x, y, z), Q),$$

$$u = u(x, y, z) = s + t, \ v = v(x, y, z) = s - t,$$

for $(x, y, z) \in \mathbf{R}^3$. Calculate

$$\frac{\partial(u, v, \phi)}{\partial(x, y, z)}$$

where ϕ is the azimuth. [Use cylindrical polar coordinates, and the chain rule.] Show that

$$\int_{\mathbf{R}^3} \frac{e^{-(s+t)}}{st} \, d(x, y, z) = 2\pi/e.$$

18.8.10 Prove a version of Theorem 18.8.3 for continuous functions taking values in a Banach space.

18.9 Differentiation under the integral sign

Suppose that A is a compact Jordan measurable subset of $\mathbf{R}^d$ and that (a, b) is an interval in $\mathbf{R}$. Suppose that f is a function on $A \times (a, b)$ taking values in a Banach space $(F, \|.\|)$, that the mapping $x \to f(x, t)$ from A to F is Riemann integrable for each $t \in (a, b)$, and that the mapping $t \to f(x, t)$ from (a, b) to F is differentiable for each $x \in A$. Let $F(t) = \int_A f(x, t) \, dx$. When is F a differentiable function of t? If it is, then when is

$$\frac{dF}{dt} = \frac{d}{dt} \left(\int_A f(x, t) \, dx \right) = \int_A \frac{\partial f}{\partial t}(x, t) \, dx?$$

In other words, when can we change the order of integration and differentiation? We give just one positive result, where we impose strong conditions on f; it is however suitable for many purposes.

Theorem 18.9.1 *Suppose that A is a compact Jordan measurable subset of $\mathbf{R}^d$ and that (a, b) is an interval in $\mathbf{R}$. Suppose that f is a continuous function on $A \times (a, b)$ taking values in a Banach space $(F, \|.\|)$, and that the partial derivative $(\partial f/\partial t)(x, t)$ exists at every point of $A \times (a, b)$ and is a continuous function on $A \times (a, b)$. Let $F(t) = \int_A f(x, t) \, dx$. Then F is a continuously differentiable function on (a, b), and*

$$\frac{dF}{dt} = \frac{d}{dt} \left(\int_A f(x, t) \, dx \right) = \int_A \frac{\partial f}{\partial t}(x, t) \, dx.$$

Proof We use Corollary 17.2.6. Suppose that $a < t < b$ and that $[t - \eta, t + \eta] \subset (a, b)$. Suppose that $\epsilon > 0$. Then $\partial f/\partial t$ is uniformly continuous on $A \times [t - \eta, t + \eta]$ and so there exists $0 < \delta < \eta$ such that if $x \in A$, $u, v \in [t - \eta, t + \eta]$ and $|u - v| < \delta$ then

$$\|(\partial f / \partial t)(x, u) - (\partial f / \partial t)(x, v)\| < \epsilon / v_d(A).$$

If $x \in A$ and $t - \delta < s < t + \delta$ then by Corollary 17.2.6

$$\left\| f(x, s) - f(x, t) - (s - t)\frac{\partial f}{\partial t}(x, t) \right\| \leq |s - t| \sup_{u \in [s,t]} \left\| \frac{\partial f}{\partial t}(x, u) - \frac{\partial f}{\partial t}(x, t) \right\|$$

$$\leq |s - t|\epsilon / v_d(A).$$

Integrating,

$$\left\| F(s) - F(t) - (s - t)\int_A \frac{\partial f}{\partial t}(x, t)\, dx \right\|$$

$$\leq \int_A \left\| f(x, s) - f(x, t) - (s - t)\frac{\partial f}{\partial t}(x, t) \right\|\, dx \leq \epsilon |s - t|,$$

which shows that F is differentiable at t, with derivative $\int_A (\partial f / \partial t)(x, t)\, dx$. The continuity of the derivative then follows from the uniform continuity of $\partial f / \partial t$ on $A \times [t - \eta, t + \eta]$. $\square$

Exercises

18.9.1 Suppose that f and g are continuous functions on $\mathbf{R}$ for which $\int_{\mathbf{R}} |f(x)|\, dx < \infty$ and $\int_{\mathbf{R}} |g(x)|\, dx < \infty$. Let

$$H(t) = \int_{A_t} f(x)g(y)\, d(x, y), \text{ where } A_t = \{(x, y) \in \mathbf{R}^2 : x + y \leq t\}.$$

Show that H is differentiable, and that $H'(t) = \int_{\mathbf{R}} f(t - x)g(x)\, dx$.

18.9.2 Suppose that f and g are continuous non-negative functions on $\mathbf{R}$ of compact support. Let

$$K(s) = \int_{B_s} f(x)g(y)\, d(x, y), \text{ where } B_s = \{(x, y) \in \mathbf{R}^2 : x \leq sy\}.$$

Show that K is differentiable, and that $K'(s) = \int_{\mathbf{R}} xf(sx)g(x)\, dx$.

18.9.3 Suppose that $f(x, y) = (1/\pi)e^{-(x^2 + y^2)/2}$, and that

$$F(s) = \int_{B_s} f(x, y)\, d(x, y), \text{ where } B_s = \{(x, y) \in \mathbf{R}^2 : x > 0, y \leq sx\}.$$

Show that F is differentiable, and that $F'(s) = \pi/(1 + s^2)$.

19

Differential manifolds in Euclidean space

19.1 Differential manifolds in Euclidean space

A *manifold* is a topological space which is locally like Euclidean space: each point has an open neighbourhood which is homeomorphic to an open subset of a Euclidean space. A *differential manifold* is one for which the homeomorphisms can be taken to be diffeomorphisms. We consider differential manifolds which are subspaces of Euclidean space.

Recall that a diffeomorphism f of an open subset W of a Euclidean space E onto a subset $f(W)$ of a Euclidean space F is a bijection of W onto $f(W)$ which is continuously differentiable, and has the property that the derivative Df_x is invertible, for each $x \in W$. If so, then $f(W)$ is open in F, and the mapping $f^{-1} : f(W) \to W$ is also a diffeomorphism. Further $\dim E = \dim F$, and Df_x has rank $\dim E$, for each $x \in E$. We split this definition into two parts.

First, suppose that W is an open subset of a Euclidean space E_d of dimension d, and that j is a continuously differentiable injective mapping of W onto a subset $j(W)$ of a Euclidean space F_{d+n} of dimension $d + n$, where $n \geq 0$. Then j is an *immersion* if the rank of Dj_x is equal to d, for each $x \in W$; that is, if Dj_x is an injective linear mapping of E into F, for each $x \in W$. If j is k-times continuously differentiable, we say that j is a $C^{(k)}$-*immersion*, and if j is a smooth mapping, we say that j is a *smooth immersion*.

Secondly, suppose that U is an open subset of a Euclidean space E_{d+n} of dimension $d + n$, where $n > 0$, and that g is a continuously differentiable mapping of U onto a subset $g(U)$ of a Euclidean space F of dimension n. Then g is a *submersion* of rank n if the rank of Dg_x is n, for all $x \in U$. Thus Dg_x is surjective, and the null-space of Dg_x has dimension d, for all $x \in U$. If g is k-times continuously differentiable, we say that g is a $C^{(k)}$-*submersion*, and if g is a smooth mapping, we say that g is a *smooth submersion*.

We use submersions to define the notion of a d-dimensional manifold M which is a subspace of a Euclidean space E_{d+n} of dimension $d + n$. A non-empty subset M of E_{d+n} is a d-*dimensional differential manifold* if for each x in M there exists an open neighbourhood U_x of x in E_{d+n}, and a submersion $g_x : U_x \to F$ of rank n such that

$$M \cap U_x = \{y \in U_x : g_x(y) = 0\};$$

locally, M is the null set of a submersion. As we shall see, the fact that the definition is a local one influences the way in which we establish properties of M. The manifold M is a $C^{(k)}$-*manifold* if each g_x can be taken to be a $C^{(k)}$-submersion, and is a *smooth manifold* if each g_x can be taken to be smooth. Differential manifolds are frequently called *differentiable manifolds*.

If $d + n = 3$ and $n = 1$, then M is a two-dimensional surface in a three-dimensional Euclidean space. More generally, if $n = 1$, then M is called a *hypersurface* in E_{d+1}. In this case, we simply require that $Dg_x \neq 0$ for each $x \in U$.

Here are three easy, but important, examples.

Example 19.1.1   The unit sphere.

Let E_{d+1} be a $(d+1)$-dimensional Euclidean space, and let $U = E_{d+1} \setminus \{0\}$. If $x \in U$, let $g(x) = \|x\|^2 - 1$. Then $Dg_x(h) = 2\langle x, h \rangle$, so that $Dg_x \neq 0$ for $x \in U$. Thus g is a submersion, and the unit sphere

$$S^d = \{x \in U : g(x) = 0\} = \{x \in E_{d+1} : \|x\| = 1\}$$

is a d-dimensional differential manifold in E_{d+1}. Since g is smooth, S^d is a smooth hypersurface in E_{d+1}.

Example 19.1.2   The graph of a differentiable function.

Suppose that f is a continuously differentiable function defined on an open subset U of a d-dimensional Euclidean space E_d, taking values in an n-dimensional Euclidean space E_n. Let $g(x, y) = f(x) - y$, for $(x, y) \in U \times E_n$. Then $Dg_{(x,y)} = (Df_x, -I)$, so that $Dg_{(x,y)}$ is a linear mapping from $E_d \times E_n$ into E_n, with rank n. Thus

$$G_f = \{(x, f(x)) : x \in U\} = \{(x, y) : g(x, y) = 0\}$$

is a d-dimensional differential manifold in $U \times E_n$.

For the next example, we need the notion of a *self-adjoint* operator. Suppose that E is a d-dimensional Euclidean space and that $T \in L(E)$. Recall that the *transpose* T' of T is the unique element of $L(E)$ for which

$\langle T(x), y \rangle = \langle x, T'(y) \rangle$, for all $x, y \in E$. T is *self-adjoint* if $T = T'$. For example, if P is an orthogonal projection of E onto a linear subspace F, and if $x, y \in E$ then

$$\langle P(x), y \rangle = \langle P(x), P(y) \rangle = \langle x, P(y) \rangle,$$

so that P is self-adjoint. If $(e_1, \ldots, e_d)$ is an orthonormal basis for E, and if T is represented by the matrix (t_{ij}), then T is self-adjoint if and only if $t_{ij} = t_{ji}$ for $1 \le i, j \le d$. Hence the set L_{sa} of self-adjoint linear operators on E is a linear subspace of $L(E)$ of dimension $d(d+1)/2$.

Example 19.1.3 The orthogonal group and special orthogonal group.

Suppose that E is a d-dimensional Euclidean space, and that $U = GL(E)$ is the group of invertible elements of $L(E)$. U is an open subset of the d^2-dimensional vector space $L(E)$. Let $g(A) = A'A - I$, for $A \in U$. Then $g(A)$ is self-adjoint, g is a smooth mapping from U into $L_{sa}(E)$ and $Dg_A(T) = A'T + T'A$. Suppose that $S \in L_{sa}(E)$. Let $T = \frac{1}{2}A'^{-1}S$. Then

$$Dg_A(T) = \tfrac{1}{2}A'A'^{-1}S + \tfrac{1}{2}SA^{-1}A = S,$$

so that Dg_A is a surjective linear mapping from $L(E)$ onto $L_{sa}(E)$. Thus the orthogonal group $O(E) = \{A \in U : g(A) = 0\}$ is a smooth manifold of dimension $d(d-1)/2$. $O(E)$ is not connected; if g^+ is the restriction of g to the open subset $U^+ = \{A \in GL(E) : \det A > 0\}$ of $L(E)$, then the special orthogonal group $SO(E)$ is equal to $\{A \in U^+ : g^+(A) = 0\}$, and is also a smooth connected manifold.

Exercises

19.1.1 Suppose that E is a differential manifold. Let $E_1 = E \times \mathbf{R}$, and let $E_0 = E \times \{0\}$. If $\alpha > 0$, let $f_\alpha(x) = \|x\|^\alpha \sin(1/\|x\|)$, for $x \in E$, and let G_α be the graph of f_α.
   (a) Is G_2 a differential manifold in E_1?
   (b) Is G_3 a differential manifold in E_1?
   (c) Is $G_3 \cap E_0$ a differential manifold in E_0?
   (d) Is $G_3 \cap (E_0 \setminus \{0\})$ a differential manifold in $E_0 \setminus \{0\}$?

19.1.2 Which of the following are manifolds in $\mathbf{R}^2$?
   (a) $\{(x, y) : y^2 = x^2 - x^4\}$.
   (b) $\{(\sin t, \sin 2t) : 0 < t < \pi\}$.

19.1.3 A real-valued function f on $\mathbf{R}^d$ is *m-homogeneous* if $f(tx) = t^m f(x)$ for all $x \in \mathbf{R}^d$ and all $t > 0$. Suppose that f is continuously differentiable and m-homogeneous and that $c \in \mathbf{R} \setminus \{0\}$. Show that if

$\{x \in \mathbf{R}^d : f(x) = c\}$ is not empty then it is a differential manifold in $\mathbf{R}^d$. What is its dimension?

19.1.4 If E is a Euclidean group then the *special linear group* $SL(E)$ is the set $\{T \in GL(E) : \det T = 1\}$. Show that $SL(E)$ is a manifold in $L(E)$. What is its dimension?

19.2 Tangent vectors

A *curve* in a Euclidean space E is a continuously differentiable mapping δ from an interval I in $\mathbf{R}$ into E (with a one-sided derivative at an end-point of I), and its *track* $[\delta]$ is the image of δ. I does not need to be a closed interval, but if I is a closed interval $[a, b]$, then δ is a continuously differentiable path from $\delta(a)$ to $\delta(b)$ (and we call δ a curve from $\delta(a)$ to $\delta(b)$). A curve is *simple* if δ is injective. If δ is a curve which is a simple closed path, we call δ a *simple closed curve*. A curve $\delta : I \to E$ is *steady* if $\|\delta'(t)\| = 1$, for each $t \in I$.

Suppose that M is a differential manifold in a Euclidean space E, and that $x \in M$. A vector h in E is a *tangent vector* to M at x if there exists an open interval $(-\delta, \delta)$ in $\mathbf{R}$ and a curve $\psi_x : (-\delta, \delta) \to E$ taking values in M, such that $\psi_x(0) = x$ and $\psi_x'(0) = h$. Let $\theta_x(t) = x + th$, for $t \in (-\delta, \delta)$. Then h is a tangent vector to M at x if and only if $\|\psi_x(t) - \theta_x(t)\| = o(|t|)$.

This definition does not involve submersions. The set of tangent vectors to M at x can be characterized in terms of submersions.

Theorem 19.2.1 *Suppose that M is a d-dimensional differential manifold in a Euclidean space E, that $x \in M$, that U_x is a neighbourhood of x in E and that $g : U_x \to F$ is a submersion for which*

$$M \cap U_x = \{y \in U_x : g(y) = 0\}.$$

If $h \in E$ then h is a tangent vector to M at x if and only if $Dg_x(h) = 0$.

Proof Suppose first that h is a tangent vector to M at x, and that $\psi : (-\delta, \delta) \to M$ satisfies the conditions of the definition. If $t \in (-\delta, \delta)$, then $g(\psi(t)) = 0$ for $t \in (-\delta, \delta)$, and so, using the chain rule,

$$Dg_x(h) = Dg_x(\psi'(0)) = (g \circ \psi)'(0) = 0.$$

The converse is harder to prove. Let us set $T_x = \{h \in E : Dg_x(h) = 0\}$: T_x is the null-space of Dg_x. Let P_x be the orthogonal projection of E onto T_x. Let $Q_x = I - P_x$, and let $N_x = Q_x(E)$; N_x is the orthogonal complement of T_x. If $y \in N_x$ and $Dg_x(y) = 0$ then $y \in T_x \cap N_x = \{0\}$, so that $y = 0$.

Thus the restriction of Dg_x to N_x is an injective linear map of N_x into F. Since $\dim N_x = \operatorname{rank}(Dg_x) = \dim F$, the restriction of Dg_x to N_x is a linear isomorphism of N_x onto F.

We now define a mapping $\widetilde{g}$ from U_x to $T_x \times F$ by setting $\widetilde{g}(y) = (P_x(y - x), g(y))$. The mapping $\widetilde{g}$ is continuously differentiable, and $D\widetilde{g}_x = (P_x, Dg_x)$. Suppose that $(h, k) \in T_x \times F$. Since g is a submersion, there exists $y \in E$ such that $Dg_x(y) = k$. Let $z = Q_x(y) + h$. Since $h - P_x(y) \in T_x$,

$$D\widetilde{g}_x(z) = (P_x(h), Dg_x(y + (h - P_x(y)))) = (h, Dg_x(y)) = (h, k).$$

Thus $D\widetilde{g}_x$ is a linear isomorphism of E onto $T_x \times F$. Since $\widetilde{g}$ is continuously differentiable, we can suppose, by replacing U_x by a smaller neighbourhood if necessary, that $D\widetilde{g}$ is invertible at each point of U_x. Applying Corollary 17.4.3, it follows that $\widetilde{g}$ is a diffeomorphism of U_x onto the open subset $\widetilde{g}(U_x)$ of $T_x \times F$.

Let $\Psi : \widetilde{g}(U_x) \to U_x$ be the inverse mapping. If $h \in T_x$, there exists $\delta > 0$ such that $(th, 0) \in \widetilde{g}(U_x)$ for $|t| < \delta$. Let $\psi(t) = \Psi(th, 0)$. Then $\psi(0) = x$. Since $\widetilde{g}(\psi(t)) = (th, 0)$, it follows that $g(\psi(t)) = 0$, so that $\psi(t) \in M$. Further, $\psi'(0) = D\Psi_{(0,0)}(h, 0) = (D\widetilde{g}_x)^{-1}(h, 0)$. Since $h \in T_x$ and $Dg_x(h) = 0$, it follows that $D\widetilde{g}_x(h) = (P_x(h), Dg_x(h)) = (h, 0)$. Thus $\psi'(0) = h$. $\qquad\square$

Corollary 19.2.2 *The set T_x of tangent vectors to M at x is a d-dimensional linear subspace of E, which depends neither on the choices of ψ in the definition of tangent vector, nor on the choice of the submersion g used to define M in a neighbourhood of x.*

The vector space T_x is called the *tangent space* of M at x, and the space N_x is called the *normal space* of M at x. Continuing with the notation of the theorem, if $y \in M \cap U_x$, let $\phi(y) = \phi_x(y) = P_x(y - x)$. Then $\widetilde{g}(y) = (\phi(y), 0)$, so that ϕ is a homeomorphism of $M \cap U_x$ onto an open subset of the tangent space T_x. The pair (U_x, ϕ) is called a *chart* of M near x. If $(e_1, \ldots, e_d)$ is an orthonormal basis for T_x, and $\phi(y) = \phi_1(y)e_1 + \cdots + \phi_d(y)e_d$, then $(\phi_1, \ldots, \phi_d)$ are *local coordinates* for M in U_x, or a *parametrization* of $M \cap U_x$. The inverse mapping $\psi = \phi_x^{-1} : \phi(U_x) \to U_x$ is an immersion.

For example, if $M = S^{n-1} = \{x \in E : q(x) = \|x\|^2 = 1\}$ is the unit sphere in an n-dimensional space and if $x \in M$ then $Dg_x(h) = 2\langle x, h \rangle$, and so $T_x = x^\perp$, the space of vectors orthogonal to x, and $N_x = \operatorname{span}(x)$.

Corollary 19.2.3 *Suppose that V is an open subset of a Euclidean space F, that $f : V \to E$ is a continuously differentiable mapping and that $f(V) \subseteq M$. Then $Df_x(F) \subseteq T_{f(x)}$ for each $x \in F$.*

Proof   Suppose that $U_{f(x)}$ is an open neighbourhood of $f(x)$ and that $g : U_{f(x)} \to G$ is a submersion for which

$$M \cap U_{f(x)} = \{y \in U_{f(x)} : g(y) = 0\}.$$

Let $V_x = f^{-1}(U_{f(x)})$. Then $g(f(y)) = 0$ for $y \in V_x$, and so $D(g \circ f)_x = Dg_{f(x)} \circ Df_x = 0$. Thus $Df_x(F) \subseteq T_{f(x)}$.   □

Corollary 19.2.4   *If $\epsilon > 0$ there exists $\delta > 0$ such that if $y \in M$ and $\|y - x\| < \delta$, then $\|(y - x) - \phi_x(y)\| < \epsilon \|\phi_x(y)\| \leq \epsilon \|y - x\|$.*

Proof   The theorem shows that $D\psi_0$ is the identity mapping on T_x. If $v \in \phi_x(U_x)$, let $\theta(v) = \psi(v) - v$. Then θ maps $\phi_x(U_x)$ into E, $\theta(0) = x$ and $D\theta_0 = 0$. Since $D\theta$ is continuous at 0, there exists $\delta > 0$ such that if $\|v\| < \delta$ then $\|D\theta_v\| < \epsilon/2$. If $\|y - x\| < \delta$ then

$$\|\phi_x(y)\| \leq \|P_x\| \cdot \|y - x\| = \|y - x\| < \delta,$$

so that

$$\|(y - x) - \phi(y)\| = \|\psi(\phi_x(y)) - \phi_x(y) - x\|$$
$$= \|\theta(\phi_x(y)) - \theta(0)\| \leq \epsilon \|\phi(y)\|.$$

by the mean value inequality.   □

Corollary 19.2.5   *If $\epsilon > 0$ there exists $\delta > 0$ such that if $y \in M$ and $\|y - x\| < \delta$, then $\|\phi_x(y)\| \leq \|y - x\| \leq (1 + \epsilon) \|\phi_x(y)\|$.*

Proof   For

$$\|\phi_x(y)\| \leq \|y - x\| \leq \|(y - x - \phi_x(y)) + \phi_x(y)\|$$
$$\leq \|y - x - \phi_x(y)\| + \|\phi_x(y)\| \leq (1 + \epsilon) \|\phi_x(y)\|.$$

□

Theorem 19.2.6   *Suppose that M is a d-dimensional differential manifold in a Euclidean space E, and that (U_x, ϕ_x) and (U_y, ϕ_y) are two charts. If $U_x \cap U_y \neq \emptyset$ then $\phi_y \circ \phi_x^{-1} : \phi_x(U_x \cap U_y) \to \phi_y(U_x \cap U_y)$ is a diffeomorphism, which is a $C^{(k)}$-diffeomorphism if M is a $C^{(k)}$-differential manifold, and is smooth if M is a smooth manifold.*

Proof   This follows directly from the corresponding properties of the mappings $\widetilde{g}$ and $\widetilde{g}^{-1}$ defined in the preceding theorem.   □

These results lead to the notion of an abstract differential manifold. This is a Hausdorff topological space (M, τ) with the property that there is a set $\mathcal{C} = \{(U, \phi)\}$ of *charts*, where

- the sets U are open subsets of M which cover M,
- if $(U, \phi) \in \mathcal{C}$, then ϕ is a homeomorphism of U onto an open subset $\phi(U)$ of $\mathbf{R}^d$,
- if (U, ϕ) and (V, ψ) are charts and $U \cap V \neq \emptyset$ then the restriction of $\psi \circ \phi^{-1}$ to $\phi(U \cap V)$ is a diffeomorphism of $\phi(U \cap V)$ onto $\psi(U \cap V)$.

Such a manifold M has a much weaker structure than a differential manifold which is a subspace of a Euclidean space (for example, there is no natural metric on M), but the study of these manifolds is the concern of differential geometry.[1] We restrict our attention to differential manifolds which are subspaces of Euclidean spaces.

Example 19.2.7   The tangent bundle of a $C^{(2)}$-differential manifold in a Euclidean space.

Suppose that M is a d-dimensional $C^{(2)}$-differential manifold in a Euclidean space E. We define the *tangent bundle* $T(M)$ of M to be the subset $\{(x, v) : x \in M, v \in T_x\}$ of $E \times E$. Let us show that $T(M)$ is a $2d$-dimensional differential manifold in $E \times E$. Suppose that $x \in M$, that U_x is an open neighhbourhood of x in E and that $g : U_x \to F$ is a submersion for which $M \cap U_x = \{y \in U_x : g(y) = 0\}$. Define $G : U_x \times E \to F \times F$ by setting $G(y, v) = (g(y), Dg_y(v))$. Then

$$T(M) \cap (U_x \times E) = \{(y, v) : G((y, v)) = 0\}.$$

We must show that G is a submersion. G is continuously differentiable, and the matrix of partial derivatives is

$$\begin{bmatrix} Dg_y & 0 \\ D^2 g_y(v, \cdot) & Dg_y \end{bmatrix}.$$

Since Dg_y has rank d, $DG_{(y,v)}$ has rank $2d$, and so G is a submersion.

We have used submersions to define differential manifolds in a Euclidean space. We can also use immersions to do this.

---

[1] For a good introduction, see Dennis Barden and Charles Thomas, *An Introduction to Differential Manifolds,* Imperial College Press, 2003.

Theorem 19.2.8 *Suppose that M is a subset of an n-dimensional Euclidean space E with the property that for each $x \in M$ there is an open neighbourhood U of x in E and an immersion j from an open subset V of a d-dimensional Euclidean space F, with $j(V) = M \cap U$. Then M is a d-dimensional differential manifold in E.*

Proof Suppose that $x \in M$. Let $z = j^{-1}(x)$ and let K be the orthogonal complement in E of $Dj_z(F)$: K is an $(n - d)$-dimensional subspace of E, so that $\dim(F \times K) = n$. Define $\tilde{j} : V \times K \to E$ by setting $\tilde{j}(y, w) = j(y) + w$. Then $\tilde{j}$ is continuously differentiable, and $D\tilde{j} = (Dj, J)$, where $J : K \to E$ is the inclusion mapping. Then $\text{rank}(D\tilde{j}_z) = n$, and so, by the inverse mapping theorem, there is a neighbourhood W of x contained in U such that $\tilde{j}$ is a diffeomorphism of $\tilde{j}^{-1}(W)$ onto W. Let $\tilde{g} = (f, g)$ be the inverse mapping. Then g is a submersion of V into W. Further, $g(y) = 0$ if and only if $\tilde{g}(y) = (f(y), 0)$, which happens if and only if $y = \tilde{j}(g(y)) = j(f(y)) \in M \cap W$. Thus M is a differential manifold. □

If the immersions are $C^{(k)}$-immersions, then M is a $C^{(k)}$ differential manifold, and if the immersions are smooth then M is a smooth manifold.

Exercise

19.2.1 Suppose that M is a d-dimensional $C^{(2)}$-differential manifold in a Euclidean space E. The *unit sphere bundle* $S(M)$ is defined to be $S(M) = \{(x, v) \in T(M) : \|v\| = 1\}$. Show that $S(M)$ is a $2d - 1$ differential manifold in $E \times E$.

19.3 One-dimensional differential manifolds

The simplest examples of differential manifolds are the one-dimensional ones. In the next theorem, we classify the connected one-dimensional manifolds contained in a Euclidean space. The results are hardly unexpected, but the proofs require some care, and illustrate the use of the results concerning paths that have been established earlier.

Theorem 19.3.1 *Suppose that M is a connected one-dimensional manifold in an open subset U of an open subset U of a Euclidean space E. There are two possibilities.*

First, M is a compact subset of U and there exists a steady simple closed curve γ in U such that $M = [\gamma]$, the track of γ.

Secondly, M is not a compact subset of U, and there exists an open interval I in $\mathbf{R}$ and a steady simple curve $\beta : I \to U$ such that $M = [\beta]$, so that β is a homeomorphism of I onto M.

Proof   We prove a series of lemmas.

Lemma 19.3.2   *Suppose that $x \in M$. There exists an open neighbourhood W_x of x in E and a chart $\chi_x : W_x \cap M \to T_x$ such that $\chi_x(W_x \cap M)$ is an interval $(-h, h)$ in T_x. If $y \in (W_x \cap M) \setminus \{x\}$ there exists a unique simple steady path $\beta : [0, L] \to M$ from x to y.*

Proof   There exist a neighbourhood U_x of x in E and a chart $\phi_x : U_x \cap M$ to the one-dimensional tangent space T_x. Since $\phi_x(U_x \cap M)$ is open in T_x, there exists an interval $(-h, h) \subseteq \phi_x(U_x \cap M)$. Let $W_x = \phi_x^{-1}(-h, h)$ and let χ_x be the restriction of ϕ_x to $W_x \cap M$. Let $\gamma(t) = t\phi_x(y)$ for $0 \leq t \leq 1$. Then $\chi_x^{-1} \circ \gamma$ is a simple curve in M from x to y; let β be its path-length parametrization. By Corollary 17.2.10, $\|\beta'(t)\| = 1$; thus β is a steady curve. It follows from Corollary 17.2.11 that β is unique.   □

We have the following uniqueness result.

Lemma 19.3.3   *Suppose that $\beta_1 : [0, L] \to M$ and $\beta_2 : [0, L] \to M$ are two steady curves in M for which $\beta_1(0) = \beta_2(0)$ and $\beta_1'(0) = \beta_2'(0)$. Then $\beta_1 = \beta_2$.*

Proof   Let

$$G = \{t \in [0, L] : \beta_1(s) = \beta_2(s) \text{ and } \beta_1'(s) = \beta_2'(s) \text{ for } 0 \leq s \leq t\}.$$

Since the functions under consideration are continuous, G is a closed interval $[0, g]$ in $[0, L]$. It follows easily from Lemma 19.3.2 that G is also an open subset of $[0, L]$; since $[0, L]$ is connected, $G = [0, L]$.   □

We now define a relation $\sim$ on M by setting $x \sim y$ if there exists a steady curve $\delta : I \to M$ for which $x, y \in [\delta]$.

Lemma 19.3.4   *The relation $\sim$ is an equivalence relation.*

Proof   Clearly $x \sim x$, and $x \sim y$ if and only if $y \sim x$. Suppose that $x \sim y$ and $y \sim z$, so that there exist a steady curve $\delta_1 : [0, L_1] \to M$ from x to y, and a steady curve $\delta_2 : [0, L_2] \to M$ from y to z. If $\delta_2'(0) = \delta_1'(L_1)$ then $\delta_1 \vee \delta_2$ is a steady curve in M from x to z. Otherwise, $\delta_2'(0) = -\delta_1'(L_1)$. Suppose that $L_1 \geq L_2$. Let $\delta_1^{\leftarrow} : [0, L_2] \to M$ be the reversal of δ_2. Then it follows from Lemma 19.3.3 that $\delta_1^{\leftarrow}(t) = \delta_2(t)$ for $0 \leq t \leq L_2$. Thus

$z = \delta_1^{\leftarrow}(L_2) = \delta_1(L_1 - L_2) \in [\delta_1]$, so that $x \sim z$. A similar argument shows that if $L_1 \leq L_2$ then $x \in [\delta_2]$, so that $x \sim z$. □

Lemma 19.3.5   *If $x, y \in M$ then $x \sim y$.*

Proof   It follows from Lemma 19.3.2 that each equivalence class is open in M. Since M is connected, M is the unique equivalence class. □

We now complete the proof of the theorem. Choose $x_0 \in M$ and choose a unit tangent vector h_0 in T_{x_0}. Let

$$I = \{t \in \mathbf{R} : \text{there exists a steady curve } \delta : [0, t] \to M$$
$$\text{with } \delta(0) = x_0 \text{ and } \delta'(0) = h_0\}.$$

(Here we allow t to be negative, or zero.) It follows from Lemma 19.3.2 that I is an open interval, from Lemma 19.3.3 that there exists a steady curve $\delta : I \to M$ with $\delta(0) = x_0$ and $\delta'(0) = h_0$ and from Lemma 19.3.5 that $[\delta] = M$.

There are now two possibilities. First, δ is a simple curve. In this case, the second possibility holds. Secondly, there exist $s, t \in I$, with $s < t$ such that $\delta(s) = \delta(t)$. It then follows from Lemma 19.3.3 that if $s + u, t + u \in I$ then $\delta(s+u) = \delta(t+u)$. Thus δ is periodic, and $I = \mathbf{R}$. Let t_0 be the fundamental period of δ: the least positive number for which $\delta(t_0) = \delta(t_0 + t)$ for all $t \in \mathbf{R}$ (see Volume I, Section 6.3, Exercise 1). Then $\delta : [0, t_0] \to M$ is a steady closed curve in M, with $[\delta] = M$. Finally, $\delta : [0, t_0] \to M$ is a simple closed curve, for if there exist $0 \leq s < t \leq t_0$ with $t - s < t_0$ for which $\delta(s) = \delta(t)$, then it follows from Lemma 19.3.3 that $t - s$ is a period of δ, contradicting the minimality of t_0. □

The leftmost trefoil in Figure 19.3a represents the track of a closed curve in the plane. It is not a manifold, since the curve intersects itself. The other two trefoils represent the track of curves in a three-dimensional space E. They are both manifolds, and are diffeomorphic to each other. On the other hand, there is no homeomorphism of E onto itself, mapping one manifold

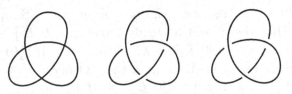

Figure 19.3a. Three trefoils.

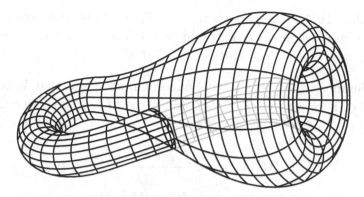

Figure 19.3b. The Klein bottle.

onto the other, since one curve is knotted and the other is not (we shall not prove this).

Similar phenomena happen in higher dimensions. The Klein bottle illustrated in Figure 19.3b is not a manifold in three-dimensional space E, since it is self-intersecting. In fact, there is no manifold in a three-dimensional space which is homeomorphic to it, since it is a one-sided surface. On the other hand, it can be represented as a manifold in a four-dimensional space: take the fourth dimension to be time, start at time 0 at the circle of self-intersection, and proceed outwards in both directions, reaching the circle of intersection again at time 1.

Exercise

19.3.1 Give an example of a one-dimensional manifold in the plane with infinitely many connected components. Can there be uncountably many connected components?

19.4 Lagrange multipliers

We begin with a result which corresponds to Rolle's theorem. We need some definitions.

Suppose that f is a real-valued function on a topological space (X, τ), and that $x \in X$. Then f has a *local maximum* (*strict local maximum*) if there is a neighbourhood V of x for which $f(y) \leq f(x)$ ($f(y) < f(x)$) for $y \in V \setminus \{x\}$. *Local minima* and *strict local minima* are defined similarly.

Suppose that f is a differentiable function on an open subset U of a normed space $(E, \|.\|)$ and that $x \in U$. Then x is a *stationary point* of f if $Df_x = 0$.

Proposition 19.4.1 *Suppose that f is a differentiable function on an open subset of a normed space $(E, \|.\|)$ and that f has a local maximum or local minimum at x. Then x is a stationary point of f.*

Proof Suppose that f has a local maximum at x, and that V is a neighbourhood of x for which $f(y) \leq f(x)$ for $y \in V$. If $h \in E$, there exists $\delta > 0$ such that $x + th \in V$ for $|t| < \delta$. Then

$$Df_x(h) = \lim_{t \searrow 0} \frac{f(x + th) - f(x)}{t} \leq 0$$

$$\text{and } Df_x(h) = \lim_{t \nearrow 0} \frac{f(x + th) - f(x)}{t} \geq 0,$$

so that $Df_x(h) = 0$. This holds for all $h \in E$, and so $Df_x = 0$. The proof for a local minimum is exactly similar. $\square$

It is important to note that the converse is not true. For example the function $f(x) = x^3$ on $\mathbf{R}$ has a stationary point at 0, but is strictly monotonic, and the function $f(x, y) = xy$ on $\mathbf{R}^2$ has a stationary point at $(0,0)$, but takes positive, negative and zero values in any neighbourhood of $(0,0)$.

Suppose now that M is a differential manifold in an open subset U of a Euclidean space E, and that f is a continuously differentiable function on U. We consider the restriction of f to M. Suppose that $x \in M$. Then f has a *constrained local maximum* (*constrained strict local maximum*) if there is a neighbourhood V of x for which $f(y) \leq f(x)(f(y) < f(x))$ for $y \in M \cap (V \setminus \{x\})$. *Constrained local minima* and *constrained strict local minima* are defined similarly.

Recall that we use the gradient ∇f_x to describe the derivative of f at x: $\nabla f_x \in E$ and $Df_x(h) = \langle h, \nabla f_x \rangle$. Now E is the orthogonal direct sum $E = T_x \oplus N_x$, where T_x is the tangent space at x and N_x is the normal space at x. We write

$$\nabla f_x = \nabla_T f_x + \nabla_N f_x, \text{ with } \nabla_T f_x \in T_x \text{ and } \nabla_N f_x \in N_x :$$

$\nabla_T f_x$ is the *tangential gradient* and $\nabla_N f_x$ is the *normal gradient* of f at x. We say that f has a *constrained stationary point* at x if $\nabla_T f_x = 0$.

Proposition 19.4.2 *Suppose that M is a differential manifold in an open subset U of a Euclidean space E, that f is a continuously differentiable function on U and that $x \in M$. If f has a constrained local maximum or minimum at x then f has a constrained stationary point at x.*

Proof   Suppose that $h \in T_x$. Then there exists $\delta > 0$ and a curve $\psi_x :$ $(-\delta, \delta) \to M$ such that $\psi_x(0) = x$ and $\psi'_x(0) = h$. Then $f \circ \psi_x$ has a local maximum or minimum at 0, so that, by the chain rule,

$$\langle h, \nabla_T f_x \rangle = \langle h, \nabla f_x \rangle = D f_x(h) = D f_x(\psi'(0)) = D(f \circ \psi_x) = 0.$$

□

Of course, $\nabla_N f_x$ need not vanish. Indeed, if U_x is a neighbourhood of x in U and $g : U_x \to F$ is a submersion for which

$$M \cap U_x = \{y \in U_x : g(y) = 0\},$$

then trivially every point y of $M \cap U_x$ is both a constrained local maximum and a constrained local minimum of g, and is therefore a constrained stationary point of g, while $\nabla_N g_y$ is a linear isomorphism of N_y onto $D g_y(E)$.

Theorem 19.4.3   *Suppose that M is a differential manifold in an open subset U of a Euclidean space E, that f is a continuously differentiable real-valued function on U and that $x \in M$. Suppose that U_x is a neighbourhood of x in U and $g : U_x \to F$ is a submersion for which $M \cap U_x = \{y \in U_x : g(y) = 0\}$. If f has a constrained local maximum or minimum at x, then there exists a unique ϕ in F for which $\nabla(f - \langle \phi, g \rangle)_x = 0$.*

Proof   Since $D g_x$ is a linear isomorphism of N_x onto F, $(D g'_x)$ is a linear isomorphism of F onto N_x. Thus there exists a unique $\phi \in F$ such that $(D g'_x(\phi)) = \nabla_N f_x$, and so

$$\langle h, \nabla_N f_x \rangle = \langle \phi, D g_x(h) \rangle \text{ for all } h \in N_x.$$

But if $h \in N_x$ then

$$\langle \phi, D g_x(h) \rangle = D(\langle \phi, g \rangle)_x(h) = \langle h, \nabla_N(\langle \phi, g \rangle)_x \rangle,$$

and so $\nabla_N(f - \langle \phi, g \rangle)_x = 0$. Since $\nabla_T(f - \langle \phi, g \rangle)_x = \nabla_T f_x - \nabla_T(\langle \phi, g \rangle)_x = 0$ it follows that $\nabla(f - \langle \phi, g \rangle)_x = 0$.

Finally, the proof shows that ϕ is unique.   □

Corollary 19.4.4   *If $F = \mathbf{R}^k$, and $g = (g_1, \ldots, g_k)$, then there exist unique $\lambda_1, \ldots, \lambda_k$ in $\mathbf{R}$ such that $\nabla(f - \sum_{j=1}^k \lambda_j g_j) = 0$.*

The quantities $\lambda_1, \ldots, \lambda_k$ are called *Lagrange multipliers*. Lagrangian multipliers provide a powerful tool for finding constrained local maxima and minima, as the following examples and Exercises 6 and 7 show.

Example 19.4.5 Diagonalizing a real quadratic form.

A *quadratic form* Q on a real vector space E is a real-valued function on E for which there exists a symmetric bilinear form b on E for which $Q(x) = b(x, x)$, for all $x \in E$. Suppose that E is a Euclidean space. Then since $Q(x + h) = Q(x) + 2b(x, h) + b(h, h)$, and since $b(h, h) = o(\|h\|)$, Q is continuously differentiable, and $DQ_x(h) = 2b(x, h)$.

Theorem 19.4.6 *Suppose that Q is a real quadratic form on a d-dimensional Euclidean space E, defined by a symmetric bilinear function b. Then there exist an orthonormal basis $(e_1, \ldots, e_d)$ of E and real numbers $\lambda_1 \geq \lambda_2 \geq \cdots \geq \lambda_d$ such that*

$$Q(x) = \lambda_1 x_1^2 + \cdots + \lambda_d x_d^2 \text{ for all } x = x_1 e_1 + \cdots + x_d e_d \in E.$$

Proof The proof is by induction on d. The result holds if $d = 1$. Suppose that it holds for $d - 1$, and that Q is a quadratic form on a d-dimensional Euclidean space E. Let S^{d-1} be the unit sphere

$$S^{d-1} = \{x \in E : g(x) = \|x\|^2 = 1\}.$$

Since Q is a continuous function on E and S^{d-1} is compact, Q attains its maximum on S^{d-1} at a point e_1 of S^{d-1}. The tangent space T_{e_1} to S^{d-1} is $e_1^\perp$. Thus $2b(e_1, h) = DQ_{e_1}(h) = 0$, for $h \in e_1^\perp$. Consequently, if $x = x_1 e_1 + h$, with $h \in e_1^\perp$, then $Q(x) = \lambda_1 x_1^2 + Q(h)$, where $\lambda_1 = Q(e_1)$.

We now apply the inductive hypothesis to the $(d-1)$-dimensional space $e_1^\perp$: there exists an orthonormal basis $(e_2, \ldots, e_d)$ of $e^\perp$ and real numbers $\lambda_2 \geq \cdots \geq \lambda_d$ such that

$$Q(y) = \lambda_2 y_2^2 + \cdots + \lambda_d y_d^2 \text{ for all } y = y_2 e_2 + \cdots + y_d e_d \in e_1^\perp.$$

Then $(e_1, \ldots, e_d)$ is an orthonormal basis of E. Since λ_1 is the supremum of Q on S^{d-1}, $\lambda_1 \geq \lambda_2 \geq \cdots \geq \lambda_d$ and

$$Q(x) = \lambda_1 x_1^2 + \cdots + \lambda_d x_d^2 \text{ for all } x = x_1 e_1 + \cdots + x_d e_d \in E.$$

$\square$

We can use this to say more about the relationship between local maxima and minima and stationary points. Suppose that f is a twice continuously differentiable function on an open subset U of a d-dimensional Euclidean space E, and that x is a stationary point of f. Then, for small h in E,

$$f(x + h) = f(x) + D^2 f_x(h, h) + r(h),$$

where $r(h) = o(\|h\|^2)$. The mapping $Q(h) = D^2 f_x(h, h)$ is a quadratic form on E, and so there exist an orthonormal basis $(e_1, \ldots, e_d)$ of E and real numbers $\lambda_1 \geq \lambda_2 \geq \cdots \geq \lambda_d$ such that

$$D^2 f_x(h, h) = \lambda_1 h_1^2 + \cdots + \lambda_d h_d^2 \text{ for all } h = h_1 e_1 + \cdots + h_d e_d \in E.$$

There are now five possibilities.

1. $\lambda_j > 0$ for $1 \leq j \leq d$. Then f has a strict local minimum at x.
2. $\lambda_j \geq 0$ for $1 \leq j \leq d$ and $\lambda_d = 0$. Then f can have a strict local minimum at x (consider the function $x_1^2 + x_2^4$ on $\mathbf{R}^2$ near $(0,0)$), a local minimum which is not strict (consider the function x_1^2 on $\mathbf{R}^2$ near $(0,0)$) or can take positive and negative values in any neighbourhood of x (consider the function $x_1^2 - x_2^4$ on $\mathbf{R}^2$ near $(0,0)$).
3. $\lambda_1 > 0$ and $\lambda_d < 0$. Then f takes positive and negative values in any neighbourhood of x. In this case x is called a *saddle point* of f: consideration of the graph of the function $f(x_1, x_2) = x_1^2 - x_2^2$ explains this terminology.
4. $\lambda_j \leq 0$ for $1 \leq j \leq d$ and $\lambda_1 = 0$. As case 2.
5. $\lambda_j < 0$ for $1 \leq j \leq d$. As case 1.

Similar results hold for constrained stationary points.

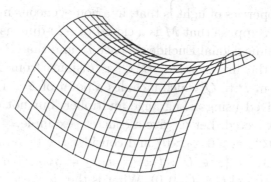

Figure 19.4a. A saddle point.

Example 19.4.7 Maximum entropy.

Suppose that $\boldsymbol{\Gamma}$ is a probability defined on a set $\{x_1, \ldots, x_n\}$ of n points, and that $\mathbf{P}(x_j) = p_j$ (so that $p_j \geq 0$ and $p_1 + \cdots + p_n = 1$). Suppose that f is a continuous strictly concave function on $[0, 1]$ which is continuously differentiable on $(0, 1)$. For what choice of $\mathbf{P}$ is $F = \sum_{j=1}^n f(p_j)$ a maximum?. First note that if we set $h(t) = f(t) - f(0)$ then $\sum_{j=1}^n h(p_j) = F - nf(0)$, and so we can suppose that $f(0) = 0$.

Consider the function $F(x) = \sum_{j=1}^n f(x_j)$ on $U = (0,1)^n$, the function $g(x) = x_1 + \cdots + x_n$, and the manifold $M = \{x \in U : g(x) = 1\}$. Suppose that $p \in M$ is a constrained stationary point of F. By Corollary 19.4.4 there exists $\lambda \in \mathbf{R}$ such that $\nabla(F - \lambda g)_p = 0$. That is,

$$f'(p_j) = \frac{\partial F}{\partial x_j}(p) = \lambda \text{ for } 1 \le j \le n.$$

Thus $f'(p_i) = f'(p_j)$ for $1 \le i, j \le n$. Since f is strictly concave, f' is strictly monotonic, and so $p_i = p_j$ for $1 \le i, j \le n$. Thus $p = (1/n, \ldots, 1/n)$, $\mathbf{P}$ is the uniform distribution on $\{x_1, \ldots, x_n\}$ and $F(x) = nf(1/n)$. It is easy to verify that this is a constrained local maximum.

We need to verify that this is the maximal value on $\overline{M}$. But if $y \in \partial M$, and $y_j \ne 0$ for k values of j, then the same argument shows that $\sum_{j=1}^n f(y_j) \le kf(1/k)$, and $kf(1/k) < nf(1/n)$, since f is strictly concave.

If $\mathbf{P}$ is a probability measure on $\{x_1, \ldots, x_n\}$, then the *Shannon entropy* of $\mathbf{P}$ is defined as $-\sum_{j=1}^n p_j \log_2 p_j$ (where $\log_2$ is the logarithm to base 2: $\log_2(t) = \log t / \log 2$, and $0. \log_2 0 = 0$). It follows that the maximum entropy occurs when $\mathbf{P}$ is the uniform distribution.

Example 19.4.8 The reflection of light.

One of the properties of light is that, in a homogeneous medium, it travels in straight lines. Suppose that M is a closed $(d-1)$-dimensional differential manifold in a d-dimensional Euclidean space E, and that P and Q are points in E for which the line segment $[P, Q]$ is disjoint from M. What is the shortest path from P to Q which includes a point of M? This is a question which can be solved using a Lagrange multiplier, but not quite in the way that might be expected. Let us give a concrete example.

Let $U = \{x \in \mathbf{R}^3 : x_1 > 0, x_2 > 0\}$ and consider the two-dimensional differential manifold $M = \{x \in U : f(x) = x_1 x_2 = 3\sqrt{2}\}$ (the 'mirror') in U. Let $P = (-1, 0, 0)$ and $Q = (1, 0, 0)$. What is the shortest path from P to Q which includes a point of M? We need some results from Euclidean geometry. Suppose that $a > 1$. Then the set $E_a = \{x \in \mathbf{R}^3 : \|x - P\| + \|x - Q\| = 2a\}$ is the ellipsoid

$$\{x \in \mathbf{R}^3 : g(x) = \frac{x_1^2}{a^2} + \frac{x_2^2 + x_3^2}{a^2 - 1} - 1 = 0\},$$

which is a compact two-dimensional differential manifold in $\mathbf{R}^3$. (See Exercise 19.4.4.) Where does the function f attain its constrained maximum on $E_a \cap U$? By Corollary 19.4.4, if x is a constrained stationary point, there exists a unique λ in $\mathbf{R}$ such that $\nabla(f - \lambda g)_x = 0$. That is,

$$x_2 - \frac{2\lambda x_1}{a^2} = 0, \qquad x_1 - \frac{2\lambda x_2}{a^2 - 1} = 0, \qquad \frac{2\lambda x_3}{a^2 - 1} = 0.$$

Thus

$$x_3 = 0 \text{ and } x_1 x_2 = \frac{2\lambda x_1^2}{a^2} = \frac{2\lambda x_2^2}{a^2 - 1},$$

so that

$$\frac{x_1^2}{a^2} = \frac{x_2^2}{a^2 - 1} = \frac{1}{2} \text{ and } x_1^2 x_2^2 = \frac{a^2(a^2 - 1)}{4}.$$

We require x to be a point of M. This happens if $a = 3$, from which it follows that

$$x_1 = 3\sqrt{2}/2, \quad x_2 = 2 \text{ and } x_3 = 0,$$

and the length of the path is 6. Note that we do not need to calculate λ.

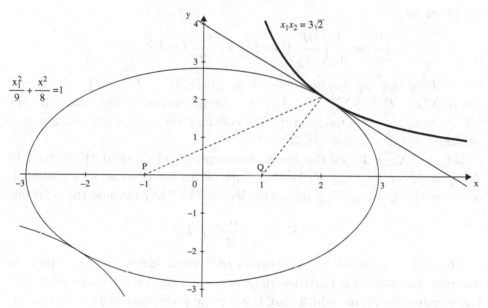

Figure 19.4b. Reflection in a hyperbolic mirror.

Why are the multipliers called Lagrange multipliers? During the second half of the eighteenth century, the Piedmontese mathematician Joseph Lagrange developed a new formulation of Newtonian mechanics. Suppose that we are considering an ensemble of N particles $P_1, \ldots P_N$, with masses $m_1, \ldots m_N$. In the Newtonian formulation, we consider the positions

$x_1, \ldots x_N$, and consider the second-order equations

$$\frac{d^2}{dt^2}(m_j x_j) = F_j, \text{ where } F_j \text{ is the force on } P_j.$$

The positions $x_1, \ldots, x_N$ are represented as a point in $\mathbf{R}^{3N}$, each particle having three coordinates. One of Lagrange's insights was that it is helpful to consider the particles' velocities, and their coordinates, on an equal footing. Thus each particle is represented by six coordinates, with three for its position and three for its velocity. Another coordinate is needed for time. Thus the configuration is in $\mathbf{R}^{6N+1} = X \times W \times T$, with basis $(e_1, \ldots, e_{3N}, f_1, \ldots, f_{3N}, h)$, where the jth particle has

$$\text{position } x^{(j)} = x_{3j-2}e_{3j-2} + x_{3j-1}e_{3j-1} + x_{3j}e_{3j}$$

$$\text{and velocity } v^{(j)} = v_{3j-2}f_{3j-2} + v_{3j-1}f_{3j-1} + v_{3j}f_{3j},$$

and h is the unit vector in the time direction. It follows from Newton's laws that if there are no constraints on the variables then the equations of motion are given by

$$\frac{\partial L}{\partial x_j} = \frac{d}{dt}\left(\frac{\partial L}{\partial v_j}\right) \text{ and } v_j = \frac{dx_j}{dt} \text{ for } 1 \leq j \leq N,$$

where L is the *Lagrangian*, defined as $L(x, v, t) = T(x, v, t) - V(x, v, t)$, where $T(x, v, t) = \frac{1}{2}\sum_{j=1}^{N} m_j v_j^2$ is the kinetic energy of the ensemble, and $V(x, v, t)$ is its potential energy. We consider these equations in an open subset $U = U_X \times U_W \times I$ of $\mathbf{R}^{6N+1}$.

Let $J : X \to W$ be the linear isomorphism of X onto W defined by $J(e_j) = f_j$ for $1 \leq j \leq N$. We can then write the equations of motion in terms of the gradients ∇_X in X and ∇_W in W. They become the equation

$$J(\nabla_X L) = \frac{d}{dt}(\nabla_W L).$$

There may however be constraints of various kinds (for example, the distance between two particles may remain constant). We consider only holonomic constraints, which only involve the positions of the particles, but neither their velocities nor time, and which are given by a submersion g of U_X into a Euclidean space F. Thus $M_X = \{x \in U_X : g(x) = 0\}$ is a manifold in U_X, and $M = M_X \times U_W \times I$ is a manifold in $\mathbf{R}^{6N+1}$. If $x \in M_X$, let T_x be the tangent space at x, let ∇_{T_x} be the corresponding gradient in T_x, and let J_x be the restriction of J to T_x. Then the equation of motion becomes

$$J_x(\nabla_{T_x} L) = \frac{d}{dt}(\nabla_W L).$$

Then, arguing as in Theorem 19.4.3, there exists a unique $\phi_x \in F$ such that

$$J_x(\nabla_X(L - \langle \phi_x, g \rangle)) = \frac{d}{dt}(\nabla_W L).$$

Suppose that $F = \mathbf{R}^k$, that $g = (g_1, \ldots, g_k)$ and that $\phi_x = (\lambda_1, \ldots, \lambda_k)$. Then the equations of motion become

$$\frac{\partial L}{\partial x_j} = \frac{d}{dt}\left(\frac{\partial L}{\partial v_j}\right) + \sum_{i=1}^{k} \lambda_i \frac{\partial g_i}{\partial x_j} \text{ for } 1 \leq j \leq N.$$

The Lagrange multipliers may vary with time, but their time derivatives do not enter into the equations. Let us give a simple example.

Example 19.4.9 The pendulum in three dimensions.

Suppose that a single particle, of mass m, is attached by a light rod of length l to a fixed point, which we take to be the origin in $\mathbf{R}^3$, and swings freely, under the influence of gravity. We take rectilinear coordinates, with x_3 in the vertical direction. Then $T(x, v) = \frac{1}{2}m(v_1^2 + v_2^2 + v_3^2)$, $V(x, v) = mgx_3$ and the constraint g is given by $x_1^2 + x_2^2 + x_3^2 = l^2$. The equations of motion then become

$$-m\frac{dv_1}{dt} = 2\lambda x_1,$$

$$-m\frac{dv_2}{dt} = 2\lambda x_2,$$

$$-mg - m\frac{dv_3}{dt} = 2\lambda x_3.$$

In fact there are solutions for which $v_3 = 0$, so that x_3 is constant. Then $\lambda = -mg/2x_3$, so that λ is constant. We require $mg < 2\lambda l$ so that $-l < x_3 < 0$. Let $\omega = \sqrt{2\lambda/m}$. Then a solution is given by $x_1 = A\cos\omega t$ and $x_2 = A\sin\omega t$, where $A^2 = l^2 - x_3^2$.

Exercises

19.4.1 Suppose that T is a self-adjoint linear operator on a d-dimensional Euclidean space E. Show that there exists an orthonormal basis $(e_1, \ldots e_d)$ and real numbers $\lambda_1 \geq \lambda_2 \geq \ldots \geq \lambda_d$ such that $T(e_j) = \lambda_j e_j$ for $1 \leq j \leq d$.

19.4.2 A linear operator on a complex inner product space F is *self-adjoint* if $\langle T(x), y \rangle = \langle x, T(y) \rangle$ for $x, y \in F$. Suppose that T is a self-adjoint linear operator on a d-dimensional complex inner product space F.

Show that $\langle T(x), x\rangle$ is real, for all $x \in F$. Show that there exist an orthonormal basis $(e_1, \ldots e_d)$ and real numbers $\lambda_1 \geq \lambda_2 \geq \ldots \geq \lambda_d$ such that $T(e_j) = \lambda_j e_j$ for $1 \leq j \leq d$.

19.4.3 Let Π be the plane $\{x \in \mathbf{R}^3 : 2\sqrt{2}x_1/3 + x_2 = 4\}$. Let $P = (1, 0, 0)$, $Q = (-1, 0, 0)$. Determine the minimum of

$$\{\|x - P\| + \|x - Q\| : x \in \Pi\}.$$

What is the geometric significance of your result?

19.4.4 Let $P = (1, 0, 0)$, $Q = (-1, 0, 0)$. Suppose that $a > 1$. Let

$$\Pi_{a^2} = \{(a^2, x_2, x_3) : x_2, x_3 \in \mathbf{R}\} \text{ and}$$

$$\Pi_{-a^2} = \{(-a^2, x_2, x_3) : x_2, x_3 \in \mathbf{R}\}.$$

Determine the sets of points

$$\{x \in \mathbf{R}^3 : a\,\|x - P\| = d(x, \Pi_{a^2})\} \text{ and}$$
$$\{x \in \mathbf{R}^3 : a\,\|x - Q\| = d(x, \Pi_{-a^2})\}.$$

Deduce that

$$\{x \in \mathbf{R}^3 : \|x - P\| + \|x - Q\| = 2a\} = \left\{x \in \mathbf{R}^3 : \frac{x_1^2}{a^2} + \frac{x_2^2 + x_3^2}{a^2 - 1} = 1\right\}.$$

19.4.5 By considering the tension T in the light rod, establish the circular movement of a pendulum as described above, using Newtonian methods. How are T and the Lagrangian multiplier λ related?

19.4.6 Suppose that $p, q > 0$ and that $1/p + 1/q = 1$.

(i) Let $S_q = \{x \in \mathbf{R}^d : \sum_{j=1}^{d} |x_j|^q = 1\}$. Show that M_q is a compact differential manifold in $\mathbf{R}^d$.

(ii) Suppose that $a_j \geq 0$ for $1 \leq j \leq d$. Let $f(x) = \sum_{j=1}^{d} a_j x_j$. Use a Lagrange multiplier to find the point $x \in S_q$ at which f attains its constrained maximum on S_q, and find the constrained maximum.

(iii) Deduce Hölder's inequality

$$\sum_{j=1}^{d} |a_j b_j| \leq \left(\sum_{j=1}^{d} |a_j|^p\right)^{1/p} \cdot \left(\sum_{j=1}^{d} |b_j|^q\right)^{1/q},$$

and show that

$$\left(\sum_{j=1}^{d} |a_j|^p\right)^{1/p} = \sup\left\{\left|\sum_{j=1}^{d} a_j b_j\right| : \sum_{j=1}^{d} |x_j|^q \leq 1\right\}.$$

(iv) Deduce Minkowski's inequality:

$$\left(\sum_{j=1}^{d} |a_j + b_j|^p\right)^{1/p} \leq \left(\sum_{j=1}^{d} |a_j|^d\right)^{1/p} + \left(\sum_{j=1}^{d} |a_j|^p\right)^{1/p}.$$

19.4.7 Suppose that $p_1, \ldots, p_d$ are positive numbers for which $\sum_{j=1}^{d} 1/p_j = 1$.

(i) Show that

$$A = \{x \in \mathbf{R}^d : \sum_{j=1}^{d} \frac{|x_j|^{p_j}}{p_j} = 1\}$$

is a compact differential manifold in $\mathbf{R}^d$.

(ii) Let $p(x) = \prod_{j=1}^{d} x_j$. Use a Lagrange multiplier to find the point $x \in A$ at which p attains its constrained maximum on S_q, and find the constrained maximum.

(iii) Establish the generalized arithmetic mean-geometric mean inequality: if $a_j \geq 0$ for $1 \leq j \leq d$, then

$$a_1^{1/p_1} \ldots a_d^{1/p_d} \leq \sum_{j=1}^{d} a_j p_j.$$

19.4.8 If R is the point of M in Example 19.4.8 for which the path PRQ has minimal length, show that PR and QR make the same angle with vectors normal to M at R. (Consider a reflection in the tangent space at R.)

19.5 Smooth partitions of unity

The definition of a differential manifold is a local one. When we turn to integration, we need to combine local results. For this, we use *smooth partitions of unity*.

Suppose that K is a compact non-empty subset of a Euclidean space E, and that $\{U_1, \ldots, U_k\}$ is a finite open cover of K. A *smooth partition of unity subordinate to the cover* is

- an open subset V such that $K \subseteq V \subseteq \cup_{j=1}^{k} U_j$,
- a sequence $(L_j)_{j=1}^{k}$ of compact sets, such that $L_j \subseteq U_j$ for $1 \leq j \leq k$, and $V \subseteq \cup_{j=1}^{k} L_j$, and

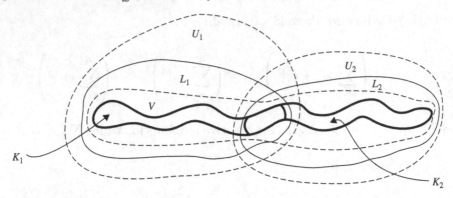

Figure 19.5a. Partition of unity.

- a sequence $(f_j)_{j=1}^k$ of smooth non-negative functions on V such that $f_j(x) = 0$ for $x \in V \setminus L_j$, and such that $\sum_{j=1}^k f_j(x) = 1$ for $x \in V$.

Theorem 19.5.1 *If K is a compact non-empty subset of a d-dimensional Euclidean space E and $\{U_1, \ldots, U_k\}$ is an open cover of K, there exists a smooth partition of unity subordinate to the cover.*

Proof We break the proof into several steps. Let $U = \cup_{j=1}^k U_j$.

Lemma 19.5.2 *Suppose that $x \in U_j$. There exists a compact subset L_x of U_j and a non-negative smooth function h_x on U such that $h_x(x) > 0$ and such that $h_x(y) = 0$ for $y \in U \setminus L_x$.*

Proof Let $(e_1, \ldots, e_d)$ be an orthonormal basis for E. There exists $\delta > 0$ such that $L_x = \{y \in E : |y_i - x_i| \le \delta \text{ for } 1 \le i \le d\} \subseteq U_j$. If $y \in U$ and $1 \le j \le k$, let

$$h_{i,x}(y) = \begin{cases} \exp(-1/(\delta^2 - |y_i - x_i|^2)) & \text{for } |y_i - x_i| < \delta, \\ 0 & \text{otherwise.} \end{cases}$$

As in Volume I, Section 7.6, $h_{i,x}$ is a smooth function on U. Let $h_x = \prod_{i=1}^d h_{i,x}$. Then L_x and h_x satisfy the requirements of the lemma. □

Lemma 19.5.3 *There exist compact sets $\{K_j : 1 \le j \le k\}$ such that $K_j \subseteq K \cap U_j$ for $1 \le j \le k$ and such that $K = \cup_{j=1}^k K_j$.*

Proof Since K is bounded, we can suppose that each U_j is bounded. Let

$$U_{j,n} = \{y \in U_j : d(y, E \setminus U_j) > 1/n\}.$$

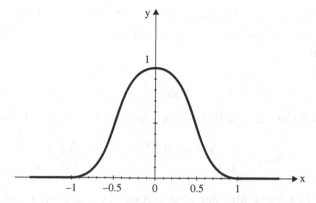

Figure 19.5b. A bump function in one dimension.

Then $\overline{U_{j,n}} \subseteq U_j$, each sequence $(U_{j,n})_{n=1}^{\infty}$ is increasing, and the sets $\{U_{j,n} : 1 \leq j \leq k, n \in \mathbf{N}\}$ form an open cover of K. There therefore exists $N \in \mathbf{N}$ such that $K \subseteq \cup_{j=1}^{k} U_{j,N}$. Let $K_j = K \cap \overline{U_{j,N}}$. Then $\{K_j : 1 \leq j \leq k\}$ satisfies the requirements of the lemma. $\qquad \square$

Lemma 19.5.4 *For each $1 \leq j \leq k$ there exists a compact set L_j such that $K_j \subseteq L_j \subseteq U_j$ and a smooth non-negative function h_j on U such that $h_j(x) > 0$ for $x \in K_j$ and $h_j(y) = 0$ for $y \in U \setminus L_j$.*

Proof For each x in K_j there exist a compact set L_x and a smooth non-negative function h_x on U which satisfy the conditions of Lemma 19.5.2. Let $U_x = \{y \in U : h_x(y) > 0\}$. Then $U_x \subseteq L_x$, and $\{U_x : x \in K_j\}$ is an open cover of K_j. There exists a finite subcover $\{U_x : x \in F_j\}$. Let $L_j = \cup_{x \in F_j} L_x$ and let $h_j = \sum_{x \in F} h_x$. Then L_j and h_j satisfy the requirements of the lemma. $\qquad \square$

We now complete the proof of the theorem. Let $h = \sum_{j=1}^{k} h_j$, and let $V = \{y \in U : h(y) > 0\}$. Then V is an open set, and $K \subseteq V \subseteq U$. Let $f_j(y) = h_j(y)/h(y)$, for $y \in V$. Then $(V, (L_j)_{j=1}^{k}, (f_j)_{j=1}^{k})$ is a partition of unity subordinate to the cover. $\qquad \square$

Suppose that K is a compact non-empty subset of a Euclidean space E, that $\{U_1, \ldots, U_k\}$ is a finite open cover of K and that $(V, (L_j)_{j=1}^{k}, (f_j)_{j=1}^{k})$ is a smooth partition of unity subordinate to the cover. Suppose further that F is a Euclidean space and that, for each $1 \leq j \leq k$, g_j is a function from U_j to F. We then define $\sum_{j=1}^{k} f_j g_j : V \to F$ by setting

$$\sum_{j=1}^{k} f_j g_j(y) = \sum \{f_j(y) g_j(y) : y \in V \cap U_j\}.$$

In other words, we set $(f_j g_j)(y) = 0$ if $f_j(y) = 0$. Then the function $\sum_{j=1}^{k} f_j g_j$ is continuous (continuously differentiable, smooth) if each of the functions g_j is.

Exercise

19.5.1 Suppose that E is a Euclidean space and that $\delta > 0$. Let

$$h(y) = \begin{cases} \exp(-1/(\delta^2 - \|y\|^2)) & \text{for } \|y\| < \delta, \\ 0 & \text{otherwise.} \end{cases}$$

Show that h is a non-negative smooth function on E. Identify the set $\{y : h(y) > 0\}$.

19.6 Integration over hypersurfaces

We now turn to integration. We restrict attention to a special case; throughout this section, we suppose that M is a connected compact d-dimensional hypersurface in a $(d + 1)$-dimensional Euclidean space E. We need one topological property of such hypersurfaces, which we state without proof.

Theorem 19.6.1 (The Jordan–Brouwer separation theorem) *Suppose that M is a connected compact d-dimensional hypersurface in a $(d+1)$-dimensional Euclidean space E. Then the open set $E \setminus M$ has two connected components: one, out[M], the outside of M, is unbounded, and the other, in[M], the inside of M, is bounded.*

When $d = 1$, so that M is the track of a simple closed curve in a two-dimensional Euclidean space, this is a special case of the Jordan curve theorem, which is proved in Volume III.

If M is a connected compact d-dimensional hypersurface and $x \in M$ then the normal space N_x is one-dimensional, and so there are two elements n_x^+ and n_x^- in N_x of norm 1. We choose n_x^+ so that $x + \lambda n_x^+$ is outside M for small positive values of λ; then $x + \lambda n_x^-$ is inside M for small positive values of λ.

Theorem 19.6.1 has the following useful consequence.

Theorem 19.6.2 *Suppose that M is a connected compact hypersurface in a Euclidean space E. There exists an open set W containing M, and a submersion $g : W \to \mathbf{R}$ such that $M = \{y \in W : g(y) = 0\}$.*

Proof Since M is compact, there exist a finite open cover $\{U_1, \ldots, U_k\}$ of M by open subsets of E and submersions $g_j : U_j \to \mathbf{R}$ such that $M \cap U_j =$

$\{y \in U_j : g_j(y) = 0\}$. We can also suppose that $M \cap U_j$ is a connected subset of M, for each j. If $x \in M \cap U_j$, let

$$\lambda_j(x) = \langle (\nabla g_j)_x, n_x^+ \rangle = \frac{\partial g_j}{\partial n_x^+}(x).$$

Then λ_j is a continuous function on $M \cap U_j$, which does not take the value 0. Replacing g_j by $-g_j$ if necessary, we can therefore suppose that $\lambda_j(x) > 0$ for all $x \in M \cap U_j$. There exists a smooth partition of unity $(V, (L_j)_{j=1}^k, (f_j)_{j=1}^k)$ subordinate to the cover. Let $g = \sum_{j=1}^k f_j g_j$. Then $M = \{y \in V : g(y) = 0\}$. If $x \in M$, there exists i such that $x \in M \cap U_i$ and $f_i(x) > 0$. Then

$$\langle \nabla g_x, n_x^+ \rangle = \sum_{j=1}^k f_j(x) \lambda_j(x) \geq f_i(x) \lambda_i(x) > 0.$$

Thus ∇g is non-zero on M. Let $W = \{y \in V : \nabla g_y \neq 0\}$. Then W is an open subset of V containing M, and the restriction of g to W is a submersion. $\square$

We now consider integration. First, let us describe the notation that will be used. If f is a Riemann integrable function on an open subset U of E, we denote the integral by $\int_U f(y) \, dv_{d+1}(y)$. On the other hand, the integral of a function g on M will be denoted by $\int_M g(x) \, d\sigma_d(x)$: thus dv_{d+1} represents a (hyper)volume integral and $d\sigma_d$ represents a (hyper)surface integral.

We begin locally. Suppose that $x \in M$. Take an orthonormal basis $(e_1, \ldots, e_{d+1})$ for E, where $(e_1, \ldots, e_d)$ is an orthonormal basis for T_x, and $e_{d+1} = n_x^+$. We shall consider cells in E defined in terms of this basis. There exists $\delta > 0$ such that if U_x is the cell

$$\{y \in E : |y_j - x_j| < \delta \text{ for } 1 \leq j \leq d+1\}$$

and $\phi_x(y) = P_x(y - x)$ for $y \in M$ then

- (U_x, ϕ_x) is a chart near x,
- $\|n_y^+ - n_x^+\| < \frac{1}{2}$ for $y \in M \cap U_x$, so that $\langle n_y^+, n_x^+ \rangle > \frac{1}{2}$, and
- $\|(y - x) - \phi_x(y)\| \leq \|\phi_x(y)\|$ for $y \in M \cap U_x$ (this is possible by Corollary 19.2.4).

Let $\psi : \phi_x(U_x) \to M \cap U_x$ be the inverse mapping, and let

$$V_x = \{v \in T_x : |v_j| < \delta \text{ for } 1 \leq j \leq d\}.$$

It follows from the third condition that if $y \in M \cap U_x$ then

$$|y_{d+1} - x_{d+1}| = \|(y - x) - \phi_x(y)\| < \max\{|y_j - x_j| : 1 \leq j \leq d\},$$

and it follows easily from this that $\phi_x(U_x) = V_x$.

We call such a cell U_x a *well-behaved neighbourhood* of x.

Suppose now that $0 < \alpha < \delta$, that

$$L_x = \{y \in E : |y_j - x_j| \le \alpha \text{ for } 1 \le j \le d+1\},$$

and that g is a continuous function on M which vanishes on $M \setminus L_x$. Suppose that $y \in M \cap L_x$ and let $\phi_x(y) = v$. We now choose a new orthonormal basis $(f_1, \ldots, f_{d+1})$ such that $(f_1, \ldots, f_d)$ is an orthonormal basis for T_x, $f_{d+1} = e_{d+1} = n_x^+$ and $n_y^+ \in \text{span}\{f_d, f_{d+1}\}$. Suppose that h is a unit vector in T_x. There exists $\eta > 0$ such that $v + th \in V_x$ for $|t| < \eta$. Then $\psi(v + th) \in M$ for $|t| < \eta$, and so $D\psi_v(h) \in T_y$. Thus the restriction of $D\psi_v$ to T_x is a linear mapping from T_x to T_y, and so the Jacobian $J\psi_v$ is defined.

We now define

$$\int_{M \cap U_x} g(y)\, d\sigma_d(y) \text{ to be } \int_{V_x} g(\psi(v)) |J\psi_v|\, dv_d(v).$$

Let us determine the value of $|J\psi_v|$. Let $C = I_1 \times \cdots \times I_d$ be a cell in T_x, defined in terms of the new basis $(f_1, \ldots, f_{d+1})$. Then $D\psi_v(C)$ is a cell in T_y with sides of lengths $l(I_1), \ldots l(I_{d-1}), l(I_d)/\langle n_y^+, n_x^+ \rangle$ (see Figure 19.6), and so $|J\psi_v| = 1/\langle n_y^+, n_x^+ \rangle$. Thus

$$\int_{M \cap U_x} g(y)\, d\sigma_d(y) = \int_{V_x} \frac{g(\psi(v))}{\left\langle n_{\psi(v)}^+, n_x^+ \right\rangle}\, dv_d(v).$$

It is now a straightforward matter to use a smooth partition of unity to define the integral of a continuous function g on M. The well-behaved neighbourhoods U_x of points x of M form an open cover of M, and so there is a finite subcover $\{U_{x_1}, \ldots, U_{x_k}\}$. Let $(V, (L_i)_{i=1}^k, (f_i)_{i=1}^k)$ be a smooth

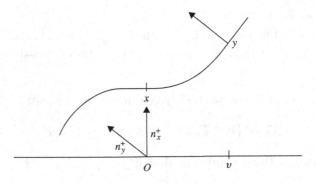

Figure 19.6. Change of variables.

partition of unity subordinate to the cover. We then define

$$\int_M g(y)\,d\sigma_d(y) \text{ to be } \sum_{i=1}^{k} \left(\int_{M\cap U_{x_i}} f_i(y)g(y)\,d\sigma_d(y) \right).$$

In particular, we write $\sigma_d(M)$ for $\int_M d\sigma_d(y)$, and write ω_d for $\sigma_d(S^d)$, where S^d is the unit sphere in a $(d+1)$-dimensional Euclidean space.

It does however remain to show that the integral does not depend on the finite subcover, nor on the choice of smooth partition of unity. This is again a fairly straightforward matter. The details are indicated in the exercises below.

Exercises

We consider the setting described above.

19.6.1 Suppose that $z \in M \cap U_x$. Show that $\langle n_y^+, n_z^+ \rangle > \frac{1}{2}$ for $y \in M \cap U_x$. Show that if we follow the procedure above, projecting onto T_z rather than T_x, then

$$\int_{M\cap U_x} g(y)\,d\sigma_d(y) = \int_{V_z} g(\psi_z(v))|(J\psi_z)_v|\,dv_d(v).$$

19.6.2 Suppose that $\{W_{x_1},\ldots,W_{x_k}\}$ is a finite cover of L_x by open cells and that $(V,(L_i)_{i=1}^{k},(f_i)_{i=1}^{k})$ is a smooth partition of unity subordinate to the cover. Show that

$$\int_{M\cap U_x} g(y)\,d\sigma_d(y) = \sum_{i=1}^{k} \left(\int_{M\cap W_{x_i}} f_i(y)g(y)\,d\sigma_d(y) \right).$$

19.6.3 Suppose that $\{W_{x_1},\ldots,W_{x_l}\}$ is a finite cover of M by open cells with the properties described above and that $(V,(L_i)_{i=1}^{k},(f_i)_{i=1}^{k})$ is a smooth partition of unity subordinate to the cover. By considering the open cover $\{U_i \cap W_j : 1 \le i \le k, 1 \le j \le l\}$, and using the previous exercises, show that $\int_M g(y)\,d\sigma_d(y)$ does not depend on the choice of cover or the choice of smooth partition of unity.

19.6.4 The following extended exercise shows how to establish a change of variables formula, in the simplest, but most useful, case. We consider the unit ball B_d and the unit sphere S^{d-1} in a d-dimensional Euclidean space E. Let A_r be the annulus $\{x : r \le \|x\| \le 1\}$, for $0 < r < 1$. Suppose that f is a continuous real-valued function on S^{d-1} and that g is a continuous function on B_d. Let $\tilde{f}(r\omega) = f(\omega)$ for $0 < r \le 1$ and $\omega \in S^{d-1}$.

(a) Show that $\tilde{f}$ is a continuous function on $B_d \setminus \{0\}$.
(b) Show that

$$(1 - r^d) \int_{S^{d-1}} f(\omega) \, d\sigma_{d-1}(\omega) = \int_{A_r} \tilde{f}(x) \, dv_d(x).$$

(c) Suppose that $\epsilon > 0$. Show that there exist $0 = r_0 < \cdots < r_k = 1$ and a function h on $B_d \setminus \{0\}$ such that
  (i) the function $h_j(\omega) = h(r_j\omega)$ is continuous on S^{d-1}, for $1 \leq j \leq k$,
  (ii) $h(t\omega) = h(r_j\omega)$ for $\omega \in S^{d-1}$ and $r_{j-1} < t \leq r_j$, for $1 \leq j \leq k$, and
  (iii) $|h(x) - g(x)| < \epsilon$ for $x \in B_d \setminus \{0\}$.
(d) Show that

$$\int_{A_r} g(x) \, dv_d(x) = \int_r^1 \left(r^{d-1} \int_{S^{d-1}} f(s\omega) \, d\sigma_{d-1}(\omega) \right) ds.$$

(e)

$$\int_{B_d} g(x) \, dv_d(x) = \int_0^1 \left(r^{d-1} \int_{S^{d-1}} f(s\omega) \, d\sigma_{d-1}(\omega) \right) ds.$$

19.6.5 Use the previous exercise to calculate ω_d.

19.6.6 Suppose that U is a bounded convex open neighbourhood of 0 in a d-dimensional Euclidean space E, that $M = \partial U$ is a $(d-1)$-dimensional differential manifold such that for each $x \in M$, $\langle x, n_x^+ \rangle > 0$. Suppose that f is a continuous real-valued function on U and that g is a continuous real-valued function on $\overline{U}$.

(a) Show that for each $\omega \in S^{d-1}$ there exists a unique $r_\omega > 0$ such that $x_\omega = r_\omega.\omega \in M$. Let $\tilde{f}(\omega) = f(x_\omega)$. (Note that $r_\omega = \|x_\omega\|$.)
(b) Show that

$$\int_M f(x) \, d\sigma_{d-1}(x) = \int_{S^{d-1}} \frac{\|x_\omega\|^d \, \tilde{f}(\omega)}{\langle x_\omega, n_{x_\omega}^+ \rangle} \, d\sigma_{d-1}(x).$$

(c) Obtain a formula for $\int_{\overline{U}} g(x) \, dv_d(x)$ corresponding to the formula in Exercise 19.6.4.

19.7 The divergence theorem

Suppose that U is an open subset of a Euclidean space E. A *vector field* F on U is a continuously differentiable mapping from U into E. For example, if f is a $C^{(2)}$ function on U, then the gradient mapping ∇f is a vector field. If F is a vector field on U for which there exists a $C^{(2)}$ function f, for which $F = \nabla f$, then F is called a *conservative vector field*, and f is a *scalar potential* for F. Another important example occurs in $\mathbf{R}^3$. Suppose that U is an open subset of $\mathbf{R}^3$ and that $f = (f_1, f_2, f_3)$ is a $C^{(2)}$ vector field on U. Let

$$(\nabla \times f)(x) = \left(\frac{\partial f_3}{\partial x_2}(x) - \frac{\partial f_2}{\partial x_3}(x), \frac{\partial f_1}{\partial x_3}(x) - \frac{\partial f_3}{\partial x_1}(x), \frac{\partial f_2}{\partial x_1}(x) - \frac{\partial f_1}{\partial x_2}(x) \right).$$

Then $\nabla \times f$ is a vector field, the *curl* of f, on U. Compare this with the cross-product of two vectors defined in Appendix C. Such vector fields occur in mathematical physics, in electromagnetic theory, in gravitation and in fluid dynamics. We consider curl further in Section 19.9.

If F is a vector field on an open subset U of a Euclidean space E and $x \in U$, then DF_x is a linear mapping of E into itself, and DF is a continuous mapping of U into $L(E)$. We define the *divergence* $\nabla.F(x)$ at x to be the trace of DF_x. If $(e_1, \ldots, e_d)$ is a basis for E, and if $F(x) = \sum_{i=1}^{d} f_i(x)e_i$, then DF is represented by the matrix $(\partial f_i/\partial x_j)$, so that

$$\nabla.F(x) = \sum_{i=1}^{d} \frac{\partial f_i}{\partial x_i}(x).$$

It is important to note that this formula does not depend upon the choice of basis (and that the basis need not be an orthonormal basis), since the trace does not depend on the choice of basis. (See Appendix B.5.) A vector field F is said to be *solenoidal* if $\nabla.F = 0$. For example, if f is a $C^{(2)}$ function on U then

$$\nabla.(\nabla \times f)(x) = \left(\frac{\partial^2 f_3}{\partial x_1 x_2}(x) - \frac{\partial^2 f_2}{\partial x_1 x_3}(x) \right)$$
$$+ \left(\frac{\partial^2 f_1}{\partial x_2 x_3}(x) - \frac{\partial^2 f_3}{\partial x_2 x_1}(x) \right)$$
$$+ \left(\frac{\partial^2 f_2}{\partial x_3 x_1}(x) - \frac{\partial^2 f_1}{\partial x_3 x_2}(x) \right) = 0,$$

so that $\nabla \times f$ is solenoidal.

Suppose that f is a $C^{(2)}$ function on U. Let $F = \nabla f$. Then f is *harmonic* if F is solenoidal; that is, if $\nabla.(\nabla f) = 0$. We write $\nabla^2 f$ for $\nabla.(\nabla f)$: $\nabla^2 f$ is the *Laplacian* of f. If $(e_1, \ldots, e_d)$ is an orthonormal basis for E, then

$$\nabla^2 f(x) = \sum_{i=1}^{d} \frac{\partial^2 f}{\partial x_i^2}(x).$$

Let us give two examples which we shall need later.

Example 19.7.1 Suppose that E is a d-dimensional Euclidean space and that $\alpha \in \mathbf{R}$.

(i) Let $\psi_\alpha(x) = \|x\|^{2-\alpha}$, for $x \in E \setminus \{0\}$. Then

$$\nabla \psi_\alpha(x) = (2 - \alpha) \frac{x}{\|x\|^\alpha} \text{ and } \nabla^2 \psi_\alpha(x) = \frac{(2 - \alpha)(d - \alpha)}{\|x\|^\alpha}.$$

(ii) Let $\phi(x) = \log \|x\|$ for $x \in E \setminus \{0\}$. Then

$$\nabla \phi(x) = \frac{x}{\|x\|^2} \text{ and } \nabla^2(x) = \frac{d - 2}{\|x\|^2}.$$

These are easy calculations. If $(e_1, \ldots, e_d)$ is an orthonormal basis for E and $x = \sum_{i=1}^{d} x_i e_i$ then $\psi_\alpha(x) = (x_1^2 + \cdots + x_d^2)^{1-\alpha/2}$, so that

$$\frac{\partial \psi_\alpha}{\partial x_i}(x) = (2 - \alpha) \frac{x_i}{\|x\|^\alpha} \text{ and } \nabla \psi_\alpha = (2 - \alpha) \frac{x}{\|x\|^\alpha}.$$

Further,

$$\frac{\partial^2 \psi}{\partial x_i^2}(x) = (2 - \alpha) \left(\frac{1}{\|x\|^\alpha} - \alpha \frac{x_i^2}{\|x\|^{\alpha+2}} \right).$$

Adding, $\nabla^2 \psi_\alpha(x) = (2 - \alpha)(d - \alpha) \|x\|^{-\alpha}$.

The calculations for (ii) are left as an exercise for the reader.

In particular, the function $x / \|x\|^d$ on $E \setminus \{0\}$ is solenoidal. If $d = 2$ then ϕ is harmonic, and if $d > 2$ then ψ_d is harmonic.

We shall consider these ideas further in the next section. First, we prove the *divergence theorem*, which is a multi-dimensional version of the fundamental theorem of calculus.

Theorem 19.7.2 (The divergence theorem) *Suppose that V is a connected bounded open subset of a $(d+1)$-dimensional Euclidean space E whose boundary M is a finite disjoint union of connected hypersurfaces $M_1, \ldots, M_k$ and that F is a vector field defined on an open set U containing $\overline{V}$. Let $B = \overline{V}$, and let*

$$\int_M \langle F(x), n_x^+ \rangle \, d\sigma_d(x) = \sum_{j=1}^{k} \left(\int_{M_j} \langle F(x), n_x^+ \rangle \, d\sigma_d(x) \right),$$

where n_x^+ is the normal at x in the direction out of V. Then

$$\int_B \nabla.F(y)dv_{d+1}(y) = \int_M \langle F(x), n_x^+ \rangle \, d\sigma_d(x).$$

Proof Note that we need to show that B is Jordan measurable, so that the first integral makes sense. This will emerge during the proof of the theorem.

We use a smooth partition of unity. For each $x \in M$ there exists a well-behaved neighbourhood U_x of x, and for each y inside M there exists an open cell C_y such that $\overline{C_y} \subseteq in[M]$. Together, these sets form an open cover of B; since B is compact, there exists a finite subcover

$$(U_{x_1}, \ldots, U_{x_k}, C_{y_{k+1}}, \ldots, C_{y_l}) = (V_1, \ldots V_l) \text{ say.}$$

We consider a smooth partition of unity $(V, (L_1, \ldots, L_l), (f_1, \ldots f_l))$ subordinate to the cover. Then

$$\int_B \nabla.F(y)dv_{d+1}(y) = \sum_{j=1}^l \int_{V_j} f_j(y)\nabla.F(y)dv_{d+1}(y)$$

and

$$\int_M \langle F(x), n_x^+ \rangle \, d\sigma_d(x) = \sum_{j=1}^k \int_{M\cap U_{x_j}} \langle f_j(x)F(x), n_x^+ \rangle \, d\sigma_d(x).$$

Now let us set $G_j = f_j F$. Since $f_j(y) = 0$ for $y \notin L_j$, we consider G_j as a function on E, setting $G_j(y) = 0$ for $y \notin L_j$. Then

$$\nabla.G_j(y) = \langle (\nabla f_j)(y), F(y) \rangle + f_j(y)\nabla.F(y)$$

so that

$$\sum_{j=1}^l \nabla.G_j = \left\langle \sum_{i=1}^l \nabla f_j, F \right\rangle + \left(\sum_{j=1}^l f_j \right) \nabla.F = \nabla.F,$$

since $\sum_{j=1}^l f_j = 1$ and $\sum_{j=1}^l \nabla f_j = \nabla 1 = 0$. It is therefore sufficient to show that

$$\int_{B\cap U_{x_j}} \nabla.G_j(y) \, dv_{d+1}(y) = \int_{M\cap U_{x_j}} \langle G_j(x), n_x^+ \rangle \, d\sigma_d(x)$$

for $1 \leq j \leq k$, and that

$$\int_{C_{y_j}} \nabla.G_j(y) \, dv_{d+1}(y) \, dy = 0,$$

for $k+1 \leq j \leq l$.

We deal with the second set of equations first. By taking a suitable orthonormal basis of E, we can suppose that

$$V_j = \{w \in E : |w_i - (y_j)_i| < \delta \text{ for } 1 \le i \le d+1\}$$

$$\text{and } L_j = \{w \in E : |w_i - (y_j)_i| \le \alpha \text{ for } 1 \le i \le d+1\}.$$

Let $G_j = (g_1, \dots g_{d+1})$. Then

$$\int_{(y_j)_i - \delta}^{(y_j)_i + \delta} \frac{\partial g_i}{\partial x_i}(w)\, dw_i =$$

$$g_i(w_1, \dots, (y_j)_i + \alpha, \dots, w_{d+1}) - g_i(w_1, \dots, (y_j)_i - \alpha, \dots, w_{d+1}) = 0;$$

integrating with respect to the other variables, we see that

$$\int_{C_{v_j}} \frac{\partial g_i}{\partial x_i}(y)\, dv_{d+1}(y) = 0.$$

Adding, it follows that

$$\int_{C_{v_j}} \nabla . G_j(y)\, dv_{d+1}(y) = 0.$$

We now turn to the first set of equations. To simplify the notation, we can suppose that $x_j = 0$. We drop the suffix j; we denote U_{x_j} by U, L_{x_j} by L, the tangent space T_{x_j} by T, the orthogonal projection onto T by P, the restriction of P to $M \cap U$ by ϕ, and the inverse mapping of $\phi(M \cap U)$ onto $M \cap U$ by ψ.

It follows from Theorem 18.5.3 that $B \cap U$ is Jordan measurable, so that the volume integral over $B \cap U$ exists. Since U is a well-behaved neighbourhood of 0, there is an orthonormal basis $(e_1, \dots, e_{d+1})$, where $(e_1, \dots, e_d)$ is an orthonormal basis for T_{x_j} and $e_{d+1} = n_{x_j}^+$. We write $G_j(y) = G(y) = \sum_{i=1}^{d+1} g_i(y)e_i$. Then

$$\int_{B \cap U} \nabla . G(y)\, dv_{d+1}(y) = \sum_{i=1}^{d+1} \int_{B \cap U} \frac{\partial g_i}{\partial x_i}(y)\, dv_{d+1}(y)$$

and

$$\int_{M \cap U} \langle G(x), n_x^+ \rangle\, d\sigma_d(x) = \sum_{i=1}^{d+1} \int_{M \cap U} g_i(x) \langle e_i, n_x^+ \rangle\, d\sigma_d(x).$$

It is therefore sufficient to show that

$$\int_{B \cap U} \frac{\partial g_i}{\partial x_i}(y) \, dv_{d+1}(y) = \int_{M \cap U} g_i(x) \langle e_i, n_x^+ \rangle \, d\sigma_d(x)$$

for $1 \leq i \leq d+1$.

We make a (non-linear) change of variables. If $y \in U$, let $\chi(y) = \psi(P(y))$ and let $\theta(y) = \chi(y) - P(y)$. If $y = z + \lambda e_{d+1}$, with $z \in T$, then

$$\theta(y) = \theta(z) = \psi(z) - z \in \operatorname{span}(e_{d+1}).$$

From this it follows that

$$\frac{\partial \theta_{d+1}}{\partial x_{d+1}} = 0 \text{ and } \frac{\partial \theta_i}{\partial x_j} = 0 \text{ for } 1 \leq i \leq d, 1 \leq j \leq d+1.$$

Now let $S(y) = y - \theta(y)$. Then $S(M \cap U) \subseteq T$, S is a diffeomorphism of $B \cap U$ onto $S(B \cap U)$ and $S(B \cap U) \subseteq T \times (-2\delta, 0)$. Further, it follows from the equations above that the Jacobian $J(S) = 1$. Let R be the inverse mapping from $S(U)$ onto U. It is then sufficient to show that

$$\int_{T \times [-2\delta, 0]} \frac{\partial g_i}{\partial x_i}(R(y)) \, dv_{d+1}(y) = \int_T g_i(\psi(w)) \frac{\left\langle e_i, n_{\psi(w)}^+ \right\rangle}{\left\langle e_{d+1}, n_{\psi(w)}^+ \right\rangle} \, d\sigma_d(w).$$

for $1 \leq i \leq d+1$.

First, let us consider the case where $i = d+1$. Then

$$\int_{T \times [-2\delta, 0]} \frac{\partial g_{d+1}}{\partial x_{d+1}}(R(y)) \, dv_{d+1}(y) =$$

$$\int_T \left(\int_{-2\delta}^0 \frac{\partial g_{d+1}}{\partial x_{d+1}}(R(y)) \, dy_{d+1} \right) dy_1, \dots, dy_d$$

$$= \int_T g_{d+1}(\psi(w)) \, d\sigma_d(w)$$

$$= \int_T g_{d+1}(\psi(w)) \frac{\left\langle e_{d+1}, n_{\psi(w)}^+ \right\rangle}{\left\langle e_{d+1}, n_{\psi(w)}^+ \right\rangle} \, d\sigma_d(w).$$

Next suppose that $1 \leq i \leq d$. Without loss of generality, we can suppose that $i = d$. First we fix all the variables y_k with $1 \leq k < d$. Suppose that $y_k = a_k$ for $1 \leq k < d$. If $y \in U$, we set $y = (a, s, t)$ and set $w = (a, s, 0)$. Let Π be the plane

$$\{y \in E : y_k = a_k \text{ for } 1 \leq k < d\},$$

and let $\widetilde{U} = U \cap \Pi$, $\widetilde{M} = (M \cap U) \cap \Pi$, $\widetilde{B} = (B \cap U) \cap \Pi$. Then

$$\widetilde{M} = \{(a, s, t) : t = \langle \psi(s), e_{d+1} \rangle\} \text{ and } \widetilde{B} \subseteq \{(a, s, t) : t \leq \psi(s)\}.$$

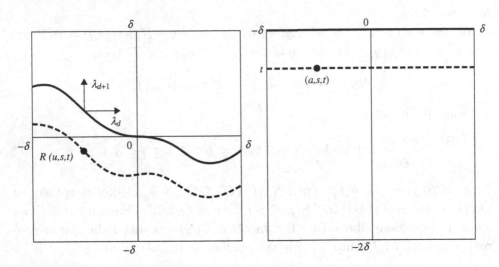

Figure 19.7. The divergence theorem.

Suppose that $y = (a, s, t) \in S(\widetilde{B})$. Let $\breve{U} = \{(s, t) : (a, s, t) \in S(\widetilde{U})\}$. If $y = (a, s, t) \in S(\widetilde{U})$, let $h(s, t) = g_d(R(y))$, and if $(s, t) \in \mathbf{R}^2 \setminus \breve{U}$ let $h(s, t) = 0$. Then the partition of unity properties imply that h is continuously differentiable, and that $h(\delta, t) = h(-\delta, t) = 0$ for $t \in \mathbf{R}$. If $(s, t) \in \breve{U}$, then

$$\frac{\partial h}{\partial s}(s, t) = \frac{\partial g_d}{\partial s}(R(y)) + \frac{\partial g_d}{\partial t}(R(y))\frac{\partial R}{\partial s}(y).$$

Now let $n^+_{\psi(w)} = \sum_{k=1}^{d+1} \lambda_k(s)e_k$, where $\lambda_k(s) = \left\langle e_k, n^+_{\psi(w)} \right\rangle$. Since U is a well-behaved neighbourhood of 0, $\lambda_{d+1}(s) > \frac{1}{2}$.

Let (u, v) be the unit tangent vector to the curve $(s, h(s, t))$, with $u > 0$. Then $(0, u, v)$ is in the tangent space of $R(T)$ at $R(y)$. But this is the same as the tangent space of M at $\psi(w)$, and so $\left\langle (0, u, v), n^+_{\psi(w)} \right\rangle = 0$; that is, $u\lambda_d(s) + v\lambda_{d+1}(s) = 0$, and so $\partial R/\partial s(y) = -\lambda_d/\lambda_{d+1}$. Now

$$\int_{-\delta}^{\delta} \frac{\partial h}{\partial s}(s, t) \, ds = h(\delta, t) - h(-\delta, t) = 0,$$

so that

$$\int_{-\delta}^{\delta} \frac{\partial g_d}{\partial s}(R(a, s, t)) \, ds = \int_{-\delta}^{\delta} \frac{\lambda_d(s)}{\lambda_{d+1}(s)} \frac{\partial g_d}{\partial t}(R(a, s, t)) \, ds.$$

Integrating with respect to t, and changing the order of integration,

$$\int_{-2\delta}^{0} \left(\int_{-\delta}^{\delta} \frac{\partial g_d}{\partial s}(R(a,s,t))\, ds \right) dt = \int_{-\delta}^{\delta} \frac{\lambda_d(s)}{\lambda_{d+1}(s)} g_d(\psi(a,s,0))\, ds$$

$$= \int_{-\delta}^{\delta} \frac{\left\langle e_d, n_{(a,s,0)}^{+} \right\rangle}{\left\langle e_{d+1}, n_{(a,s,0)}^{+} \right\rangle} g_d(\psi(a,s,0))\, ds.$$

Integrating over the remaining variables,

$$\int_{S(B \cap U)} \frac{\partial g_d}{\partial x_d}(y)\, dv_{d+1}(y) = \int_{T} g_d(\psi(w)) \frac{\left\langle e_d, n_{\psi(w)}^{+} \right\rangle}{\left\langle e_{d+1}, n_{\psi(w)}^{+} \right\rangle}\, d\sigma_d(w).$$

$\square$

There are many consequences of the divergence theorem.

Corollary 19.7.3 *Suppose that M is a d-dimensional connected compact hypersurface in a $(d+1)$-dimensional Euclidean space E, that $B = \overline{in}[M] = M \cup in[M]$ and that F is a vector field defined on an open set U containing B. Then*

$$\int_{B} \nabla . F(y)\, dv_{d+1}(y) = \int_{M} \left\langle F(x), n_x^{+} \right\rangle\, d\sigma_d(x).$$

Proof   This is a consequence of the Jordan–Brouwer separation theorem: M is the boundary of $in(M)$. $\square$

Corollary 19.7.4 *If f is a $C^{(2)}$-function defined on U then*

$$\int_{B} \nabla^2 f(y)\, dv_{d+1}(y) = \int_{M} \frac{\partial f}{\partial n_x^{+}}(x)\, d\sigma_d(x).$$

Proof   Apply the theorem to ∇f. $\square$

Corollary 19.7.5 $(d+1)v_{d+1}(B) = \int_M \left\langle x, n_x^{+} \right\rangle\, d\sigma_d(x).$

Proof   Take $F(x) = x$. Then $\nabla . F = d + 1$. $\square$

Corollary 19.7.6 $(d+1)v_{d+1}(B_r(y)) = r\sigma_d(S_r(y)) = r^d \omega_d.$

Proof   For $\langle x - y, n_x^{+} \rangle = r$. $\square$

Corollary 19.7.7 *If $y_0 \in V$ then*

$$\int_{M} \frac{\langle x, n_x^{+} \rangle}{\|x - y_0\|^{d+1}}\, d\sigma_d(x) = \omega_d.$$

If $y_0 \in E \setminus \overline{V}$ then

$$\int_M \frac{\langle x - y_0, n_x^+ \rangle}{\|x - y_0\|^{d+1}} \, d\sigma_d = 0.$$

Proof If $y_0 \in E \setminus \overline{V}$ then $(y - y_0)/\|y - y_0\|^{d+1}$ is solenoidal on V, and the result follows from the divergence theorem.

If $y_0 \in V$, there exists $r_0 > 0$ such that if $0 < r < r_0$ then $B_r(y_0) \subseteq V$. For such r, let $V_r = V \setminus B_r(y_0)$, so that $\partial V_r = M \cup S_r(y_0)$. Since $(y - y_0)/\|y - y_0\|^{d+1}$ is solenoidal on V_r, it follows that

$$\int_M \frac{\langle x - y_0, n_x^+ \rangle}{\|x - y_0\|^{d+1}} \, d\sigma_d(x) = -\int_{S_r(y_0)} \frac{\langle x - y_0, n_x^+ \rangle}{\|x - y_0\|^{d+1}} \, d\sigma_d(x).$$

Bearing in mind that in this equation n_x^+ is pointing towards y_0 on $S_r(y_0)$, so that $\langle x - y_0, n_x^+ \rangle = -\|x - y_0\|$, it follows that

$$-\int_{S_r(y_0)} \frac{\langle x - y_0, n_x^+ \rangle}{\|x - y_0\|^{d+1}} \, d\sigma_d(x) = \frac{1}{r^d} \sigma_d(S_r(y_0)) = \omega_d$$

$\square$

Corollary 19.7.8 *If $y_0 \in V$ then*

$$\frac{d+1}{r} \left(\frac{1}{\sigma_d(S_r(y_0))} \int_{S_r(y_0)} \langle F(x), n_x^+ \rangle \, d\sigma_d(x) \right) \to \nabla . F(y)$$

as $r \searrow 0$.

Proof There exists r_0 such that $B_r(y_0) \subseteq V$ for $0 < r < r_0$. Applying the divergence theorem to the open ball $N_r(y_0)$, and using Corollary 19.7.6,

$$\frac{1}{v_{d+1}(B_r(y_0))} \int_{B_r(y_0)} \nabla . F(y) dv_{d+1}(y)$$

$$= \frac{(d+1)}{r \sigma_d(S_r(y_0))} \left(\int_{S_r(y_0)} \langle F(x), n_x^+ \rangle \, d\sigma_d(x) \right).$$

But

$$\frac{1}{v_{d+1}(B_r(y_0))} \int_{B_r(y_0)} \nabla . F(y) dv_{d+1}(y) \to \nabla . F(y_0) \text{ as } r \searrow 0,$$

and so the result follows.

$\square$

Note that

$$\frac{1}{\sigma_d(S_r(y_0))} \int_{S_r(y_0)} \langle F(x), n_x^+ \rangle \, d\sigma_d(x) = \frac{1}{r^d \omega_d} \int_{S_r(y_0)} \langle F(x), n_x^+ \rangle \, d\sigma_d(x)$$

is the average value of $\langle F(x), n_x^+ \rangle$ on $S_r(y_0)$.

Corollary 19.7.9 (Green's formulae) *Suppose that f and g are $C^{(2)}$ functions on U. Then*

$$\int_M f(x) \frac{\partial g}{\partial n_x^+}(x) \, d\sigma_d(x) = \int_B \Big(\langle \nabla f(y), \nabla g(y) \rangle + f(y) \nabla^2 g(y) \Big) \, dv_{d+1}(y),$$

and

$$\int_M \left(f(x) \frac{\partial g}{\partial n_x^+}(x) - g(x) \frac{\partial f}{\partial n_x^+}(x) \right) \, d\sigma_d(x)$$
$$= \int_B \Big(f(y) \nabla^2 g(y) - g(y) \nabla^2 f(y) \Big) \, dv_{d+1}(y).$$

Proof Apply the theorem to $H = f\nabla g$ and $K = f\nabla g - g\nabla f$:

$$\nabla.H = \langle \nabla f, \nabla g \rangle + f\nabla^2 g \text{ and } \nabla.K = f\nabla^2 g - g\nabla^2 f.$$

$\square$

Corollary 19.7.10 *If U is an open subset of $\mathbf{R}^3$ with boundary M consisting of a finite disjoint union of 2-manifolds, and if f is a $C^{(2)}$-function defined on an open set U containing B then*

$$\int_M \langle (\nabla \times f)(x), n_x^+ \rangle \, d\sigma_2(x) = 0.$$

Proof Apply the theorem to the solenoidal vector field $\nabla \times f$. $\square$

Exercises

19.7.1 Establish the formulae in Example 19.7.1 (ii).

19.7.2 Suppose that G is a vector field on an open set U and that f is a continuously differentiable function on U. Show that

$$\nabla(fG) = \langle \nabla f, G \rangle + f(\nabla.G).$$

19.7.3 Suppose that f is a $C^{(2)}$ function on an open subset U of $\mathbf{R}^3$. Show that $\nabla \times (\nabla f) = 0$.

19.7.4 Suppose that F is a vector field on an open subset of $\mathbf{R}^3$ for which there exists a vector field A (a *vector potential*) such that $F = \nabla \times A$. Show that F is solenoidal.

19.8 Harmonic functions

In the previous sections, we considered functions defined on an open subspace U of a $(d+1)$-dimensional Euclidean space, and hypersurfaces of dimension d. In this section **we change the dimension by** 1; we consider functions defined on an open subset U of a d-dimensional space.

Recall that a real-valued function f defined on an open subset U of a Euclidean space E is *harmonic* if it is twice continuously differentiable, and $\nabla^2 f = 0$.

Harmonic functions have good averaging properties. Let us introduce some notation. If $x \in S_r(x_0)$ we denote by n_x the unit normal vector in the direction away from x_0. If $0 < s < r$ we denote by $A_{s,r}(x_0)$ the annular set $\{x \in E : s \le \|x - x_0\| \le r\}$: it has boundary $S_s(x_0) \cup S_r(x_0)$.

Proposition 19.8.1 *If f is a harmonic function on an open subset U of a d-dimensional Euclidean space E and $B_r(x_0) \subseteq U$ then*

$$\int_{S_r(x_0)} \frac{\partial f}{\partial n_x}(x)\, d\sigma(x) = 0.$$

Proof   Apply the divergence theorem to ∇f.   □

Theorem 19.8.2 *Suppose that f is harmonic on an open subset U of a d-dimensional Euclidean space E and that $B_r(x_0) \subseteq U$. Then*

$$f(x_0) = \frac{1}{\sigma_{d-1}(S_r(x_0))} \int_{S_r(x_0)} f(x)\, d\sigma_{d-1}(x)$$

$$= \frac{1}{r^{d-1}\omega_{d-1}} \int_{S_r(x_0)} f(x)\, d\sigma_{d-1}(x).$$

Proof   We deal with the case where $d > 2$; the proof for $d = 2$ is essentially the same. Let $0 < s < r$. Applying Green's formula to f and $\psi_d(x - x_0) = \|x - x_0\|^{2-d}$,

$$0 = \int_{A_{s,r}} \psi_d(y - x_0)\nabla^2 f(y) - f(y)\nabla^2 \psi_d(y)\, dv_d(y)$$

$$= I_r - I_s,$$

where

$$I_r = \int_{S_r(x_0)} \left(\psi_d(x - x_0) \frac{\partial f}{\partial n_x}(x) - f(x) \frac{\partial \psi_d}{\partial n_x}(x - x_0) \right) d\sigma_{d-1}(x)$$

$$= \frac{1}{r^{d-1}} \int_{S_r(x_0)} \left(r \frac{\partial f}{\partial n_x}(x) - (2 - d)f(x) \right) d\sigma_{d-1}(x)$$

$$= \frac{d-2}{r^{d-1}} \int_{S_r(x_0)} f(x) \, d\sigma_{d-1}(x),$$

using Proposition 19.8.1 and the equation

$$\frac{\partial \psi_d}{\partial n_x} = \langle \nabla \psi_d, n_x \rangle = \frac{(2-d)r}{r^d} = \frac{2-d}{r^{d-1}}.$$

Similarly.

$$I_s = \int_{S_s(x_0)} \left(\psi_d(x - x_0) \frac{\partial f}{\partial n_x}(x) - f(x) \frac{\partial \psi_d}{\partial n_x}(x - x_0) \right) d\sigma_{d-1}(x)$$

$$= \frac{d-2}{s^{d-1}} \int_{S_s(x_0)} f(x) \, d\sigma_{d-1}(x).$$

Since $I_s \to (d-2)f(x_0)\sigma_{d-1}(S^{d-1})$ as $s \searrow 0$, the result follows. $\qquad\square$

This theorem has the following consequence.

Theorem 19.8.3 *Suppose that f is a non-constant harmonic function on a connected open subset U of a d-dimensional Euclidean space E. Then f has no local maximum or minimum.*

Proof Suppose that f has a local maximum at y_0 and that $f(y_0) = a$. Let $F = \{y \in U : f(y) = a\}$. Since f is continuous, F is a closed subset of U. We show that F is open. Suppose that $y \in F$. There exists $r > 0$ such that $B_r(y) \subseteq U$. Suppose that $z \in B_r(y)$ and that $\|z - y\| = s < r$. Then $f(y)$ is the average value of f on $S_s(y)$. Since f is continuous and $f(y) \geq f(w)$ for $w \in S_s(y)$, it follows that $f(w) = f(y)$ for all $w \in S_s(y)$. In particular, $f(z) = f(y) = a$. Thus $B_r(y) \subseteq F$, and so F is open. Since U is connected, $F = U$, and f is constant. $\qquad\square$

We now consider the simplest form of the Dirichlet problem: if f is a continuous function on the unit sphere S^{d-1} in $\mathbf{R}^d$, is there a continuous function u on the closed unit ball B_d which is harmonic on the open unit ball U_d and is equal to f on S^{d-1}? For this, we need the *Poisson kernel*. This is defined as

$$P_x(y) = \frac{1 - \|y\|^2}{\omega_{d-1} \|y - x\|^d} \text{ for } x \in S^{d-1}, y \in U_d.$$

Note that $P_x(0) = 1/\omega_{d-1}$.

Proposition 19.8.4 $P_x(y)$ *is a harmonic function of y in U_d.*

Proof We prove this by a direct calculation. Let $v(y) = 1 - \|y\|^2$ and let $w(y) = 1/\|x - y\|^d$. Then

$$\frac{\partial v}{\partial y_i}(y) = -2y_i,$$

$$\nabla^2 v(y) = -2d,$$

$$\frac{\partial w}{\partial y_i}(y) = \frac{-d(y_i - x_i)}{\|y - x\|^{d+2}}$$

$$\frac{\partial^2 w}{\partial y_i^2}(y) = \frac{-d}{\|y - x\|^{d+2}} + \frac{d(d+2)(y_i - x_i)^2}{\|y - x\|^{d+4}}$$

$$\text{and } \nabla^2 w(y) = \frac{-d^2}{\|y - x\|^{d+2}} + \frac{d(d+2)}{\|y - x\|^{d+2}} = \frac{2d}{\|y - x\|^{d+2}}.$$

Hence

$$\omega_{d-1} \nabla^2 P_x(y) = w(y)\nabla^2 v(y) + 2\langle \nabla v(y), \nabla w(y) \rangle + v(y)\nabla^2 w(y)$$

$$= \frac{-2d}{\|y - x\|^d} + \frac{4d \langle y, y - x \rangle}{\|y - x\|^{d+2}} + \frac{2d(1 - \|y\|^2)}{\|y - x\|^{d+2}}$$

$$= \frac{2d}{\|y - x\|^{d+2}} \left(-\|y - x\|^2 + 2\langle y, y - x \rangle + (1 - \|y\|^2) \right) = 0.$$

$\square$

Theorem 19.8.5 (i) $P_x(y) > 0$ for $x \in S^{d-1}$, $y \in U_d$.
   (ii) *If $x \in S_{d-1}$ and $\delta > 0$ then $P_x(rz) \to 0$ uniformly on $\{z \in S^{d-1} : \|z - x\| > \delta\}$ as $r \nearrow 1$.*
   (iii) $\int_{S^{d-1}} P_x(y) \, d\sigma_{d-1}(x) = 1$ *for $y \in U_d$.*

Proof   (i) and (ii) follow from inspection of the definition of the Poisson kernel. Suppose that $y \in U$, and that $y = rz$, with $z \in S^{d-1}$. Since $P_x(y)$ is a harmonic function of y,

$$1 = \omega_{d-1} P_z(0) = \int_{S^{d-1}} P_z(rx) \, d\sigma_{d-1}(x).$$

But $\|rx - z\| = \|x - rz\|$, so that $P_z(rx) = P_x(rz) = P_x(y)$, and so

$$1 = \int_{S^{d-1}} P_x(y) \, d\sigma_{d-1}(x).$$

$\square$

Theorem 19.8.6 (Solution of the Dirichlet problem)   *Suppose that f is a continuous function on the unit sphere S^{d-1} in $\mathbf{R}^d$. There exists a continuous function u on the closed unit ball B_d which is harmonic on the open unit ball U_d and is equal to f on S^{d-1}.*

Proof   Let

$$u(y) = \begin{cases} \int_{S^{d-1}} f(x) P_x(y) \, d\sigma_{d-1}(x) & \text{for } y \in U, \\ f(x) & \text{for } x \in S^{d-1}. \end{cases}$$

Differentiating twice under the integral sign, we see that

$$\nabla^2 u(y) = \int_{S^{d-1}} f(x) \nabla^2 P_x(y) \, d\sigma_{d-1}(x) = 0, \text{ for } y \in U,$$

so that u is harmonic. It remains to show that u is continuous on B. It is certainly continuous on U. Let $u_r(x) = u(rx)$ for $x \in S^{d-1}$ and $0 < r < 1$. Since each u_r is continuous, it is sufficient, by the general principal of uniform convergence, to show that $u_r \to f$ uniformly on S^{d-1}. Suppose that $\epsilon > 0$. Since S^{d-1} is compact, f is uniformly continuous on S^{d-1}, and so there exists $\delta > 0$ such that if $\|z - x\| < \delta$ then $|f(z) - f(x)| < \epsilon/3$. By Theorem 19.8.5 there exists $0 < r_0 < 1$ such that $|P_x(rz)| < \epsilon/(3\|f\|_\infty \omega_{d-1})$ for $r_0 < r < 1$ and $\|z - x\| \geq \delta$. If $x \in S^{d-1}$ and $r_0 < r < 1$ then, using the results of Theorem 19.8.5,

$$|f(x) - u_r(x)| = \left| \int_{S^{d-1}} (f(x) - f(z)) P_z(rx) \, d\sigma_{d-1}(z) \right|$$

$$\leq \int_{\|z-x\|<\delta} |f(x) - f(z)| P_z(rx) \, d\sigma_{d-1}(z)$$

$$+ \int_{\|z-x\|\geq\delta} (|f(x)| + |f(z)|) P_{rx}(z) \, d\sigma_{d-1}(z)$$

$$\leq \epsilon/3 + 2\epsilon/3 = \epsilon.$$

$\square$

Of course, we can prove a corresponding result for every ball $B_r(x)$. We now have a converse to Theorem 19.8.2.

Theorem 19.8.7 *Suppose that f is a continuous function on an open subset U of a Euclidean space E with the property that for every $x_0 \in U$ and $\epsilon > 0$ there exists $0 < r < \epsilon$ such that $B_r(x_0) \subseteq U$ and*

$$f(x_0) = \frac{1}{\sigma_{d-1}(S_r(x_0))} \int_{S_r(x_0)} f(x) \, d\sigma_{d-1}(x) \qquad (\star)$$

Then f is harmonic on U.

Proof Suppose that $x_0 \in U$, and let $r > 0$ be chosen so that $B_r(x_0) \subseteq U$ and $(\star)$ holds. Let u be the solution to the Dirichlet problem for $B_r(x_0)$, and let $v = f - u$. Then $v(x) = 0$ for $x \in S_r(x_0)$, and, since harmonicity is a local property, it is sufficient to show that $v(y) = 0$ for $y \in B_r(x_0)$. Suppose not, and suppose that $v(x) > 0$ for some $x \in B_r(x_0)$. Let $a = \sup\{v(y) : y \in B_r(x_0)\}$, and let $F = \{y \in B_r(x_0) : f(y) = a\}$. Since $B_r(x_0)$ is compact, F is a non-empty closed set in $B_r(x_0)$, and $F \cap S_r(x_0)$ is empty. Similarly, there exists $y_0 \in F$ such that $\|y_0 - x_0\| = s = \sup\{\|y - x_0\| : y \in F\}$, and $0 \leq s < r$. By hypothesis, there exists $0 < t < r - s$ such that $(\star)$ holds. But then

$$a = v(y_0) = \frac{1}{\sigma_{d-1}(S_t(y_0))} \int_{S_t(x_0)} v(x) \, d\sigma_{d-1}(x),$$

which is not possible, since $v(x) \leq a$ for $x \in S_t(y_0)$, and there are points x in $S_t(y_0)$ for which $v(x) < a$. $\qquad \square$

Note that the condition on f does not involve derivatives, but ensures that f is twice continuously differentiable.

Corollary 19.8.8 *A harmonic function f on U is infinitely differentiable.*

Proof Take an orthonormal basis $(e_1, \ldots, e_d)$ for E. An inductive argument shows that it is sufficient to show that $\partial f/\partial x_j$ is harmonic, for $1 \leq j \leq d$. Suppose that $x_0 \in U$ and that $B_r(x_0) \subseteq U$. Then

$$f(x_0) = \frac{1}{\sigma_{d-1}(S_r(x_0))} \int_{S_r(x_0)} f(x) \, d\sigma_{d-1}(x),$$

by Theorem 19.8.2. Differentiating under the integral sign,

$$\frac{\partial f}{\partial x_j}(x_0) = \frac{1}{\sigma_{d-1}(S_r(x_0))} \int_{S_r(x_0)} \frac{\partial f}{\partial x_j}(x) \, d\sigma_{d-1}(x),$$

and so $\partial f/\partial x_j$ is harmonic. $\qquad \square$

Corollary 19.8.9 *Suppose that $(f_n)_{n=1}^{\infty}$ is a sequence of harmonic functions on U which converges locally uniformly to a function f. Then f is harmonic.*

Proof For if $B_r(x_0) \subseteq U$ then, using Theorem 19.8.2,

$$f(x_0) = \lim_{n \to \infty} f_n(x_0)$$

$$= \lim_{n \to \infty} \frac{1}{\sigma_{d-1}(S_r(x_0))} \int_{S_r(x_0)} f_n(x)\, d\sigma_{d-1}(x)$$

$$= \frac{1}{\sigma_{d-1}(S_r(x_0))} \int_{S_r(x_0)} (\lim_{n \to \infty} f_n(x))\, d\sigma_{d-1}(x)$$

$$= \frac{1}{\sigma_{d-1}(S_r(x_0))} \int_{S_r(x_0)} f(x)\, d\sigma_{d-1}(x),$$

and so the result follows from Theorem 19.8.7. $\qquad\square$

We shall consider harmonic functions further in Volume III.

Exercises

19.8.1 What are the harmonic functions on an open interval of the real line?

19.8.2 Suppose that f and f^2 are harmonic on an open connected subset of $\mathbf{R}^d$. Show that f is constant.

19.8.3 Suppose that f is a harmonic function on an open subset U of a d-dimensional Euclidean space E. Show that if $B_r(x) \subseteq U$ then

$$f(x) = \frac{1}{v_d(B_r(x))} \int_{B_r(x)} f(y)\, dV_d(y). \qquad (*)$$

Show conversely that if f is continuous on U, and that if $(*)$ holds for all $B_r(x)$ contained in U, then f is harmonic.

19.9 Curl

We now study the operator 'curl' in more detail. This requires knowledge of the material in Appendix C. Throughout this section, we suppose that $F = (f_1, f_2, f_3)$ is a vector field defined on an open subset V of $\mathbf{R}^3$.

Proposition 19.9.1 *Suppose that v is a unit vector in $\mathbf{R}^3$. Then*

$$\langle \nabla \times F, v \rangle = \nabla.(F \times v).$$

Proof   If $\tau(1) = 2$, $\tau(2) = 3$ and $\tau(3) = 1$, then $\epsilon_\tau = 1$. Hence

$$\langle \nabla \times F, v \rangle = \left\langle \sum_{\sigma \in \Sigma_3} \epsilon_\sigma \frac{\partial f_{\sigma(3)}}{\partial x_{\sigma(2)}} e_{\sigma(1)}, v \right\rangle = \sum_{\sigma \in \Sigma_3} \epsilon_\sigma \frac{\partial f_{\sigma(3)}}{\partial x_{\sigma(2)}} v_{\sigma(1)}$$

$$= \sum_{\sigma \in \Sigma_3} \epsilon_\sigma \frac{\partial f_{\sigma(2)}}{\partial x_{\sigma(1)}} v_{\sigma(3)} = \nabla.(F \times v).$$

$\square$

This clearly corresponds to properties of the scalar triple product.
We now apply the divergence theorem.

Theorem 19.9.2   *Suppose that U is an open subset of $\mathbf{R}^3$ with $B = \overline{U} \subseteq V$, and with boundary M consisting of a finite disjoint union of 2-manifolds. Then*

$$\int_B (\nabla \times F)(y)\, dv_3(y) = - \int_M (F(x) \times n_x^+)\, d\sigma_2(x).$$

Proof   Let v be a unit vector in $\mathbf{R}^3$. Using the preceding proposition and the divergence theorem,

$$\left\langle \int_B (\nabla \times F)(y)\, dv_3(y), v \right\rangle = \int_B \langle (\nabla \times F)(y), v \rangle\, dv_3(y)$$

$$= \int_B \nabla.(F(y) \times v)\, dv_3(y)$$

$$= \int_M \langle F(x) \times v, n_x^+ \rangle\, d\sigma_2(x)$$

$$= - \int_M \langle F(x) \times n_x^+, v \rangle\, d\sigma_2(x)$$

$$= - \left\langle \int_M (F(x) \times n_x^+)\, d\sigma_2(x), v \right\rangle.$$

This holds for all unit vectors v, and so the result follows. $\square$

Suppose that $x_0 \in V$. We apply the theorem to balls $B_r(x_0) \subseteq V$. Suppose that $B_r(x_0) \subseteq V$ and that G is a continuous vector field on V. We set

$$A_r(G)(x_0) = \frac{1}{v_3(B_r(x_0))} \int_{B_r(x_0)} G(y)\, dv_3(y) = \frac{3}{4\pi r^3} \int_{B_r(x_0)} G(y)\, dv_3(y),$$

$$a_r(G)(x_0) = \frac{1}{\sigma_2(S_r(x_0))} \int_{S_r(x_0)} G(x)\, d\sigma_2(x) = \frac{1}{4\pi r^2} \int_{S_r(x_0)} G(x)\, d\sigma_2(x).$$

$A_r(G)$ is the average value of G on the ball $B_r(x_0)$ and $a_r(G)$ is the average value of G on the sphere $S_r(x_0)$.

Corollary 19.9.3 $-3a_r(F \times n_x^+)(x_0)/r \to (\nabla \times F)(x_0)$ *as* $r \to 0$.

Proof   For $A_r(\nabla \times F)(x_0) = -3a_r(F \times n_x^+)(x_0)/r$ and $A_r(\nabla \times F)(x_0) \to (\nabla \times F)(x_0)$ as $r \to 0$. $\square$

This expresses $\nabla \times F$ as a limit of volume integrals.

It is more informative to express the components of $\nabla \times F$ as a limit of planar line integrals. In order to simplify notation, let us suppose that $x_0 = 0$, and let us consider the component of $\nabla \times F$ in the e_3 direction. By Proposition 19.9.1,

$$\langle \nabla \times F, e_3 \rangle = \nabla.(F \times e_3) = \nabla.(f_2 e_1 - f_1 e_2) = \frac{\partial f_2}{\partial x_1} - \frac{\partial f_1}{\partial x_2}.$$

If $y = y_1 e_1 + y_2 e_2 \in U \cap e_3^{\perp}$, let $G(y) = f_2(y)e_1 - f_1(y)e_2$. The two-dimensional divergence of G in the plane satisfies $\nabla.G = \langle \nabla \times F, e_3 \rangle$. Suppose that $N_{r_0}(0) \subseteq U$. If $0 < r < r_0$, let $\gamma_r : [0, 2\pi] \to E$ be the circular path $\gamma_r(s) = r(\cos s \, e_1 + \sin s \, e_2)$ in $U \cap e_3^{\perp}$, and let B_r be the disc $\{w : w \in e_3^{\perp}, \|w\| \le r\}$. If $\gamma_r(s) \in [\gamma_r]$, then $n_{\gamma_r(s)}^+ = \cos(s) \, e_1 + \sin(s) \, e_2$. Applying the two-dimensional divergence theorem, it follows that

$$\int_{B_r} \langle (\nabla \times F)(y), e_3 \rangle \, dv_2(y) = \int_{B_r} \nabla.G(y) \, dv_2(y)$$

$$= r \int_0^{2\pi} (f_2(\gamma_r(s)) - f_1(\gamma_r(s))) \, ds.$$

Now $\gamma_r'(s) = r(-\sin s \, e_1 + \cos s \, e_2)$, and the unit tangent $t(\gamma_r(s))$ at $\gamma_r(s)$ is $-\sin s \, e_1 + \cos s \, e_2$, so that

$$\int_{B_r} \langle (\nabla \times F)(y), e_3 \rangle \, dv_2(y) = \int_0^{2\pi} \langle F(\gamma_r(s)), \gamma_r'(s) \rangle \, ds$$

$$= r \int_0^{2\pi} \langle F(\gamma_r(s)), t(\gamma_r(s)) \rangle \, ds.$$

Theorem 19.9.4

$$\langle (\nabla \times F)(0), e_3 \rangle = \lim_{r \to 0} \frac{1}{\pi r^2} \int_0^{2\pi} \langle F(\gamma_r(s)), \gamma_r'(s) \rangle \, ds$$

$$= \lim_{r \to 0} \frac{1}{\pi r} \int_0^{2\pi} \langle F(\gamma_r(s)), t(\gamma(s)) \rangle \, ds.$$

Proof   For

$$A_r = \frac{1}{\pi r^2} \int_{B_r} \langle (\nabla \times F)(y), e_3 \rangle \, dv_2(y) \to \langle (\nabla \times F)(0), e_3 \rangle$$

as $r \to 0$.   □

Exercise

19.9.1 Suppose that F and G are vector fields on an open set U in $\mathbf{R}^3$ and
that f is a continuously differentiable function on U.
(i) Show that $\nabla \times (fG) = \nabla f \times G + f(\nabla.G)$.
(ii) Let

$$G.\nabla = g_1 \frac{\partial}{\partial x_1} + g_2 \frac{\partial}{\partial x_2} + g_3 \frac{\partial}{\partial x_3}.$$

Show that $\nabla \times (F \times G) = (\nabla.G)F - (\nabla.F)G + (G.\nabla)F - (F.\nabla)G$.

Appendix B

Linear algebra

B.1 Finite-dimensional vector spaces

We are concerned with real vector spaces, but the results extend readily to complex vector spaces, as well. We describe briefly the ideas and results that we need[1].

Let K denote either the field $\mathbf{R}$ of real numbers or the field $\mathbf{C}$ of complex numbers. A *vector space* E over K is an abelian additive group $(E, +)$, together with a mapping *(scalar multiplication)* $(\lambda, x) \to \lambda x$ of $K \times E$ into E which satisfies

- $1.x = x$,
- $(\lambda + \mu)x = \lambda x + \mu x$,
- $\lambda(\mu x) = (\lambda\mu)x$,
- $\lambda(x + y) = \lambda x + \lambda y$,

for $\lambda, \mu \in K$ and $x, y \in E$. The elements of E are called *vectors* and the elements of K are called *scalars*.

It then follows that $0.x = 0$ and $\lambda.0 = 0$ for $x \in E$ and $\lambda \in K$. (Note that the same symbol 0 is used for the additive identity element in E and the zero element in K.)

A non-empty subset F of a vector space E is a *linear subspace* if it is a subgroup of E and if $\lambda x \in F$ whenever $\lambda \in K$ and $x \in F$. A linear subspace is then a vector space, with the operations inherited from E. If A is a subset of E then the intersection of all the linear subspaces containing A is a linear subspace, the subspace span (A) *spanned* by A. If A is empty,

---

[1] For a fuller account, see, for example, Alan F. Beardon, *Algebra and Geometry*, Cambridge University Press, 2005.

then $E_A = \{0\}$; otherwise

$$E_A = \{\lambda_1 a_1 + \cdots + \lambda_n a_n : n \in \mathbf{N}, \lambda_i \in K, a_i \in A \text{ for } 1 \le i \le n\}.$$

A subset B of E is *linearly independent* if whenever $b_1, \ldots, b_k$ are distinct elements of B and $\lambda_1, \ldots, \lambda_k$ are scalars for which

$$\lambda_1 b_1 + \cdots + \lambda_k b_k = 0$$

then $\lambda_1 = \cdots = \lambda_k = 0$. A subset B of E which is linearly independent and which spans E is called a *basis* for E.

A vector space E is *finite-dimensional* if it is spanned by a finite set, Every finite-dimensional vector space E has a basis.

Proposition B.1.1 *If A is a linearly independent subset of E contained in a finite subset C of E which spans E then there is a basis B for E with $A \subseteq B \subseteq C$.*

Proof   Consider a maximal linearly independent subset of C which contains A, or a minimal spanning subset of C which contains A.   □

Corollary B.1.2   *Every finite-dimensional space E has a basis.*

Proof   Take $A = \emptyset$, C a finite spanning set.   □

When we list a basis as $(b_1, \ldots b_d)$, we shall always suppose that the elements are distinct.

Proposition B.1.3   *If $B = (b_1, \ldots, b_d)$ is a basis for E, then any element x of E can be written uniquely as $x = x_1 b_1 + \cdots + x_d b_d$ (where $x_1, \ldots, x_d$ are scalars).*

Proof   Since B spans E, x can be written as $x = x_1 b_1 + \cdots + x_d b_d$. If $x = x_1' b_1 + \cdots + x_d' b_d$ then $(x_1 - x_1')b_1 + \cdots + (x_d - x_d')b_d = 0$, so that $x_i - x_i' = 0$ for $1 \le i \le d$, by linear independence.   □

Proposition B.1.4   *If $B = (b_1, \ldots, b_k)$ and $C = (c_1, \ldots. c_l)$ are finite bases for E then $k = l$.*

Proof   For $1 \le i \le k$ we can write $b_i = \sum_{j=1}^{l} \gamma_{ji} c_j$, and for $1 \le j \le l$ we can write $c_j = \sum_{m=1}^{k} \beta_{mj} b_m$. Then

$$b_i = \sum_{j=1}^{l} \gamma_{ji} \left(\sum_{m=1}^{k} \beta_{mj} b_m \right) = \sum_{m=1}^{k} \left(\sum_{j=1}^{l} \beta_{mj} \gamma_{ji} \right) b_m.$$

Since $(b_1, \ldots, b_k)$ is a basis, the expression for b_i is unique, and so $1 = \sum_{j=1}^{l} \beta_{ij} \gamma_{ji}$. Consequently, $k = \sum_{i=1}^{k} (\sum_{j=1}^{l} \beta_{ij} \gamma_{ji})$. Similarly, $l = \sum_{j=1}^{l} (\sum_{i=1}^{k} \gamma_{ji} \beta_{ij})$, and so $k = l$. $\qquad\square$

Corollary B.1.5 *If $B = \{b_1, \ldots, b_d\}$ is a basis for E, and A is a linearly independent subset of E then A is a finite set, and $|A| \le |B|$.*

Proof   Suppose that F is a finite subset of A. By Proposition B.1.1, there exists a finite basis G of E with $F \subseteq G \subseteq F \cup B$. Then $|F| \le |G| = |B|$. Since this holds for all finite subsets of A, A is finite, and $|A| \le |B|$. $\qquad\square$

Thus any two bases have the same number of elements; this number is the *dimension* $\dim E$ of E. If $\dim E = d$, we say that E is *d-dimensional*.

Corollary B.1.6 *If C is a spanning subset of a k-dimensional vector space E and $|C| = k$ then C is a basis for E.*

Proof   For C contains a subset B which is a basis, and $|B| = k = |C|$, so that $C = B$. $\qquad\square$

As an example, let $E = K^d$, the product of d copies of K, with addition defined coordinatewise, and with scalar multiplication

$$\lambda(x_1, \ldots, x_d) = (\lambda x_1, \ldots, \lambda x_d).$$

Let $e_j = (0, \ldots, 0, 1, 0, \ldots, 0)$, with 1 in the jth position. Then K^d is a vector space, and $(e_1, \ldots, e_d)$ is a basis for K^d, the *standard basis*. As another example, let $M_{d,k} = M_{d,k}(K)$ denote the set of all K-valued functions on $\{1, \ldots, d\} \times \{1, \ldots, k\}$. $M_{d,k}$ becomes a vector space over K when addition and scalar multiplication are defined coordinatewise. The elements of $M_{d,k}$ are called *matrices*. We denote the matrix taking the value 1 at (i, j) and 0 elsewhere by E_{ij}. Then the set of matrices $\{E_{ij} : 1 \le i \le d, 1 \le j \le k\}$ forms a basis for $M_{d,k}$, so that $M_{d,k}$ has dimension dk. A matrix t in $M_{d,k}$ is denoted by an array

$$\begin{bmatrix} t_{11} & \cdots & t_{1k} \\ \vdots & \ddots & \vdots \\ t_{d1} & \cdots & t_{dk} \end{bmatrix}.$$

If $1 \le i \le d$ and $1 \le j \le k$, let

$$r_i = [t_{i1}, \ldots, t_{ik}] \text{ and let } c_j = \begin{bmatrix} t_{1j} \\ \vdots \\ t_{dj} \end{bmatrix}.$$

r_i is the *ith row* of t and c_j is the *jth column* of t.

We denote $M_{d,d}$ by M_d: M_d is the vector space of *square matrices*. We define the identity matrix I to be $I = \sum_{j=1}^d E_{jj}$.

B.2 Linear mappings and matrices

A mapping $T : E \to F$, where E and F are vector spaces over the same field K, is *linear* if

$$T(x + y) = T(x) + T(y) \quad \text{and} \quad T(\lambda x) = \lambda T(x) \quad \text{for all } \lambda \in K, x, y \in E.$$

The *image* $T(E)$ is a linear subspace of F and the *null-space* $N(T) = \{x \in E : T(x) = 0\}$ is a linear subspace of E. If E is finite-dimensional, then the dimension of $T(E)$ is the *rank* of T and the dimension of $N(T)$ is the *nullity* $n(T)$ of T.

Theorem B.2.1 (The rank-nullity formula) *If $T : E \to F$ is a linear mapping and if E is finite-dimensional then*

$$\operatorname{rank}(T) + n(T) = \dim E.$$

Proof   Let B be a basis for E. Then $T(B)$ spans $T(E)$, and so $T(E)$ is finite-dimensional. Let $(y_1, \ldots, y_r)$ be a basis for $T(E)$ and let $(x_1, \ldots, x_n)$ be a basis for $N(T)$. For each $1 \le j \le r$ there exists $z_j \in E$ such that $T(z_j) = y_j$. We show that $(z_1, \ldots, z_r, x_1, \ldots, x_n)$ is a basis for E, so that $\operatorname{rank}(T) + n(T) = \dim E$.

Suppose that $x \in E$ and that $T(x) = \lambda_1 y_1 + \cdots + \lambda_r y_r$. Let $v = \lambda_1 z_1 + \cdots + \lambda_r z_r$. Then $T(v) = T(x)$, so that

$$u = x - v \in N(T) = \operatorname{span}(x_1, \ldots, x_n).$$

Thus $x = u + v \in \operatorname{span}(z_1, \ldots, z_r, x_1, \ldots, x_n)$, and $(z_1, \ldots, z_r, x_1, \ldots, x_n)$ spans E. If

$$x = (\lambda_1 z_1 + \cdots + \lambda_r z_r) + (\mu_1 x_1 + \cdots + \mu_n x_n) = 0$$

then $T(x) = \lambda_1 y_1 + \cdots + \lambda_r y_r = 0$, so that $\lambda_i = 0$ for $1 \le i \le r$, and $x = \mu_1 x_1 + \cdots + \mu_n x_n = 0$. Hence $\mu_j = 0$ for $1 \le j \le n$. Thus $(z_1, \ldots, z_r, x_1, \ldots, x_n)$ is linearly independent. $\qquad\square$

A bijective linear mapping $J : E \to F$ is called an *isomorphism*. A linear mapping $J : E \to F$ is an isomorphism if and only if $J(E) = F$ and $N(J) = \{0\}$. If J is an isomorphism, then $\dim E = \dim F$. For example, if $(f_1, \ldots, f_d)$ is a basis for F then the linear mapping $J : K^d \to F$ defined by $J(\lambda_1, \ldots, \lambda_d) = \lambda_1 f_1 + \cdots + \lambda_d f_d$ is an isomorphism of K^d onto F.

The set $L(E, F)$ of linear mappings from E to F is a vector space, when we define

$$(S + T)(x) = S(x) + T(x) \quad \text{and} \quad (\lambda S)(x) = \lambda(S(x))$$

for $S, T \in L(E, F)$, $x \in E$, $\lambda \in K$. If $T \in L(E, F)$ and $S \in L(F, G)$ then the composition $ST = S \circ T$ is in $L(E, G)$. We write $L(E)$ for $L(E, E)$; elements of $L(E)$ are called *endomorphisms* of E.

An element T of $L(E, F)$ is *invertible* if there exists an element $T^{-1} \in L(F, E)$, the *inverse* of T, such that $T^{-1} \circ T = I_E$, the identity on E, and $T \circ T^{-1} = I_F$, the identity mapping on F. T is invertible if and only if it is a linear isomorphism of E onto F.

The set of invertible elements of $L(E)$ is a group under composition, with identity element I_E. It is called the *general linear group* $GL(E)$. When $E = K^d$, it is denoted by $GL_d(K)$. It follows from the rank-nullity formula that if $T \in L(E)$, then T is invertible if and only if it has a left inverse, and if and only if it has a right inverse.

Suppose that E and F are finite-dimensional vector spaces over K, and that $(e_1, \ldots, e_k)$ is a basis for E, $(f_1, \ldots, f_d)$ a basis for F and that $T \in L(E, F)$. Let $T(e_j) = \sum_{i=1}^d t_{ij} f_i$.

$$\text{If } x = \sum_{j=1}^k x_j e_j \text{ then } T(x) = \sum_{i=1}^d \left(\sum_{j=1}^k t_{ij} x_j \right) f_i. \qquad (*)$$

Proposition B.2.2 *The mapping $T \to (t_{ij})$ is then an isomorphism of $L(E, F)$ onto $M_{d,k}$, so that $\dim L(E, F) = dk = \dim E . \dim F$.*

Proof The mapping $T \to (t_{ij})$ is clearly linear and injective. On the other hand, if $(t_{ij}) \in M_{d,k}$ then the formula $(*)$ defines an element $T \in L(E, F)$ whose image is (t_{ij}), and so the mapping is also surjective. $\qquad \square$

We say that T is *represented by* the matrix (t_{ij}). If $(g_1, \ldots, g_l)$ is a basis for G, and $S \in L(F, G)$ is represented by the matrix (s_{hi}) then the product $R = ST \in L(E, G)$ is represented by the matrix (r_{hj}), where $r_{hj} = \sum_{i=1}^d s_{hi} t_{ij}$. This expression defines matrix multiplication.

A matrix t in M_d is *invertible* if the element $T \in L(K^d)$ which it defines is invertible. This is so if and only if there is a matrix t^{-1} in M_d such that $tt^{-1} = t^{-1}t = I$. The matrix t^{-1} is then unique: it is the *inverse* of t.

As an example, suppose that $(e_1, \ldots, e_d)$ and $(f_1, \ldots, f_d)$ are bases of E. Then the identity mapping $I : (E, (f_1, \ldots, f_d)) \to (E, (e_1, \ldots, e_d))$ is represented by a matrix b. I is invertible, and so therefore is b. Then b^{-1} represents the mapping $I : (E, (e_1, \ldots, e_d)) \to (E, (f_1, \ldots, f_d))$. Suppose now that $T \in L(E)$ and that T is represented by the matrix t with respect to the basis $(e_1, \ldots, e_d)$. Then, considering the composite mapping

$$(E, (f_1, \ldots, f_d)) \xrightarrow{I} (E, (e_1, \ldots, e_d)) \xrightarrow{T} (E, (e_1, \ldots, e_d)) \xrightarrow{I} (E, (f_1, \ldots, f_d)),$$

we see that T is represented by the matrix $b^{-1}tb$ with respect to the basis $(f_1, \ldots, f_d)$.

If $T \in L(K^d)$ and T is represented by the matrix $t = (t_{ij})$, then t can be written as a finite product of matrices of a particularly simple form.

A matrix of the form $I + \lambda E_{ij}$, where λ is a scalar and $i \neq j$, is called an *elementary shear matrix*. Such a matrix is invertible, with inverse $I - \lambda E_{ij}$. The corresponding element of $L(K^d)$ is called an *elementary shear operator*. The matrix product $(I + \lambda E_{ij})t$ is the matrix obtained by adding λ times the jth row of t to the ith row, and leaving the other rows unchanged. This multiplication is call a *row operation*. Similarly, the matrix product $t(I + \lambda E_{ij})$ is the matrix obtained by adding λ times the ith column of t to the jth column, and leaving the other rows unchanged. This multiplication is call a *column operation*.

A matrix of the form $\lambda_1 E_{11} + \cdots + \lambda_d E_{dd}$, where $\lambda_1, \ldots \lambda_d$ are scalars, is called a *diagonal matrix* and is denoted by $\mathrm{diag}(\lambda_1, \cdots \lambda_d)$. If it is invertible, the corresponding element of $L(K^d)$ is called a *scaling operator*. The matrix product $\mathrm{diag}(\lambda_1, \cdots \lambda_d)t$ is obtained by multiplying the ith row of t by λ_i, for $1 \leq i \leq d$, and the matrix product $t\,\mathrm{diag}(\lambda_1, \cdots \lambda_d)$ is obtained by multiplying the jth column of t by λ_j, for $1 \leq j \leq d$. The matrix $\mathrm{diag}(\lambda_1, \cdots \lambda_d)$ is invertible if and only if each λ_j is non-zero, and the inverse is then $\mathrm{diag}(\lambda_1^{-1}, \cdots \lambda_d^{-1})$.

Theorem B.2.3   *If $t \in M_d$ then $t = p\lambda q$, where p and q are finite products of elementary shear matrices, and λ is a diagonal matrix.*

Proof   We show that there exist finite products $\widetilde{p}$ and $\widetilde{q}$ for which $\widetilde{p}t\widetilde{q} = d$, a diagonal matrix. Then $\widetilde{p}$ and $\widetilde{q}$ are invertible, their inverses p and q are finite products of elementary shear operations, and $t = pdq$.

If $t = 0$ there is nothing to prove. Otherwise, by using a row operation and a column operation if necessary, we obtain a matrix t' for which $t'_{11} \neq 0$. By using row operations and column operations, we obtain a matrix t'' for which $t''_{1j} = 0$ and $t''_{i1} = 0$ for $2 \leq i, j \leq d$. Now repeat the procedure, to obtain a matrix t''' for which $t'''_{ij} = 0$ for $i = 1, 2$, for $j = 1, 2$ and $i \neq j$, and then iterate.   □

B.3 Determinants

If $\sigma \in \Sigma_d$, the group of permutations of the set $\{1, \ldots, d\}$, we define the *signature* ϵ_σ to be

$$\epsilon_\sigma = \prod_{1 \leq i < j \leq d} \left(\frac{\sigma(j) - \sigma(i)}{j - i} \right).$$

Then $\epsilon_\sigma = \pm 1$. If $\sigma, \tau \in \Sigma_d$ then

$$\epsilon_{\sigma\tau} = \prod_{1 \leq i < j \leq d} \left(\frac{\sigma(\tau(j)) - \sigma(\tau(i))}{j - i} \right)$$

$$= \prod_{1 \leq i < j \leq d} \left(\frac{\sigma(\tau(j)) - \sigma(\tau(i))}{\tau(j) - \tau(i)} \right) \prod_{1 \leq i < j \leq d} \left(\frac{\tau(j) - \tau(i)}{j - i} \right) = \epsilon_\sigma . \epsilon_\tau.$$

Thus the mapping $\sigma \to \epsilon_\sigma$ is a homomorphism of the group Σ_d into the multiplicative group $D_2 = \{1, -1\}$. The kernel

$$A_d = \{\sigma \in \Sigma_d : \epsilon_\sigma = 1\}$$

is the *alternating group*. If σ is a transposition, then $\epsilon_\sigma = -1$. Any permutation can be written (in many ways) as a product of transpositions; it follows that the number of transpositions is always even if $\epsilon_\sigma = 1$ and is always odd if $\epsilon_\sigma = -1$.

Suppose now that $t \in M_{d,d}$. The *determinant* $\det t$ is defined as

$$\det t = \sum_{\sigma \in \Sigma_d} \epsilon_\sigma t_{1,\sigma(1)} \cdots t_{d,\sigma(d)}.$$

Note that, since $\epsilon_\sigma = \epsilon_{\sigma^{-1}}$,

$$\det t = \sum_{\sigma \in \Sigma_d} \epsilon_\sigma t_{\sigma^{-1}(1),1} \cdots t_{\sigma^{-1}(d),d} = \sum_{\sigma \in \Sigma_d} \epsilon_\sigma t_{\sigma(1),1} \cdots t_{\sigma(d),d}.$$

Here are some basic properties of the determinant function.

Theorem B.3.1 *Suppose that* $t, u \in M_{d,d}$, *that* $\lambda = \mathrm{diag}(\lambda_1, \ldots, \lambda_d)$ *is a diagonal matrix and that* $s = I + \mu E_{ij}$ *is an elementary shear matrix.*

(i) $\det \lambda = \lambda_1 \ldots \lambda_d$ *and* $\det t\lambda = \det t . \det \lambda$.

(ii) *If* t *has two equal columns, then* $\det t = 0$.

(iii) $\det s = 1$ *and* $\det ts = \det t$.

(iv) $\det tu = \det t . \det u$.

Proof   (i) It follows from the definition that $\det \lambda = \lambda_1 \ldots \lambda_d$.

$$\det t\lambda = \sum_{\sigma \in \Sigma_d} \epsilon_\sigma t_{1,\sigma(1)} \lambda_{\sigma(1)} \cdots t_{d,\sigma(d)} \lambda_{\sigma(d)}$$

$$= \lambda_1 \ldots \lambda_d \sum_{\sigma \in \Sigma_d} \epsilon_\sigma t_{1,\sigma(1)} \cdots t_{d,\sigma(d)} = \det t . \det \lambda.$$

(ii) Suppose that the kth and the lth columns are equal. Let τ be the transposition (k,l) ; then $t_{ij} = t_{i\tau(j)}$, for $1 \leq i, j \leq d$. Then

$$\det t = \sum_{\sigma \in A_d} \epsilon_\sigma t_{1,\sigma(1)} \cdots t_{d,\sigma(d)} + \sum_{\sigma \in A_d} \epsilon_{\tau\sigma} t_{1,\tau(\sigma(1))} \cdots t_{d,\tau(\sigma(d))}$$

$$= \sum_{\sigma \in A_d} \epsilon_\sigma t_{1,\sigma(1)} \cdots t_{d,\sigma(d)} + \epsilon_\tau \sum_{\sigma \in A_d} \epsilon_\sigma t_{1,\sigma(1)} \cdots t_{d,\sigma(d)} = 0.$$

(iii) It follows from the definition that $\det s = 1$.

$$\det ts = \sum_{\sigma \in \Sigma_d} \epsilon_\sigma t_{\sigma(1),1} \cdots (t_{\sigma(j),j} + \mu t_{\sigma(j),i}) \cdots t_{d,\sigma(d)}$$

$$= \det t + \mu \sum_{\sigma \in \Sigma_d} \epsilon_\sigma t_{\sigma(1),1} \cdots t_{\sigma(i),i} \cdots t_{\sigma(j),i} \cdots t_{\sigma(d),d} = \det t,$$

since the second sum is the determinant of a matrix with two equal columns.

(iv) By Theorem B.2.3, we can write $u = p\lambda q$ where p and q are products of elementary shear matrices, and λ is a diagonal matrix. Then

$$\det u = \det p\lambda q = \det p\lambda = \det p . \det \lambda = \det \lambda$$

and

$$\det tu = \det tp\lambda q = \det tp\lambda = \det tp . \det \lambda = \det t . \det u.$$

$\square$

The determinant determines whether or not a matrix is invertible.

Corollary B.3.2   *A matrix u in $M_{d,d}$ is invertible if and only if its determinant is non-zero.*

Proof   By Theorem B.2.3, $u = p\lambda q$, where p and q are products of elementary shear matrices and $\lambda = \mathrm{diag}(\lambda_1, \ldots, \lambda_d)$ is a diagonal matrix. Thus $\det u = \det p . \det \lambda \det q = \det \lambda = \lambda_1 \ldots \lambda_d$. Since elementary shear matrices are invertible, u is invertible if and only if d is. But d is invertible if and only if $\lambda_1 \ldots \lambda_d \neq 0$.

$\square$

If u is invertible then $u.u^{-1} = I$, so that $1 = \det I = \det u.u^{-1} = \det u.\det u^{-1}$, and $\det u^{-1} = (\det u)^{-1}$.

Suppose now that T is an endomorphism of E which is represented by a matrix t with respect to a basis $(e_1, \ldots, e_d)$ and by a matrix s with respect to a basis $(f_1, \ldots, f_d)$. Then there is an invertible matrix b such that $s = b^{-1}tb$, and so $\det s = \det b^{-1}.\det t.\det b = \det t$. This means that we can define the determinant $\det T$ of the endomorphism T to be $\det t$: the definition does not depend on the choice of basis.

B.4 Cramer's rule

Suppose that $u \in M_{d,d}$ is invertible. How can we calculate its inverse?

Suppose that $1 \le i, j \le d$. Define the matrix $u^{(i,j)} \in M_{d,d}$ by setting

$$u_{kl}^{(i,j)} = \begin{cases} 1 & \text{if } k = i \text{ and } l = j, \\ 0 & \text{if } k \ne i \text{ and } l = j, \\ u_{kl} & \text{otherwise.} \end{cases}$$

The matrix $u^{(i,j)}$ is obtained by changing the terms in the jth column of u. Thus

$$u^{(i,j)} = \begin{bmatrix} u_{11} & \cdots & u_{1,j-1} & 0 & u_{1,j+1} & \cdots & u_{1d} \\ \vdots & \ddots & \vdots & \vdots & \vdots & \ddots & \vdots \\ u_{i-1,1} & \cdots & u_{i-1,j-1} & 0 & u_{i-1,j+1} & \cdots & u_{i-1,d} \\ u_{i,1} & \cdots & u_{i,j-1} & 1 & u_{i,j+1} & \cdots & u_{i,d} \\ u_{i+1,1} & \cdots & u_{i+1,j-1} & 0 & u_{i+1,j+1} & \cdots & u_{i+1,d} \\ \vdots & \ddots & \vdots & \vdots & \vdots & \ddots & \vdots \\ u_{d1} & \cdots & u_{d,j-1} & 0 & u_{d,j+1} & \cdots & u_{dd} \end{bmatrix}.$$

Let $U_{ji} = \det u^{(i,j)}$. Note the change in the order of the coefficients. The matrix U is called the *adjugate* of the matrix u. It follows from the definition of the determinant that $\det u = \sum_{i=1}^{d} U_{ji}u_{ij}$. On the other hand, if $j \ne k$, replace the jth column of u by the kth, to give the matrix $u^{[j,k]}$. Then $u^{[j,k]}$ has two columns equal, so that

$$\det u^{[j,k]} = \sum_{i=1}^{d} U_{ji}u_{ik} = 0.$$

Thus $U.u = (\det u)I$, and so $u^{-1} = (1/\det u)U$. This formula is known as *Cramer's rule*.

Let us give one explicit example.

Example B.4.1 Suppose that u is in $SO(3)$, the group of orthogonal matrices with determinant 1. Then

$$u^{-1} = U = \begin{bmatrix} u_{22}u_{33} - u_{23}u_{32} & u_{13}u_{32} - u_{12}u_{33} & u_{12}u_{23} - u_{13}u_{22} \\ u_{23}u_{31} - u_{21}u_{33} & u_{11}u_{33} - u_{13}u_{31} & u_{13}u_{21} - u_{11}u_{23} \\ u_{21}u_{32} - u_{22}u_{31} & u_{12}u_{31} - u_{11}u_{32} & u_{11}u_{22} - u_{12}u_{21} \end{bmatrix}$$

B.5 The trace

If $t \in M_d$, we define the *trace* $\mathrm{tr}(t)$ of t to be the sum of the diagonal terms: $\mathrm{tr}(t) = \sum_{i=1}^{d} t_{ii}$. If $s, t \in M_d$ then

$$\mathrm{tr}(st) = \sum_{i=1}^{d}\left(\sum_{j=1}^{d} s_{ij}t_{ji}\right) = \sum_{j=1}^{d}\left(\sum_{i=1}^{d} t_{ji}s_{ij}\right) = \mathrm{tr}(ts).$$

Suppose now that T is an endomorphism of E which is represented by a matrix t with respect to a basis $(e_1, \ldots, e_d)$ and by a matrix s with respect to a basis $(f_1, \ldots, f_d)$. Then there is an invertible matrix b such that $s = b^{-1}tb$, and so $\mathrm{tr}(s) = \mathrm{tr}(b^{-1}tb) = \mathrm{tr}(tbb^{-1}) = \mathrm{tr}(t)$. This means that we can define the trace $\mathrm{tr}(T)$ of the endomorphism T to be $\mathrm{tr}(t)$: the definition does not depend on the choice of basis.

Appendix C

Exterior algebras and the cross product

C.1 Exterior algebras

Suppose that E is a real vector space. An element of E, a vector, can be considered to have magnitude and direction. In the same way, if x and y are two vectors in E then they somehow relate to an area in span (x, y). If we wish to make this more specific, we certainly require that the area should be zero if and only if x and y are linearly dependent. A similar remark applies to higher dimensions. We wish to develop these ideas algebraically.

A finite-dimensional (associative) real *algebra* $(A, \circ)$ is a finite-dimensional real vector space equipped with a law of composition: that is, a mapping (*multiplication*) $(a, b) \to a \circ b$ from $A \times A$ into A which satisfies

- $(a \circ b) \circ c = a \circ (b \circ c)$ (associativity),
- $a \circ (b + c) = a \circ b + a \circ c$,
- $(a + b) \circ c = a \circ c + b \circ c$,
- $\lambda(a \circ b) = (\lambda a) \circ b = a \circ (\lambda b)$,

for $\lambda \in \mathbf{R}$ and $a, b, c \in A$. (As usual, multiplication is carried out before addition).

An algebra A is *unital* if there exists $1 \in A$, the *identity element*, such that $1 \circ a = a \circ 1 = a$ for all $a \in A$. For example, $M_d(\mathbf{R})$ and $M_d(\mathbf{C})$ are unital algebras. Both $\mathbf{R}$ and $\mathbf{C}$ can be considered as finite-dimensional real algebras: $\mathbf{R}$ has real dimension 1 and $\mathbf{C}$ has real dimension 2.

Suppose now that E is a finite-dimensional real vector space. An *exterior algebra* for E is a unital algebra $(\bigwedge^*(E), \wedge)$, together with an injective linear mapping $j : E \to \bigwedge^*(E)$ with the following properties.

(†) $j(x_1) \wedge \cdots \wedge j(x_k) = 0$ if and only if $x_1, \ldots, x_k$ are linearly dependent elements in E.

(‡) $j(E)$ generates $\bigwedge^*(E)$: any element of $\bigwedge^*(E)$ can be written as a linear combination of the identity and products of the form $j(x_1) \wedge \cdots \wedge j(x_k)$.

We identify E and $j(E)$, and so we write x for $j(x)$.

Proposition C.1.1   *If* $(\bigwedge^*(E), \wedge)$ *is an exterior algebra for E and $x, y \in E$ then $x \wedge y = -y \wedge x$. More generally, if $x_1, \ldots x_k \in E$ and σ is a permutation of $\{1, \ldots, k\}$, then*

$$x_{\sigma(1)} \wedge \cdots \wedge x_{\sigma(k)} = \epsilon_\sigma x_1 \wedge \cdots \wedge x_k.$$

Proof   For

$$0 = (x + y) \wedge (x + y) = x \wedge x + x \wedge y + y \wedge x + y \wedge y = x \wedge y + y \wedge x.$$

The second statement follows from this, since σ can be written as a product of transpositions which transpose two adjacent elements of $\{1, \ldots, k\}$.   □

This shows first that any area that $x \wedge y$ might represent is a signed area, so that the value may be positive or negative, and secondly that the order of the terms in a product is all-important.

We must show that an exterior algebra exists, and that it is essentially unique. It is possible to define the exterior algebra in a coordinate free way, but it is probably simpler to use a basis $(e_1, \ldots e_d)$ of E. We set $\Omega = \{1, \ldots, d\}$. We consider a 2^d-dimensional space $\bigwedge^*(E)$ with basis $\{e_A : A \subset \Omega\}$ indexed by the subsets of E; thus an element x of $\bigwedge^*(E)$ can be written uniquely as $x = \sum_{A \subseteq \Omega} x_A e_A$. We define the mapping j by setting $j(\sum_{i=1}^d x_i e_i) = \sum_{i=1}^d x_i e_{\{i\}}$, and define multiplication in in the following way.

- $e_\emptyset$ is the identity element of $\bigwedge^*(E)$;
- If $A \cap B \neq \emptyset$, then $e_A \wedge e_B = 0$;
- If A and B are disjoint, and if $A \cup B = C$, then we can write

$$A = \{i_1 < \ldots < i_{|A|}\}$$
$$B = \{j_1 < \ldots < j_{|B|}\}$$
$$C = \{k_1 < \ldots < k_{|C|}\}.$$

Here the order of the terms is all-important. If $A, B \subseteq \Omega$, we denote the sequence $(i_1, \ldots, i_{|A|}, j_1, \ldots, j_{|B|})$, by $A \# B$. Let σ be the permutation of C which arranges the sequence $A \# B$ in increasing order. We define $e_A \wedge e_B$ to be $\epsilon_\sigma e_C$.

- We extend multiplication by linearity. If $x = \sum_{A \subseteq \Omega} x_A e_A$ and $y = \sum_{B \subseteq \Omega} y_B e_B$, then

$$x \wedge y = \sum \{ x_A y_B e_A \wedge e_B : A, B \subseteq \Omega \}.$$

First we show that this multiplication is associative. Suppose that A, B, C are subsets of Ω. If A, B and C are not pairwise disjoint, it is easy to see, be considering various cases, that $(e_A \wedge e_B) \wedge e_C = 0$ and $e_A \wedge (e_B \wedge e_C) = 0$. Suppose that A, B and C are pairwise disjoint, and that $D = A \cup B \cup C$. Then $(e_A \wedge e_B) \wedge e_C = \epsilon_\sigma e_D$, where σ is the permutation obtained by first arranging $A \# B$ in increasing order, and then arranging $(A \cup B) \# C$ in increasing order. Similarly $e_A \wedge (e_B \wedge e_C) = \epsilon_\tau e_D$, where τ is the permutation obtained by first arranging $B \# C$ in increasing order, and then arranging $A \# (B \cup C)$ in increasing order. Clearly $\sigma = \tau$, so that $(e_A \wedge e_B) \wedge e_C = e_A \wedge (e_B \wedge e_C)$. The associativity of multiplication follows from this: $\bigwedge^*(E)$ is a unital algebra.

Next we show that condition (†) is satisfied. First suppose that $x_1, \ldots, x_k$ are not linearly independent. Then there exist $\lambda_1, \ldots, \lambda_k$, not all zero, such that $\sum_{j=1}^k \lambda_j x_j = 0$. Without loss of generality, $\lambda_k \neq 0$. Then

$$0 = x_1 \wedge \cdots \wedge x_{k-1} \wedge \left(\sum_{j=1}^k \lambda_j x_j \right) = \lambda_k (x_1 \wedge \cdots \wedge x_k),$$

so that $x_1 \wedge \cdots \wedge x_k = 0$.

Secondly, suppose that $x_1, \ldots, x_k$ are linearly independent. By Proposition B.1.1, there exist $x_{k+1}, \ldots, x_d$, so that $(x_1, \ldots, x_d)$ is a basis for E. Let $x_j = \sum_{i=1}^d x_{ij} e_i$. Expanding the terms in the product, we see that

$$x_1 \wedge \cdots \wedge x_d = \left(\sum_{\sigma \in \Sigma_d} \epsilon_\sigma x_{\sigma(1),1} \cdots x_{\sigma(d),d} \right) e_\Omega = (\det X) e_\Omega,$$

where X is the matrix (x_{ij}). But X is the matrix of the endomorphism $T \in L(E)$ which maps e_j to x_j, for $1 \leq j \leq d$. This is invertible, and so $\det X \neq 0$. Thus $x_1 \wedge \cdots \wedge x_d \neq 0$, and so $x_1 \wedge \cdots \wedge x_k \neq 0$

Finally it follows from the construction that the condition (‡) is satisfied.

It is easy to see that the exterior algebra is essentially unique. Suppose that $(F, \wedge')$ is an exterior algebra for E, with mapping $j' : E \to F$. If $A = \{i_1 < \ldots < i_{|A|}\} \subseteq \Omega$, let $\pi(e_A) = j'(e_{i_1}) \wedge' \cdots \wedge' j'(e_{i_{|A|}})$. Then it follows from (†) that $\pi(e_A) \neq 0$, and from the construction of $\bigwedge^*(E)$ that $\pi(e_A) \wedge' \pi(e_B) = \pi(e_A \wedge e_B)$. Thus π extends to an injective algebra homomorphism of $\bigwedge^*(E)$ into F. Finally, condition (‡) ensures that π is surjective.

As a result, we refer to 'the exterior algebra' $\bigwedge^*(E)$ of E, rather than 'an exterior algebra'.

The exterior algebra $\bigwedge^*(E)$ is a *graded algebra*; we can write

$$\bigwedge{}^*(E) = \mathbf{R}.1 \oplus E \oplus (E \wedge E) \oplus \cdots \oplus \bigwedge{}^k(E) \oplus \cdots \oplus \bigwedge{}^d(E),$$

where $\bigwedge^k(E) = \mathrm{span}\,\{x_1 \wedge \cdots \wedge x_k : x_1, \ldots x_k \in E\}$. That is to say, if $y \in \bigwedge^*(E)$ then y can be written uniquely as $y = y_0.1 + \sum_{k=1}^d y_k$, where $y_k \in \bigwedge^k(E)$ for $1 \le k \le d$.

If $(e_1, \ldots, e_d)$ is a basis for E as above, then expanding the products, it follows that $\bigwedge^k(E) = \mathrm{span}\,\{e_A : |A| = k\}$, so that $\bigwedge^k(E)$ has dimension $\binom{d}{k}$. In particular, $\bigwedge^d(E)$ is one-dimensional, and in this setting, is the span of e_Ω; e_Ω is called the *unit volume element*. As we have seen, if we consider a different basis $(f_1, \ldots, f_d)$, then $f_\Omega = (\det T)e_\Omega$, where T is the linear mapping which sends e_j to f_j, for $1 \le j \le d$. In particular, if E is a Euclidean space and $(e_1, \ldots, e_d)$ and $(f_1, \ldots, f_d)$ are orthonormal bases, then $f_\Omega = e_\Omega$ if $T \in SO(E)$, and $f_\Omega = -e_\Omega$ otherwise. In the former case, the bases have the same *chirality*, or *handedness*, and in the latter, opposite chirality, or handedness.

Exercise

C.1.1 The elements of $E \wedge E$ are called *bivectors*, and elements of the form $x \wedge y$ are called *simple bivectors*. Suppose that $(e_1, \ldots e_4)$ is a basis for a four-dimensional space E. Calculate

$$(e_1 \wedge e_2 + e_3 \wedge e_4) \wedge (e_1 \wedge e_2 + e_3 \wedge e_4),$$

and conclude that $e_1 \wedge e_2 + e_3 \wedge e_4$ is a bivector, but not a simple bivector.

C.2 The cross product

We now restrict attention to the case where E is a three-dimensional Euclidean space, with orthonormal basis (e_1, e_2, e_3). Then $\bigwedge^*(E)$ is an eight-dimensional unital algebra, and

$$\bigwedge{}^*(E) = \mathbf{R}.1 \oplus E \oplus (E \wedge E) \oplus \mathbf{R}.e_\Omega.$$

Let $f_1 = e_2 \wedge e_3$, $f_2 = e_3 \wedge e_1$ and $f_3 = e_1 \wedge e_2$. Then (f_1, f_2, f_3) is a basis for $E \wedge E$. If $u = u_1 e_1 + u_2 e_2 + u_3 e_3$ and $v = v_1 e_1 + v_2 e_2 + v_3 e_3$ then

$$u \wedge v = (u_2 v_3 - u_3 v_2)f_1 + (u_3 v_1 - u_1 v_3)f_2 + (u_1 v_2 - u_2 v_1)f_3.$$

Proposition C.2.1 *Every element of $E \wedge E$ is a simple bivector.*

Proof Suppose that $y = y_1 f_1 + y_2 f_2 + y_3 f_3 \in E \wedge E$. If $y_1 = 0$ then $y = e_1 \wedge (y_3 e_2 - y_2 e_3)$, and if $y_1 \neq 0$ then

$$y = (y_2 e_1 - y_1 e_2) \wedge (y_3 e_1 - y_1 e_3)/y_1.$$

$\square$

Suppose that $x, y, z \in E$. Then $x \wedge y \wedge z = v(x, y, z)e_\Omega$, where $v(x, y, z)$ is the signed volume of the parallelepiped defined by x, y and z. Let $\phi_{y,z}(x) = v(x, y, z)$. Then $\phi_{y,z}$ is a linear functional on E. By the Fréchet-Riesz representation theorem (Theorem 14.3.7), there exists a unique element of E, which we denote by $y \times z$, such that

$$v(x, y, z) = \phi_{y,z}(x) = \langle x, y \times z \rangle.$$

The vector $y \times z$ is called the *cross product* of y and z. Since $\langle y, y \times z \rangle = y \wedge y \wedge z = 0$ and $\langle z, y \times z \rangle = z \wedge y \wedge z = 0$, $y \times z$ is a vector which is orthogonal to span (y, z).

The quantity $\langle x, y \times z \rangle$ is called the *scalar triple product* of x, y and z. Since $x \wedge y \wedge z = \langle x, y \times z \rangle \, e_\Omega$,

$$\langle x, y \times z \rangle = \langle y, z \times x \rangle = \langle z, x \times y \rangle = \det T,$$

where T is the endomorphism of E which maps e_1 to x, e_2 to y and e_3 to z.

The mapping $(y, z) \to y \times z$ is a bilinear mapping of $E \times E$ onto E, which satisfies $y \times z = -z \times y$. Similarly, the mapping $y \wedge z \to y \times z$ is a bijective linear mapping of $E \wedge E$ onto E.

It is important to note that the cross product is not an associative product. For example, $e_1 \times (e_1 \times e_2) = e_1 \times e_3 = -e_2$, while $(e_1 \times e_1) \times e_2 = 0 \times e_2 = 0$.

Let us consider the cross product in more detail. Suppose that $x, y, z \in E$. If y and z are linearly dependent, then $y \times z = 0$. Otherwise, let $g_1 = y/\|y\|$, let $w = z - \langle z, g_1 \rangle g_1$ and let $g_2 = w/\|w\|$. Then $y \times z = y \times w$, and g_1 and g_2 are orthogonal unit vectors. Let g_3 be the unit vector orthogonal to g_1 and g_2 for which $g_1 \wedge g_2 \wedge g_3 = e_\Omega$. Then

$$\langle g_1, g_1 \times g_2 \rangle = 0, \quad \langle g_2, g_1 \times g_2 \rangle = 0 \text{ and } \langle g_3, g_1 \times g_2 \rangle = 1,$$

so that $g_1 \times g_2 = g_3$. Consequently,

$$y \times z = y \times w = (\|y\| \cdot \|w\|)g_1 \times g_2 = (\|y\| \cdot \|w\|)g_3.$$

Now

$$\|w\|^2 = \|z\|^2 - 2 \langle z, g_1 \rangle^2 + \langle z, g_1 \rangle^2 = \|z\|^2 - \langle z, y \rangle^2 / \|y\|^2,$$

and so
$$\|y \times z\|^2 = \|y\|^2 \|z\|^2 - \langle z, y \rangle^2$$

What can we say about the *vector triple product* $x \times (y \times z)$?

Proposition C.2.2 *Suppose that* $x, y, z \in E$.
(i) $x \times (y \times z) = \langle x, z \rangle y - \langle x, y \rangle z$.
(ii) $x \times (y \times z) + y \times (z \times x) + z \times (x \times y) = 0$.

Proof   (i) If y and z are linearly dependent, both sides of the equation are zero. Otherwise, let us use the notation above. Since

$$x = \langle x, g_1 \rangle g_1 + \langle x, g_2 \rangle g_2 + \langle x, g_3 \rangle g_3,$$

$$
\begin{aligned}
x \times (y \times z) &= (\|y\| \cdot \|w\|)(\langle x, g_1 \rangle (g_1 \times g_3) + \langle x, g_2 \rangle (g_2 \times g_3)) \\
&= (\|y\| \cdot \|w\|)(- \langle x, g_1 \rangle g_2 + \langle x, g_2 \rangle g_1) \\
&= - \langle x, y \rangle w + \langle x, w \rangle y.
\end{aligned}
$$

But

$$- \langle x, y \rangle w = - \langle x, y \rangle z + \langle x, y \rangle \langle z, g_1 \rangle g_1$$
$$\text{and } \langle x, w \rangle y = \langle x, z \rangle y - \langle z, g_1 \rangle \langle x, g_1 \rangle y = \langle x, z \rangle y - \langle z, g_1 \rangle \langle x, y \rangle g_1.$$

Adding, we obtain the result.
 (ii) follows by adding the formulae for each of the three terms.   □

Appendix D

Tychonoff's theorem

We prove Tychonoff's theorem, that the topological product of compact topological spaces is compact. The key idea is that of a filter. This generalizes the notion of a sequence in a way which allows the axiom of choice to be applied easily.

A collection $\mathcal{F}$ of subsets of a set S is a *filter* if

F1 if $F \in \mathcal{F}$ and $G \supseteq F$ then $G \in \mathcal{F}$,
F2 if $F \in \mathcal{F}$ and $G \in \mathcal{F}$ then $F \cap G \in \mathcal{F}$,
F3 $\emptyset \notin \mathcal{F}$.

Here are three examples.

- If A is a non-empty subset of S then $\{F : A \subseteq F\}$ is a filter.
- Suppose that (X, τ) is a topological space, and that $x \in X$. The collection $\mathcal{N}_x$ of neighbourhoods of x is a filter.
- If (s_n) is a sequence in S then

$$\{F : \text{there exists } N \text{ such that } s_n \in F \text{ for } n \geq N\}$$

is a filter.

Filters can be ordered. We say that $\mathcal{G}$ *refines* $\mathcal{F}$, and write $\mathcal{G} \geq \mathcal{F}$, if $\mathcal{G} \supseteq \mathcal{F}$.

We now consider a topological space (X, τ). We say that a filter $\mathcal{F}$ *converges* to a *limit* x (and write $\mathcal{F} \to x$) if $\mathcal{F}$ refines $\mathcal{N}_x$. Clearly if $\mathcal{G}$ refines $\mathcal{F}$ and $\mathcal{F} \to x$ then $\mathcal{G} \to x$.

The Hausdorff property can be characterized in terms of convergent filters.

Proposition D.0.6 (X, τ) *is Hausdorff if and only if whenever* $\mathcal{F} \to x$ *and* $\mathcal{F} \to y$ *then* $x = y$.

Proof   If (X, τ) is Hausdorff, if $\mathcal{F} \to x$, and if $x \neq y$ then there exist disjoint open sets U and V with $x \in U$ and $y \in V$. Since $U \in \mathcal{N}_x$, $U \in \mathcal{F}$. But then $V \notin \mathcal{F}$, since $U \cap V = \emptyset$. Thus $\mathcal{F}$ does not refine $\mathcal{N}_y$, and so $\mathcal{F} \not\to y$.

Conversely if (X, τ) is not Hausdorff there exist distinct x and y such that if $U \in \mathcal{N}_x$ and $V \in \mathcal{N}_y$ then $U \cap V$ is not empty. Let

$$\mathcal{F} = \{F : F \supseteq U \cap V \text{ for some } U \in \mathcal{N}_x \text{ and } V \in \mathcal{N}_y\}.$$

Then $\mathcal{F}$ is a filter which converges to both x and y. $\square$

We say that x is *adherent* to a filter $\mathcal{F}$ if $x \in \overline{F}$ for each $F \in \mathcal{F}$.

Proposition D.0.7   *If $\mathcal{F} \to x$ then x is adherent to $\mathcal{F}$.*

Proof   For if $F \in \mathcal{F}$ and $N \in \mathcal{N}_x$ then $N \in \mathcal{F}$ and so $F \cap N \in \mathcal{F}$. Thus $F \cap N$ is not empty, and so $x \in \overline{F}$. $\square$

Proposition D.0.8   *If x is adherent to $\mathcal{F}$ then there is a refinement $\mathcal{G}$ such that $\mathcal{G} \to x$.*

Proof   Let

$$\mathcal{G} = \{G : G \supseteq F \cap N \text{ for some } F \in \mathcal{F}, N \in \mathcal{N}_x\}.$$

$\mathcal{G}$ is a filter which refines both $\mathcal{F}$ and $\mathcal{N}_x$. $\square$

Suppose that S and T are sets, and that f is a mapping from S to T. If $\mathcal{F}$ is a filter on S the image filter $f(\mathcal{F})$ on T is defined by

$$f(\mathcal{F}) = \{H \subseteq T : f^{-1}(H) \in \mathcal{F}\}.$$

It is easy to check that this is a filter.

Proposition D.0.9   *Suppose that X and Y are topological spaces, that f is a mapping from X to Y and that $x \in X$. Then f is continuous at x if and only if whenever $\mathcal{F} \to x$ then $f(\mathcal{F}) \to f(x)$.*

Proof   If f is continuous at x and $\mathcal{F} \to x$ then if $N \in \mathcal{N}_{f(x)}$ then $f^{-1}(N) \in \mathcal{N}_x \subseteq \mathcal{F}$, so that $N \in f(\mathcal{F})$. Thus $f(\mathcal{F})$ refines $\mathcal{N}_{f(x)}$ and $f(\mathcal{F}) \to f(x)$.

Conversely if the condition is satisfied, then since $\mathcal{N}_x \to x$, $f(\mathcal{N}_x) \to f(x)$, so that if $N \in \mathcal{N}_{f(x)}$ then $N \in f(\mathcal{N}_x)$, and so $f^{-1}(N) \in \mathcal{N}_x$. Thus f is continuous at x. $\square$

Proposition D.0.10   *A topological space (X, τ) is compact if and only if every filter on X has an adherent point.*

Proof Suppose that (X, τ) is compact, and that $\mathcal{F}$ is a filter on X. The collection of closed sets $\{\overline{F} : F \in \mathcal{F}\}$ has the finite intersection property, and so has a non-empty intersection. Any point of the intersection is adherent to $\mathcal{F}$.

Conversely, suppose that the condition is satisfied, and that $\mathcal{C}$ is a collection of closed sets with the finite intersection property. Let $\mathcal{D}$ be the set of finite intersections of sets in $\mathcal{C}$ and let

$$\mathcal{F} = \{F : F \supseteq D, \text{ for some } D \in \mathcal{D}\}.$$

$\mathcal{F}$ is a filter: if x is an adherent point then

$$x \in \cap\{\overline{C} : C \in \mathcal{C}\} = \cap\{C : C \in \mathcal{C}\}.$$

$\square$

Proposition D.0.11 *A topological space (X, τ) is compact if and only if every filter on X has a convergent refinement.*

Proof Propositions D.0.7, D.0.8 and D.0.10. $\square$

Compare this with the Bolzano–Weierstrass theorem.

Recall that if $(X_\alpha)_{\alpha \in A}$ is a family of topological spaces, we give the product $\prod_{\alpha \in A} X_\alpha$ the product topology, taking as basis of open sets the sets of the form $\cap_{i=1}^n \pi_{a_i}^{-1}(O_{\alpha_i})$, where O_{α_i} is open in X_{α_i}. This means that $N \in \mathcal{N}_x$ if and only if $N \supseteq \cap_{i=1}^n \pi_{\alpha_i}^{-1}(N_{\alpha_i})$, where $N_{\alpha_i} \in \mathcal{N}_{\pi_{\alpha_i}(x)}$ for some $\alpha_1, \ldots, \alpha_n$ in A.

Proposition D.0.12 *Suppose that $\mathcal{F}$ is a filter on $\prod_{\alpha \in A} X_\alpha$. Then $\mathcal{F} \to x$ if and only if $\pi_\alpha(\mathcal{F}) \to \pi_\alpha(x)$ for each $\alpha \in A$.*

Proof If $\mathcal{F} \to x$, then $\pi_\alpha(\mathcal{F}) \to \pi_\alpha(x)$, since π_α is continuous. Conversely suppose that $\pi_\alpha(\mathcal{F}) \to \pi_\alpha(x)$, for each α. Then if $N \in \mathcal{N}_x$, $N \supseteq \cap_{i=1}^n \pi_{\alpha_i}^{-1}(N_{\alpha_i})$, for suitable n, α_i and N_{α_i}. But $N_{\alpha_i} \in \pi_{\alpha_i}(\mathcal{F})$, since $\pi_{\alpha_i}(\mathcal{F}) \to \pi_{\alpha_i}(x)$, and so $\pi_{\alpha_i}^{-1}(N_{\alpha_i}) \in \mathcal{F}$. Thus $N \in \mathcal{F}$, and so $\mathcal{F} \to x$. $\square$

One major virtue of filters is that they allow the axiom of choice, in the form of Zorn's Lemma, to be applied easily. The filters on a set S are ordered by refinement. If $\mathcal{C}$ is a chain of filters refining $\mathcal{F}$ then $\{G : G \in \mathcal{G} \text{ for some } \mathcal{G} \in \mathcal{C}\}$ is a filter which is an upper bound for $\mathcal{C}$. Thus by Zorn's Lemma, every filter has a maximal refinement. A maximal filter is called an *ultrafilter*.

Proposition D.0.13 *Every filter has an ultrafilter refinement.*

Proposition D.0.14 *If $\mathcal{U}$ is an ultrafilter on S and $A \subseteq S$ then exactly one of A and $C(A)$ is in $\mathcal{U}$. Conversely if $\mathcal{F}$ is a filter on S with the property that if $A \subseteq S$ then either $A \in \mathcal{F}$ or $C(A) \in \mathcal{F}$ then $\mathcal{F}$ is an ultrafilter.*

Proof   Since $A \cap C(A) = \emptyset$, at most one of A and $C(A)$ can belong to $\mathcal{U}$. Suppose that $A \cap U$ is non-empty, for each $U \in \mathcal{U}$. Then the sets $\{V : V \supseteq A \cap U$ for some $U \in \mathcal{U}\}$ form a filter which refines $\mathcal{U}$, and so is equal to $\mathcal{U}$, by maximality. Thus $A \in \mathcal{U}$. Otherwise, there exists $U_0 \in \mathcal{U}$ such that $A \cap U_0$ is empty. Then $C(A) \supseteq U_0$, so that $C(A) \in \mathcal{U}$.

   Conversely, let $\mathcal{G}$ be a refinement of $\mathcal{F}$, and suppose that $G \in \mathcal{G}$. If $C(G) \in \mathcal{F}$, then $C(G) \in \mathcal{G}$, giving a contradiction. So $G \in \mathcal{F}$, and $\mathcal{F}$ is maximal.   □

Proposition D.0.15 *If $\mathcal{U}$ is an ultrafilter on S and f is a mapping from S to T then $f(\mathcal{U})$ is an ultrafilter on T.*

Proof   If $B \subseteq T$ then $f^{-1}(C(B)) = C(f^{-1}(B))$, so that either $f^{-1}(B) \in \mathcal{U}$ or $f^{-1}(C(B)) \in \mathcal{U}$. Thus either B or $C(B)$ is in $f(\mathcal{U})$.   □

Proposition D.0.16 *A topological space (X, τ) is compact if and only if every ultrafilter on X converges.*

Proof   If (X, τ) is compact and $\mathcal{U}$ is an ultrafilter on X, then $\mathcal{U}$ has a convergent refinement, by Proposition D.0.11. But any refinement of $\mathcal{U}$ is $\mathcal{U}$ itself.

   Conversely, if the condition is satisfied, and $\mathcal{F}$ is a filter on X, then $\mathcal{F}$ has an ultrafilter refinement, by Proposition D.0.13. This converges, by hypothesis, and so (X, τ) is compact, by Proposition D.0.11.   □

Theorem D.0.17 (Tychonoff's theorem)   *If $(X_\alpha, \tau_\alpha)_{\alpha \in A}$ is a family of compact topological spaces, then $\prod_{\alpha \in A} X_\alpha$ is compact in the product topology.*

Proof   Let $\mathcal{U}$ be an ultrafilter on $\prod_{\alpha \in A} X_\alpha$. Then for each α, $\pi_\alpha(\mathcal{U})$ is an ultrafilter on X_α, and so it converges, to x_α, say. Let $x = (x_\alpha)$, so that $x_\alpha = \pi_\alpha(x)$. Then $\mathcal{U} \to x$ in the product topology, by Proposition D.0.12. This gives the result.   □

Exercises

D.0.1 Suppose that $\mathcal{F}$ is a filter on S, and that f is a mapping from S to T. When is $\{f(F) : F \in \mathcal{F}\}$ a filter?

D.0.2 Let (x_n) be a sequence in a topological space (X, τ), and let $\mathcal{F}$ be the filter $\{A : \text{there exists } N \text{ such that } x_n \in A \text{ for all } n \geq N\}$. Show that $x_n \to x$ if and only if $\mathcal{F} \to x$.

D.0.3 A collection $\mathcal{B}$ of non-empty subsets of S is a *filter base* if whenever $B_1, B_2 \in \mathcal{B}$ then there exists $B_3 \in \mathcal{B}$ with $B_3 \subseteq B_1 \cap B_2$. Show that

$$\mathcal{F} = \{F : F \supseteq B \text{ for some } B \in \mathcal{B}\}$$

is a filter.

D.0.4 A filter base $\mathcal{B}$ is *free* if $\cap\{B : B \in \mathcal{B}\} = \emptyset$. Characterize compactness in terms of the non-existence of certain free filter bases.

D.0.5 Suppose that for each x in a set X there is given a filter $\mathcal{N}_x$ with the following properties:

(a) $x \in N$ for each $N \in \mathcal{N}_x$;

(b) if $N \in \mathcal{N}_x$ there exists $M \in \mathcal{N}_x$ with $M \subset N$ such that $M \in \mathcal{N}_y$ for each $y \in M$.

Show that the collection of sets $\{U : U \in N_y \text{ for all } y \in U\}$ is a topology on X and that the filters $\mathcal{N}_x$ are the neighbourhood filters for this topology.

Index

Contents

Volume I

Volume III

Printed in the United States
by Booktango

Printed in the United States
By Bookmasters